Systems Analysis & Design

Systems Analysis & Design

Perry Edwards

Mitchell McGRAW-HILL

New York St. Louis San Francisco Auckland Bogotá Caracas
Lisbon London Madrid Mexico Milan Montreal New Delhi Paris
San Juan Singapore Sydney Tokyo Toronto Watsonville

Mitchell **McGRAW-HILL**
Watsonville, CA 95076

Systems Analysis & Design

1 2 3 4 5 6 7 8 9 0 DOH DOH 9 0 9 8 7 6 5 4 3 2

ISBN 0-07-019573-0

Sponsoring editor: Erika Berg
Editorial assistant: Jennifer Gilliland
Technical reviewer: Efrem Mallach
Director of production: Jane Somers
Production assistant: Leslie Austin
Project manager: Jane Granoff
Cover and interior designer: Gary Head, Publishing Principals
Cover photo: Gary Head, Publishing Principals
Copy editor: Karen Richardson
Compositor: Barbara Gelfand
Illustrator: Sean Scullion
Printer and binder: R. R. Donnelley and Sons

Library of Congress Card Catalog No. 92-64037

Contents in Brief

Contents

Chapter 11	*Network Design* *363*

Chapter 12   *Software Design*   390

Management expects information systems to satisfy their information needs to solve their business problems. Systems are expected to be delivered on time, within budget, with features promised, free of errors, as well as meeting users' needs. Besides demanding clients, today's systems analysts face ever-changing development methodologies and technologies, and resistance to change.

This book is designed for introductory systems analysis and design courses that address such varied issues. *Systems Analysis and Design* provides a solid foundation of systems principles and an understanding of how businesses function, while heightening students' sensitivity to the people issues analysts face daily.

The goal of *Systems Analysis and Design* is to help launch the careers of successful systems analysts—or of users assuming an active role in building systems that satisfy their organization's information needs.

Here is how *Systems Analysis and Design* meets this challenge.

PROVIDES A REAL-WORLD CONTEXT

Most students taking a systems analysis and design course have little—if any—business experience. In addition to Goals, Preview, Key Terms, Review Questions, Discussion Questions, Application Questions, and Research Questions, each chapter includes the following learning aids:

Real-World Case Studies

The two integrated cases and variety of mini-cases reflect the experiences of actual businesses. By providing a real-world context, they illustrate the systems principles discussed throughout the text and motivate student understanding.

- Fleet Feet, a large shoe store franchise. Used throughout the text, this integrated case study traces the development of a new accounts payable system.

■ View Video, a small videotape rental store. While most cases assume that all companies build systems from scratch, View Video reflects the growing majority of "new" systems that are actually modifications of existing systems.

■ Mini-cases. These vignettes reflect the issues that practicing analysts commonly face and are expected to resolve.

These cases demonstrate that in order for an information system to be successful, it must promote the effectiveness and efficiency of an organization. In other words, systems are designed and developed for two primary reasons: to satisfy users' information needs and to help solve problems.

Boxed Inserts

The following inserts are featured throughout the text, as illustrated in the detailed table of contents. They are intended to balance the technical issues with the people issues that analysts commonly confront when developing systems.

■ "Working with People" inserts highlight people issues commonly faced by analysts: coping with fearful users, keeping pace with state-of-the-art technology, retraining users, appreciating ergonomics, handling change, and more. These inserts emphasize why user participation is critical to the development of successful systems.

■ "Working with Technology" inserts offer practical tips, techniques, and checklists for designing screens, reports, and databases; testing procedures; conducting walk-throughs; writing reports; and organizing presentations to management.

GIVES CURRENT COVERAGE

The following table outlines the book's coverage of the topics that reviewers believe are increasingly important to future systems analysts.

Topic	Text Coverage	Benefit
Computer-Aided Software Engineering (CASE)	Integrated into every chapter.	Illustrates use of CASE at each phase of the life cycle.
User Participation	"Working with People" boxed inserts—one per chapter. See the Detailed Table of Contents.	Emphasizes the benefits of involving users in development—and the costs of excluding them.

Topic	Text Coverage	Benefit
Prototyping	Chapter 6.	Illustrates how this emerging methodology enhances the traditional life cycle.
Quality Assurance	A theme of Chapters 13 and 14.	Encourages students to strive for error-free systems.
Managing Costs and Schedules	Chapters 4, 5, 13, 14, and 15.	Stresses the need to make trade-offs to meet budget and schedule.
Downsizing and Networking Issues	Chapter 11.	Prepares students to deal effectively with issues of increasing concern to business.
Object-Oriented Analysis and Design Concepts	Chapters 6, 12, and 15.	Clarifies the difference between structured and object-oriented approaches to development.
International Issues	International examples appear in every chapter.	Heightens students' sensitivity to cultural differences that impact the development of systems.

INSTRUCTOR SUPPORT

Because CASE concepts are integrated throughout the entire text, *Systems Analysis and Design* provides a conceptual foundation for the use of any CASE tool. In addition, the text is supported by:

■ **Instructor's Manual with Transparency Masters**

Includes Summary, Teaching Tips and Techniques, Key Terms, Answers to Questions, detailed outline and course syllabus, 40 term projects, three quizzes, and transparency masters for each chapter. Also contains an optional case study, complete with narrative describing the problem, data flow diagrams, a relational database design, sample reports, data-entry screens, and problem specifications.

■ **Computerized Test Bank**

The computerized test bank set includes over 1400 multiple choice and true/false questions. Included in the set are the questions on disk, printed questions, and RHTest test-generator for the IBM PC and compatibles (3 1/2"). The printed test questions are also available separately.

■ **Supplementary Casebooks**

Casebook for Systems Analysis and Design: FSS, Inc., by Robert Marble, takes a process-oriented approach to developing systems. Mitchell McGRAW-HILL, 1992, ISBN: 0-07-040190-X.

Casebook for Systems Analysis and Design: JPS, Inc., Robert Marble, takes a data-oriented approach to developing systems. Mitchell McGRAW-HILL, 1993, ISBN: 0-07-040193-4.

ACKNOWLEDGMENTS

Many people contributed to this text's development, each in a unique way. The manuscript was class-tested by Clifford Burns and me. I thank Cliff and our students for their invaluable suggestions. Reviewers from around the country helped develop each successive draft of the manuscript. My special thanks to:

Fred Augustine—Stetson University

C.T. Cadenhead—Richmond College

John Eatman—University of North Carolina, Greensboro

Karen-Ann Kievit—Loyola-Marymount University

Peeter Kirs—Florida International University

Constance Knapp—Pace University

Sue Krimm—Pierce College

Efrem Mallach—University of Massachusetts, Lowell

Ted Sturynt—Stetson University

James Westfall—University of Evansville

The staff at Mitchell McGRAW-HILL is wonderful and I appreciate all of their efforts. Many thanks to Erika Berg, Jennifer Gilliland, Jane Somers, and Leslie Austin. I'd also like to thank my copyeditor, Karen Richardson, and project manager, Jane Granoff.

Many family activities were missed or delayed due to this project. I thank Kathleen, Jennifer, Sarah, and Benjamin for their understanding of my writing these past few years.

Perry Edwards

An Introduction to the Systems Process

The Systems Development Life Cycle

GOALS

After reading this chapter, you should be able to do the following:

- Define the word *system* and give at least three examples of common systems
- Explain how computer-based information systems benefit businesses
- Describe the relationship between the design of an organization and the design of the business systems that support that organization
- Explain the job roles of users and analysts as they relate to business systems
- Describe communication skills that are important to analysts
- State the four primary steps in the systems life cycle
- Know why it is necessary to design a "logical" version of a system before designing the "physical" version
- Describe how business systems have traditionally developed
- List four advantages of the CASE method of software development
- Describe modern methods for developing business systems, and explain how and why the new methods are better than the old ones

PREVIEW

The use of computers in our society has become so widespread that we find ourselves in the middle of changes as dramatic as those that attended the Industrial Revolution in the nineteenth century. Computers no longer benefit just large corporations and government agencies but—in the form of personal, portable, and laptop computers—have begun to help everyone, including students and their families, traveling business people, and small business owners.

What does the computer revolution mean? In a word, it means change. And change brings with it new fields of study, new ways of getting work done, and sometimes also fear of the unknown. Prior to the advent of the computer, organizations and individuals processed data manually or mechanically. As a result, information often arrived too late for it to help people make the right decisions. Even worse, manually generated information often contained the types of errors that cause costly mistakes. Modern computers not only process data faster, thus ensuring more timely information, but they can do so with extreme accuracy, giving people the precise facts that they need to operate efficiently. Computers also free busy managers to spend more time on increasing sales, planning, or personnel problems.

This text will teach you to analyze the ways in which organizations process data so that you can better understand and improve on those ways. In this first chapter, you will learn the basic concepts of systems analysis, design, implementation, and maintenance—beginning with a definition of systems.

WHAT IS A SYSTEM?

Systems. We see, hear, or read this word almost every day. We have learned to read, write, and calculate within an educational system. The human body is a biological system. Nightly news reporters analyze our political system, often examining the way it interacts with the political systems of our allies and adversaries. As consumers, employees, or employers, we struggle with a worldwide economic system, governed by the laws of supply and demand, that seems to grow more complex every day. In the field of business, organizations rely on many systems, including accounting or finance, sales, personnel, order-entry, and word processing systems. As we move to a global market, we must work with a currency system. A **system** is a combination of resources working together to convert inputs to usable outputs.

All these systems exhibit certain characteristics. First, they consist of interrelated and interdependent elements. Our own human biological system, for example, contains bones and organs, many different types of cells, and an assortment of organic and inorganic compounds that function together according to certain biological "rules." Similarly, business systems involve organizational structures, people, and all types of equipment (such as paper clips, fax machines, voice mail systems, and computers), all of which work together under certain policies and procedures.

Within any system, the individual pieces coordinate to accomplish specific tasks, jobs, or functions. All the accountants and clerks in a company, for example, follow prescribed guidelines for using their machines to convert data about transactions into useful decision-making information that takes the form of various reports such as balance sheets, income statements, and lists of suppliers.

To study and understand systems better, we can subdivide them into smaller systems, each of which contains its own interacting elements. For example, we can subdivide the American political system into national, state, and local government subsystems, which consist of various agencies that contain further subsystems. Similarly, within an organization's information system reside accounting, production, logistics, and other systems. These in turn contain certain subsystems. Accounting systems, for example, contain subsystems for general ledger, accounts receivable, accounts payable, inventory control, and payroll.

COMPUTER-BASED BUSINESS SYSTEMS

In this text, we will concentrate on business systems with the ultimate goal of learning how we can improve them with the use of computers. Despite the fact that not all business systems require a computer, we will emphasize computer-based systems throughout our discussion. Bear in mind that careful analysis of an existing manual system may lead to retaining that system; using a computer is not always the best solution to a problem. For example, an analyst may decide that it makes more sense to record sales contacts on 3"-by-5" cards or on a Rolodex than use a personal computer and a database.

A computer-based business system involves six interdependent elements: hardware (machines), software (instructions or programs), people (programmers, managers, or users), procedures (rules), data, and information. All six elements interact to convert data into information.

The word **data** means accumulated but unorganized facts (such as a list of the type of shoes worn by each American at a given time). **Information** implies usefully organized and reported facts (such as the most popular type of running shoe worn by women aged 25 to 35). However, one person's information is another's data. In the case of the popular type of shoe, the manufacturer needs to know the bestselling shoe for women between the ages of 25 and 35 so that they can make more shoes of that particular style and color.

Figure 1.1a illustrates the conversion of data into information. At the input stage, data enter the system for processing, where they are organized and tabulated. At the output stage, information leaves the system. Obviously, processing vast amounts of data into information by hand is costly and time-consuming. Since computers can convert data to information much more easily and inexpensively, business decision-makers have naturally become increasingly dependent upon them.

As Figure 1.1b also shows, a computer-based system allows for adjustment. A manager will study the information produced by the system, comparing it with expected results. If the results seem to require some fine-tuning, the manager may decide to alter the original data. We call this comparison and use of information in adjusting the system feedback. For example, if a computer-based inventory system shows a particular style of running shoe as a poor seller, the manager may decide to lower the price. By entering the new price back into the system, the company can see how it might affect sales. We can experience feedback in the form of a necessary adjustment, confirming that the system is producing the desired output.

To see how computer-based systems improve a business's activity, consider the following situation. Mountain Motors, an automobile-parts chain with five stores, 55 employees, and an appointments secretary, prides itself on its preventive auto-care practice. Customers periodically receive reminders that their cars require regular battery, tire, oil, alignment, and air-conditioning checkups. The appointments secretary uses a computer to record the date of each customer's last visit and the services performed. Each month the computer scans customer files, alerting the secretary to customers who have not brought their vehicles to the shop for 6 months. Since these customers should receive reminders, the computer's printer produces notices that the secretary can mail to them.

Figure 1.1 A system converts input data into output information.

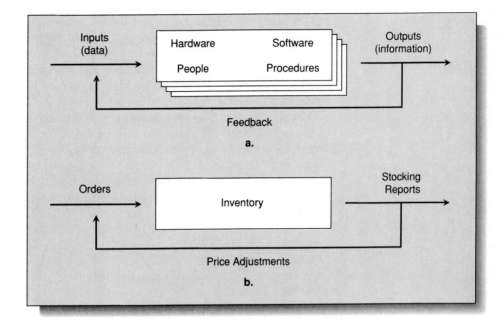

Mountain Motors' method of reminding customers exhibits all the traits of a system. In this example, the input is the record of the customer's last visit. The output is the reminder notice. This system includes four components: people (the appointments secretary, her boss, and customers), hardware (the computer), software (programs designed to extract the customers' names and addresses), and procedures (6-month rule for recall). Feedback occurs when customers fail to respond to their reminders and need a follow-up telephone call from the appointments secretary.

Advantages and Disadvantages of Business Systems

The ever-expanding use of computers by American businesses has led to a broadening of their applications and a stepped-up search for their tangible benefits. In the 1960s, when the use of computers in business was still fairly new, the information processing operation in a company was usually regarded as a "back-office" function and almost always had an accounting flavor. Now, companies are so thoroughly computerized that they can barely function on a day-to-day basis without their computer-based systems. Nearly all of the functional areas of businesses—accounting, logistics, human resources, sales, production, and others—are fully reliant on their information systems.

Today, a company would not consider trying to produce an electric utility bill for 4 million customers without a computer. Banks could not prepare checking account statements without information systems. The airlines would not function without SABRE, Apollo, and other reservation systems. Production lines in auto assembly plants at General Motors,

Ford, Chrysler, Honda, and others would halt without the computer systems that ensure that the right parts and subassemblies arrive at the right vehicle on the production line at the right time. Universities would take weeks instead of a few days to compile, print, and mail end-of-term grade reports. Indeed, even the tiniest organization can become paralyzed without a computer system. Information processing is definitely no longer a back-office operation.

A company's degree of dependence on its information systems is assessed by using the concept of information intensity. Some organizations are inherently low in information intensity, such as a sand and gravel quarry company, an auto body shop, or a logging company. Others are inherently high in information intensity, such as banks, airlines, insurance companies, and the Social Security Administration. In these kinds of organizations, the product or service is based on information itself. In a bank, the money on deposit by a customer does not exist as actual currency or in any other physical form, but simply as data stored on computer disks. Obviously, a loss of information systems and stored data in a bank would not only halt daily operation, but would cause a loss of assets that could lead to the failure of the bank.

Universities and consulting firms have conducted several studies of how long different types of organizations could continue to operate if they suddenly lost all computer support. These studies have uniformly found that the most information-intensive firms could survive only a couple of days, and the least information-intensive firms could survive only about ten days. These results vividly illustrate this heavy dependence of companies on their information systems. The depth of this computer dependence is enormous: for most business organizations of other than the smallest size or the lowest information intensity, loss of their information systems would mean corporate death!

Computer dependence has taken on still another dimension in recent years: organizations have discovered that information systems can serve as competitive weapons. American Hospital Supply Corporation found, for example, that by putting their order-entry terminals in the hospitals that buy from them, their business volume and market share increased markedly. It was simply easier for the hospital purchasing agents to order from AHS than from the competition. McKesson Corporation gave retail druggists small bar code readers that enabled them to inventory a store very quickly and then to order more shelf stock automatically by plugging the bar code reader into a telephone modem and dialing the McKesson computer. Naturally, druggists preferred to order from McKesson, and their profitability and market share improved greatly.

While many of the benefits of computerization are quite tangible—such as more accurate account sales forecasting and tighter financial controls, which positively affect the bottom line—other benefits of computerization are intangible. Greater worker enjoyment, relief from boring work, and improved customer satisfaction can also result from well-designed information systems and have an effect on the bottom line. In the future, we may see some surprising new uses for computers added to the list—perhaps in the areas of artificial intelligence, expert systems, and desktop publishing—especially when applications produce direct benefits. Regardless of specific future uses and benefits, businesses will surely continue to make computers an integral part of their daily activities.

Information Systems in Business

Before initiating a computerized system, the analyst must understand the firm's objectives and the way it has organized itself. Regardless of its type or size, an organization normally develops a goal, a structure that identifies its purpose, and lines of authority, responsibility, and accountability. In fact, some organizations have a printed and published corporate mission, which outlines the organization's purpose and philosophy. See Figure 1.2 for an example of a cellular communications company's goals and values. Today, many companies in America, large and small, public and private, have adopted corporate missions similar to the one in this figure.

Figure 1.2 Many of today's companies have published corporate missions, goals, and values. This one comes from McCaw Cellular Communications, Inc., an organization that provides cellular telephone and paging systems across the United States.

McCaw's Goals and Values

Our goal is to establish our company as the premier conveniences communications company in the world. To do this, we must earn the continuing loyalty of customers by providing them with network service systems that they acknowledge to be of superior value in a way that is profitable to us, thus creating long-term rewards for our shareholders and employees. Therefore, we will:

1. **Hire and develop great people** (it's the most important thing we do). Decentralize and empower them to make decisions, but balance this to take advantage of our strengths.

2. **Stay close to our customers.** Listen to them and care for them beyond their expectations.

3. **Provide superior network service systems** of the best quality, as defined by the customer.

4. **Pursue excellence in all we do.** It helps make customers happy and gives real meaning to life.

5. **Keep it simple.** Focus on results (satisfying customers), not on form (administrative processes). This will be especially important as we grow.

6. **Run lean** (but spend wisely to achieve our goals and values).

7. **Be humble.** It helps keep an open mind, a caring attitude, and respect for others.

8. **Be a team player.** Teams are more powerful than individuals.

9. **Employ good judgment.** It makes empowerment work.

10. **Keep our promises.** It builds precious credibility.

11. **Consider the future** (with an eye on the customer). Be flexible and open to new ideas and change. Be respectfully irreverent, questioning established ways, the "impossible," and things that conflict with our goals and values.

During the 1970s, the role of the data processing department expanded to provide a wider variety of services to the production, marketing, and sales departments. This growing role of computer-based information systems forced the department onto an equal footing with other departments, and it brought about a new title for many such departments, management information systems, often abbreviated MIS. The following example of an organization chart shows you how an MIS department fits into a typical company.

An **organization chart** depicts lines of authority and assigns responsibility between individuals and departments. Smaller companies with relatively few employees generally prefer a line structure. A line structure assigns overall responsibility and authority to managers at the top, and limited authority at each lower level. Managers on the same level enjoy equal authority and maintain their departments as independent units. Each vice president (V.P.) reports to the president. When a V.P. needs the assistance of another department, he or she may seek out their counterpart in that department.

As companies grow, they usually add a staff structure to line responsibilities. With a staff structure, certain departments, such as human resources (personnel) and management information systems (MIS) serve all the other departments. Figure 1.3 shows such a line and staff structure at Dynamic Systems, Inc. Notice how each vice president enjoys direct access to the service departments of management information systems and human resources. Each V.P. also retains direct communication with the president. However, the manager of management information systems and the human resources director work on a different level than that of the vice presidents, and their departments serve the V.P.s in an advisory capacity only. Departments that operate in a staff capacity usually perform a function for the entire organization, such as hiring new employees or providing computer services.

In some organizations, however, each department maintains its own MIS department. In this environment, users are responsible for their own software, operations, maintenance, and so on. The net effect is a shift in responsibility from a central group to the actual end-user group—those that care the most about their specific application. As the cost of hardware and software drops, this pattern will likely take over, with many different departments doing their own data processing in a decentralized fashion. The need for a central computer may not disappear, but its function may change from the sole processor of an organization's data to a central depository for all the organization's data. In this new role, the central computer may serve as a large corporate database server in which individual departments can fetch and compare their performance with the organization as a whole.

Analysts and Users

The success of systems projects depends on careful planning and thoughtful preparation, a job that falls on the shoulders of the **systems analyst**. The systems analyst is the person (or persons) who guides the analysis, design, implementation, and maintenance of a given system. In performing these four tasks, the analyst must always match the information system objectives with the goals of the organization.

Most systems analysts are college graduates with majors in computer science, engineering, business administration, information systems, or economics. Many analysts have

Figure 1.3 Line and staff organization chart for Dynamic Systems, Inc. The two staff departments are management information systems (MIS) and human resources (personnel).

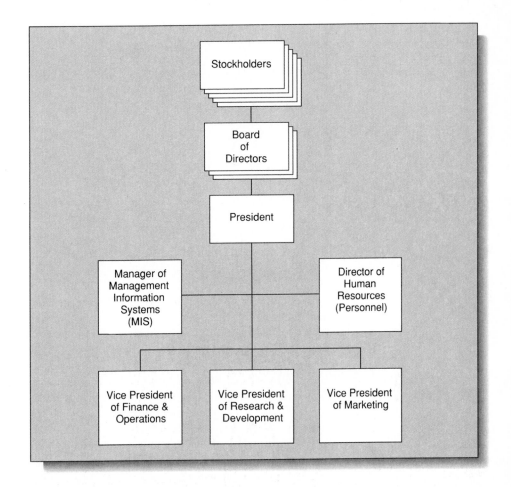

an advanced degree, such as an M.B.A. (Masters of Business Administration). Most analysts have worked for a few years as applications programmers before turning to analysis. All analysts must possess a basic understanding of the organization's business functions.

The career path for analysts may lead to supervision or management. Analysts must have the ability to listen; assess situations and draw conclusions; possess a strong general business orientation; speak and write effectively; prepare clear, concise reports; and work well with others to gain their respect and confidence.

Regardless of a business's particular structure, its MIS department will exhibit an internal structure of its own. In Figure 1.4, we see a typical organization chart for an MIS department. Although not the only structure in use today, it does reflect the common practice of separating the various MIS processing functions into small teams (programming, operations, documentation, and quality assurance)—each with specific charges. New to many organizations is the quality assurance team. People working on this team certify that the software works as intended, hence ensuring that the software meets stringent tests for accuracy before releasing it for customer use.

Figure 1.4 Organization chart for the management information systems (MIS) department at U.S. Computer Systems.

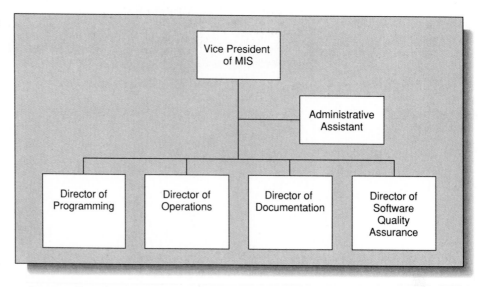

In some firms, an analyst may officially serve the MIS department, but also assume a staff responsibility with respect to a particular department of the organization. In a small West Coast county office of education, for example, each analyst reports in a line fashion to the MIS manager, but also takes responsibility for one specific area, such as elementary schools, middle schools, or high schools.

If a systems analyst comes into contact with many departments and subdepartments throughout a company, the analyst must thoroughly understand the company's organization and how the MIS department fits into the overall structure. Clear lines of authority must exist for the analyst to deal effectively with individuals at all levels in all areas. In the morning, an analyst may work with clerical employees, examining their requirements for increased productivity by speeding the data-entry process for payroll; a half-hour later, the same analyst may talk with the vice president of sales, analyzing the need for more detailed reports showing sales by individual, territory, and region. A week later, the analyst may present both needs to the company president. Periodically, the analyst may talk to customers from all over the world about their specific needs or errors that they have found. From the analyst's perspective, the clerical workers, vice president, president, and customers are all **users** of new or existing systems.

When designing any new system, the analyst must remain acutely sensitive to user needs. From interviews with users, the analyst acquires the first-hand knowledge necessary to achieve the goal of a successful system, but such interviews require a great deal of tact because users or employees may feel threatened by the analyst's appearance in their offices. They may even fear losing their jobs. A successful analyst must remain aware of company needs, but he or she must also deal with a wide range of nontechnical issues. Nothing can replace rapport with users, and good analysts strive to establish this rapport from the very beginning.

Oakes Crafts, which suffered a serious morale problem when it automated its organization, illustrates this point. Makers of innovative modern furniture, everyone at Oakes took great pride in old-fashioned hand-made quality, and Dorothy, the bookkeeper, loved

PERSONAL TRAITS OF AN ANALYST

Ben Morjig nervously waited for his interview for a position as a systems analyst with General Products, an international manufacturing conglomerate. Ben received his bachelor of science degree in mathematics in June two years ago from Collingsworth College. He easily passed all his mathematics, physics, chemistry, history courses and three programming classes. These past two years, Ben had spent in traveling around Europe, treating himself to an extended vacation after four years at Collingsworth. Ben felt confident that his background and international travel would impress the recruiter. He also felt that because he had put himself through college by working alone nights and weekends in an electronics store, potential employers would realize that he not only had an affinity for technology, but was a motivated, hard-working individual. When he heard his name called, Ben walked into the room where the General Products representative, Debbie Ferris, waited.

The interview seemed to go smoothly and Ben was pleased with his performance. Anticipating questions even before Ms. Ferris asked them, he had taken the initiative throughout. Every step of the way, he asked what General Products could do for him. Understandably, the rejection letter he received a few days later shocked him. What had gone wrong? Had he made some technical blunders? Had he said something to irritate the interviewer? Hoping to learn from any mistakes he might have made, Ben called General Products to find out why he had failed. Unable to recall his interviewer's name, Ben asked for General Products' human resources department, which located his file and put him in touch with Ms. Ferris. From his conversation with her and from later talks with a classmate who eventually got the job, Ben realized where he'd gone wrong.

First, an analyst must enjoy working with people. Ben's description of his sales job indicated that he liked to work by himself, with little or no supervision. Also, an analyst must communicate with people who do not understand, and perhaps even fear, computers, but Ben's technical lingo had gone right over Ms. Ferris's head. General Products' analysts, she felt, are agents of change, and since change threatens most people, an analyst must remain sensitive to such emotions and demonstrate how easy new systems are to use and how they can improve the user's job performance. Ben, she thought, saw computers as ivory-towered gadgets that only the initiated could really understand and use. Unfortunately, Ben had tried so hard to impress her with his mastery of computer wizardry that he failed to ask one penetrating question about General Products. And he continued

to seek out what General could do for him and not what Ben could do to help General Products, a fatal mistake.

Effective communication skills include listening, and Ms. Ferris rated Ben as a talker, not a listener. Analysts must pick out subtle nuances in how they talk with users, who may downplay a vital concern or overdramatize a trivial one. A good listener means letting the user answer questions rather than leading the user to the answer that the analyst expects. Listening means taking the time to hear the users' complaints and empathize with them. Ben's lack of listening skills made Ms. Ferris feel that he couldn't function in a team environment, working with business people where the "big picture" was very important.

When Ben pointed out his strong technical background and his programming skills in BASIC, Pascal, and Modula-2, Ms. Ferris suggested that an analyst could get as much mileage out of a degree in business or economics as one in computer science or information systems. Furthermore, she told Ben that the commercial business world used COBOL and C along with object-oriented languages and that Ben had not taken courses in these languages. She also related that General had moved into SQL (Structured Query Language) and relational databases and Ben had skipped these classes as well.

Unlike his classmate, Ben hadn't taken courses in English, speech, economics, composition, and communications while in college. Because analysts write many reports and documents and make frequent oral presentations to users, as well as to management and other personnel, they must learn to write and speak clearly. Poor writing and speaking skills can weaken the effectiveness of an otherwise knowledgeable and skillful analyst.

Analysts usually possess a bachelor's degree, and sometimes a master's degree, in business, computer science, information systems, or a related subject. Although a degree cannot guarantee a good analyst, the college experience provides a means of acquiring both technical and interpersonal skills.

Finally, an analyst must gain business experience, usually by working as a programmer. Ben had yet to program outside of a college environment. Not only does programming for an organization familiarize the analyst with information principles and procedures, it facilitates the building of interpersonal and problem-solving skills as well. So, although Ben was not hired at General Products, he learned a valuable lesson there: A successful analyst must be a technical expert, skillful communicator, businessperson, psychologist, and problem-solver, all wrapped up in one.

to hand-letter each bill. When that became impossible and the company decided to automate its billing system, Dorothy threatened to quit, saying, "It will ruin our image as a company that cares."

The analyst that Oakes hired, sympathizing with her point of view, showed Dorothy how a special printer could maintain the custom look but generate bills 10 times faster. As the months passed without any evidence of customer dissatisfaction, Dorothy's skepticism lessened, until she found herself actually enjoying the computer.

If an analyst fails to establish rapport and trust, the user may cause the system to fail through lack of employee cooperation. Had Oakes' owner so valued Dorothy that he felt willing to jeopardize profitability in order to keep her happy, no new system would have come on-line, and if the analyst hadn't patiently shown Dorothy the benefits of the printer, she might have willingly or unwittingly subverted the new billing system.

From this case, we can see that an analyst is much more than a technical expert. Rather, a good analyst is part problem-solver and part psychologist, a person who identifies with more than technical problems and proposes solutions that take more than technology into account. The best analysts communicate closely with users, write clear reports, and use state-of-the-art technical knowledge of hardware and software to forge a productive bond between users and software developers.

THE SYSTEMS LIFE CYCLE

Systems projects involve a sequence of phases called a **systems life cycle**, which consists of four distinct phases: analysis, design, implementation, and maintenance (Figure 1.5).

Systems analysis involves studying the ways an organization currently retrieves and processes data to produce information with the goal of determining how to make it work better. To do so, the systems analyst may develop alternative systems and evaluate each in terms of cost/benefit and feasibility.

Systems design requires the analyst to decide on the formats of the reports that the system will produce, define data-storage methods, plan how to collect and input data, and define the necessary programs. Design is the second phase in the systems life cycle.

Systems implementation includes actual programming, testing, training, and use of the new system. Upon completion of the system, the analyst, users, and management evaluate the system to ensure that it fulfills all its goals. Implementation is the third phase of the systems process.

Systems maintenance is the fourth and final phase in the systems life cycle. Systems maintenance includes repairing the system when errors are detected, enhancing the system when the user needs new functions, or changing the system to meet new laws or changes in the organization's purpose or goals.

Logical Versus Physical

Information systems develop first at the logical level, then at the physical level. This distinction between logical and physical system activities is one of the fundamental concepts

Figure 1.5 The systems process or life cycle in diagram form.

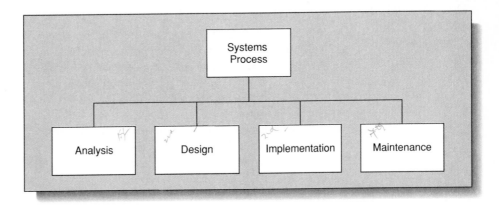

of the systems profession. In the logical phase, we decide on the number of the rooms, usage patterns, and so on. In the physical phase, we draw the blueprint showing bathrooms off corridors, electrical outlets, placement of doors, and other factors.

Within the logical and physical levels, we have two further subdivisions: current and new. The current logical represents how a current system logically operates, whereas the new logical is how the new system is supposed to operate. The current physical represents how the system in use really operates and the new physical is how a new system is supposed to operate.

Logical design means defining the system, but not actually building the system. The result of a logical system design effort is a complete set of system capabilities, but not a system that actually functions.

The physical part of system activity converts the logical definition into reality—including blueprints, programs, databases, and data collection screens—so that the system comes to life.

There are many reasons for separating the logical and physical phases of systems work, and we will cover them in detail later in the book. Before constructing a house, blueprints must be completed; before constructing a multi-million dollar information system, computer designs must be done. Putting logical design before physical design, and clearly separating the two, simply reflects a common-sense sequence of steps in constructing an information system—just as a successful house-builder would plan the house before building it.

Phase 1: Systems Analysis

Systems analysis includes both a preliminary and a detailed stage. During preliminary analysis, the analyst takes a quick look at what is wanted and whether its costs and benefits justify the perceived want. Detailed analysis includes an in-depth look at what is wanted and contains more refined cost and benefit studies.

The preliminary analysis stage begins when someone perceives a problem, desires a modification to an existing system, needs a repair to an existing system, or demands an entirely new system. A sales manager, for example, might feel that it takes too long for a sales report to reach her desk or that the sales reports lack vital information, such as sales by region or sales by salesperson. If so, the manager can submit a request for a study to the department responsible for the firm's information processing. If the study leads to a system that will better benefit the company, an information processing manager will approve the request and assign an analyst to conduct a preliminary investigation to determine exactly what the need involves and whether overall benefits will exceed the cost of fulfilling it. In this example, the person initiating the request for a preliminary analysis, the sales manager, is a user.

During this problem-definition stage, the analyst first concentrates on answering the question: "What is the problem?" The analyst answers this tough question by examining the current system (if there is one), identifying any problem areas, and considering how to best satisfy the user's perceived need.

Eventually, the analyst summarizes the information gleaned, including personnel requirements, basic costs, and potential benefits of the new system, in a formal report we will call the preliminary report. In the preliminary report, the analyst defines the problem, summarizes interviews and meetings with users and management, and presents reviews of relevant documents.

If the analyst urges further study, and management approves continuation of the project, detailed analysis commences. Detailed analysis expands the preliminary effort to include a complete analysis of all possible alternative solutions to the problem and a complete explanation of what appears to be the most practical solution.

The report that results from the detailed analysis becomes a document called a feasibility study or a requirements document. This report further expands the systems study, outlining in far greater detail:

- Problem definition
- Scope and objectives of a "new" system
- Alternative solutions
- Cost and benefit estimates derived from alternatives
- Potential organizational or policy changes
- Description of the new system's major outputs
- Recommended alternative and course of action

The requirements document (or feasibility study) becomes the centerpiece of a formal presentation to users and managers. If the president of the organization thinks the systems need is extremely important, he or she may play a role in final approval. However, in large organizations, vice presidents or their subordinates often handle routine systems decisions that primarily affect their departments. The sales vice president and controller, for example, may approve a new expense account reporting system for a company's sales staff.

Some feasibility studies lead to a decision to stop the systems study and continue with the existing system. Such decisions to terminate the systems process may come about when cost factors outweigh expected benefits, when management perceives that it will take too long to implement the new system, or when the overall goals of the organization change.

Phase 2: Systems Design

After decision-makers and the user have approved the requirements document, the analyst begins the second step in the systems life cycle: systems design. During this phase, the analyst schedules design activities, works with the user to determine the various data inputs to the system, plans how data will flow through the system, designs required outputs, and writes program specifications. Again, the analyst's activities focus on solving a user's problem in logical terms.

During this second step, analysts employ a variety of tools such as data flow diagrams, entity-relationship diagrams, data dictionaries (lists of terms and their definitions), and the Gantt chart (a graph depicting design events, personnel assignments, and time schedules).

A new tool, **prototyping software**, has evolved recently. Prototyping software allows the analyst to generate quickly an initial version of the system for the user to review. The prototyping software includes facilities that automatically generate screens, printed reports, and essential logic functions for collecting and storing data. Essentially, the user and the analyst use prototyping to experiment with the new system. Prototyping enables the analyst to move quickly from a preliminary logical design for the system to a physical realization of that design. After experimenting with the prototype, changes are made to the original logical design, which is then re-prototyped. This prototype-review-change process continues until the user and the analyst are both satisfied with the system's design.

The system's design converts the theoretical solution introduced by the feasibility study into a logical reality. During design, the analyst:

- Draws a model of the new system, using data flow and entity-relationship diagrams
- Devises formats for all the reports that the system will generate
- Develops a method for collecting and/or inputting data
- Defines the detailed data requirements with a data dictionary
- Writes program specifications
- Specifies control techniques for the system's outputs, database, and inputs
- Identifies and orders any hardware or software that the system will need

By the end of the logical design phase, the analyst has prepared complete **systems specifications** in the form of a detailed report, with step-by-step instructions that describe the proposed system. After the specifications are complete and approved by the users and management, the analyst writes a set of programming and database specifications that contain sufficient technical detail for programmers to begin physically creating the system. Program specifications include the description of the output, input, control, program definitions, and file designs for the system.

The program specifications, like the requirements document, undergo user and management review. If everyone finds the specifications satisfactory, they will probably permit implementation of the system.

Phase 3: Systems Implementation

During the implementation phase, the system actually takes physical shape. As in the other two stages, the analyst, his or her associates, and the user perform many tasks, including:

- Writing, testing, debugging, and documenting programs
- Converting data from the old to the new system
- Training the system's users
- Ordering and installing any new hardware required by the system
- Developing operating procedures for the computer center staff
- Establishing a maintenance procedure to repair and enhance the system
- Completing system documentation
- Evaluating the final system to make sure that it is fulfilling original needs and that it began operation on time and within budget

The analyst's involvement in each of these activities varies from organization to organization. For a small organization, the analyst may perform all the phases and tasks. In large organizations, specialists may work on different phases and tasks, such as training, ordering equipment, converting data from the old method to the new, or certifying the correctness of the system.

The implementation phase ends with an evaluation of the system after placing it into operation for a period of time. By then, most program errors will have shown up and most costs will have become clear. To make sure that the system performs as expected, however, a final system audit or review takes place. The system audit is a last check or review of a system to ensure that it meets design criteria. Evaluation forms the feedback part of the cycle that keeps implementation going as long as the system continues in operation.

Phase 4: Systems Maintenance

Like housework, dirty clothes, and weeds, systems work never seems to end; users almost always want changes or encounter problems. Thus, the systems maintenance part of the systems process deserves special attention. It is during systems maintenance that the analyst:

- Resolves necessary changes
- Corrects errors
- Enhances or modifies the system
- Assigns staff to perform maintenance activities
- Provides for scheduled maintenance

Most systems spend the bulk of their time in the maintenance phase, with constant enhancements and repairs. Studies show that more money is spent in this fourth phase than in all of the others combined. Writing systems that require as little maintenance as possible is one of the primary goals—as well as one of the benefits—of today's modern methodology of software development.

THE SYSTEMS DEVELOPMENT LIFE CYCLE: AN EXAMPLE

Returning to Mountain Motors, the auto-care group that uses a computer to prepare reminder notices for customers, let's suppose the company's controller decides she needs more timely information about the company's cash position. Perhaps, she thinks, a weekly report on all unpaid invoices or bills would show her which invoices require immediate attention, thus enabling her to determine the amount of cash needed at any given time. Accountants call this type of planning cash flow analysis or cash flow projection.

The controller could describe her need for a cash flow projection to a systems analyst, thus initiating the systems cycle. Part of her needs statement is the rationale for the report. In this case, she says that the report will allow her to see, in summary form, the changes in financial position of Mountain Motors over the time period of the report. Also, she can tell where Mountain acquired financial resources and where it spent them.

The analyst studies the company's general ledger system carefully to discover how it currently produces balance sheets. Using experience as a guide, the analyst estimates the new costs Mountain Motors might incur in generating the desired cash flow report. With costs, benefits, and a schedule estimated, the user (in this case, the controller) can decide whether or not to continue the systems study.

If she asks the analyst to proceed with design, the analyst focuses on a format for the contents of the cash flow report. (To see a sample cash flow report, refer to Figure 1.6.) The controller may suggest changes in the format, such as adding a line for totals by cash flow area and identifying it with the caption "NET CASH FLOWS, OPERATIONS." The analyst modifies the report format to include the new specification and then resubmits the report for approval.

Once the user and the analyst agree on the report format, the analyst can determine where and how to acquire the pertinent data. Since Mountain Motors already uses a system to generate a balance sheet, some of the data already exists. However, this report requires a comparison of two balance sheets, which means that the system must store the prior balance sheet's data, as well as the current one.

Finally, the analyst determines whether to store and maintain data in conventional files or in a database management system. Since Mountain Motors uses a relational database manager for its computer-based payroll and accounts receivable systems, the analyst would probably choose the existing database manager as the data-storage method.

Some systems will require the analyst to specify additional hardware. Here, the analyst must determine exactly what to order, arrange for payment, and schedule delivery, installation, and operator training. Because Mountain Motors can use its existing hardware for the new report, it does not need additional hardware. Since their printer can print 10 pages per minute or 600 pages an hour, a small report such as this new one would consume little additional printer time.

Figure 1.6 The statement of cash flow summarizes the significant financial actions that took place during the period and identifies major changes in the financial position of the company. It does not show how the increase or decrease in cash and cash equivalents occurred.

June 21, 1993 8:21 A.M.	**Mountain Motors** Statement of Cash Flow		Page 1

CASH FLOWS, OPERATIONS:	June 92	June 91	Inc / (Dec)
Year-to-Date-Earnings	$10,975.75	$8,573.43	$2,402.30
Adjustments to Year-to-Date-Earnings:			
Trade Receivables	<16,662.02>	<14,777.96>	<1,884.06>
Employee Receivables	<100.00>	<200.00>	100.00
Inventory	<17,680.49>	<13,876.12>	<3,804.37>
Prepayments	<250.00>	<500.00>	250.00
Trade Payables	11,531.18	9,360.16	2,171.02
Inventory Purchase Receiving	257.35	230.14	27.21
Sales Tax Payable	1,943.57	1,629.67	313.90
Federal Tax Withheld	1,717.00	833.00	884.00
Social Security Withheld	722.16	350.35	371.81
Social Security, Employer	371.80	350.35	21.45
Federal Unemployment	96.10	44.10	52.00
State Tax Withheld	252.50	122.50	130.00
State Disability	91.88	45.08	46.80
State Unemployment	323.30	156.80	166.40
Other Employee Withholdings	250.00	150.00	100.00
Union Dues	50.50	24.90	25.60
NET CASH FLOWS, OPERATIONS			1,374.06
CASH FLOWS, FINANCE and INVESTING:			
Accum. Deprec., Equipment	4,360.08	2,170.04	2,190.04
Accum. Deprec., Buildings	4,999.92	2,716.62	2,83.30
Notes Payable, Mortgage	65,031.59	76,109.34	<11,077.75>
Retained Earnings	40,668.33	12,692.91	27,975.42
NET CASH FLOWS, FINANCING and INVESTING			21,371.01
Net Increase (Decrease) in CASH and CASH EQUIVALENTS			22,745.07
CASH and CASH EQUIVALENTS Beginning of the period			30,305.33
CASH and CASH EQUIVALENTS Current			$3,050.40
CASH and CASH EQUIVALENTS:			
Petty Cash			100.00
Cash in Checking			17,950.00
Cash in Savings			35,000.00
TOTAL CASH and CASH EQUIVALENTS			53,050.40

At last, the analyst can draft a "program specification," which is a narrative describing the main functions of the needed programs. The specifications will include a schedule for completing and testing the programs and directions for locating test data. Once the programs are running smoothly, the users begin training. Later, a final audit will help verify that the system works correctly.

After operating the system for awhile, the user may discover another application for the data, even though the system is now in a maintenance mode. For example, Mountain Motors may request that the report appear in a second way: alphabetically by adjustments to year-to-date-earnings. This would allow the user to quickly locate an adjustment category. Such an eventual by-product of a system may not surface during analysis or design, but with experience, a good analyst can learn to anticipate future needs and even plan for them.

This is an example of the system life cycle as it would unfold for a relatively small project in a small organization. At the other extreme, systems such as the United Airlines Revenue Accounting System require several years to complete, cost millions of dollars, involve hundreds of programmers and analysts, and consist of several million lines of computer code. In both small and enormous systems projects, though, the basic steps in the systems life cycle remain the same.

SYSTEMS DEVELOPMENT: THEN AND NOW

Then: Art and Chaos

Until recent years, the process of creating information systems was more like an art form than an exact science. This means that different analysts took different approaches to working with users, writing system specifications, and designing the details of system operation. Different programmers wrote program code to implement those systems in very different ways. The situation resulted not from any lack of professionalism or intelligence on the part of analysts and programmers, but simply because the information systems development process was new and young. Standardized techniques for system implementation were rare. Some companies managed to institute uniform approaches to systems work for their own internal use, but one company's approach still differed from another company's approach, and there were no industry-wide standards. The result was chaos.

This chaos had all kinds of disadvantages. One obvious disadvantage was that analysts could seldom fully understand each other's work. One systems analyst might choose to describe the features users desired in a particular system by writing up those "system requirements" in the form of normal English language paragraphs. Another analyst might represent the system's requirements using flowcharts containing symbols unknown to the first analyst. A third analyst might prefer to use a combination of segments of program code and logic diagrams to describe users' needs. The problem with using these different approaches is that, years later, an analyst who was given the job of adding some new capabilities to the system might not understand the system's documentation because it was in a format that he or she had never used. The result was that the system modification job would take much longer and cost more than it should because the analyst would first have to take the time to figure out how to read the documentation before he or she could learn how the system worked, and then apply the modification. Even worse, some system modification projects were undertaken with absolutely no system documentation because the company's procedures did not require that the analyst create documentation when the system was originally developed.

WORKING WITH TECHNOLOGY

THE DEVELOPERS OF STRUCTURED PROGRAMMING

The year is 1965, and Edgar Dijkstra has just set the programming world on its ear with the publication of his landmark paper, "Programming Considered as a Human Activity." Dijkstra said, "I am of the opinion that it is worthwhile to investigate to what extent the needs of Man and Machine go hand in hand and to see what techniques we can devise for the benefit of all of us." Just what was this obscure Dutchman proposing?

Clarity and simplicity were on Dijkstra's mind. He was the first to question how programs were written and to propose a more orderly approach. Up until this time, programs were mathematically elegant, but often looked like Einstein's blackboard by the time they were completed. Sometimes there were mistakes, or the programmers wanted part of the program to repeat. If so, programmers could patch up the code quickly (although sloppily) with the GO TO statement. Dijkstra wrote a letter to the editor of *Communications of the ACM* insisting, "The GO TO statement as it stands is just too primitive; it is too much an invitation to make a mess of one's program," and the critics howled with rage. People are still publishing articles today that argue the pros and cons of the GO TO statement.

Two Italians, C. Bohm and G. Jacopini, theoretically proved that the structured method could work in 1966; IBM tested it in "The Super-Programmer Project." Dr. Harlan Mills, as chief programmer, undertook to use it on a project that until then would have taken about 30 human years to complete; he finished it single-handedly in 6 human years.

This experience led to IBM's developing *The New York Times* information bank, utilizing the chief programmer team concept and structured techniques. The project was not only successful, but programmer productivity using structured methodology improved four to six times over previous approaches.

Oddly, most people were amazed. Apparently, programmers assumed they had to write mathematically elegant, complex programs to make the computer work efficiently. Now they were learning that it was far more efficient to organize, or structure, the programming to create greater reliability.

This was an exciting development. Now programmers could write programs that were very easy to understand and could be read from top to bottom, without any GO TO's branching back to some earlier point. Of course, many people wanted to convert old unstructured programs to structured ones, but they quickly encountered problems. A major stumbling block that had nothing to do with design was that the programming languages then in use did not allow writing programs that would eliminate GO TO statements. A great deal of code was rewritten, flexibility and maintenance suffered, and people realized that the proverbial cart was somewhat before the horse. In an article entitled "The Revolution in Programming," published in *Datamation* in 1973, Daniel D. McCracken wrote, "Structured programming is a major intellectual invention, one that will come to be ranked with the subroutine concept or even the stored program concept."

Another disadvantage to the chaotic early approaches to systems work was poor skill transfer. You might become a very proficient systems analyst after three years of experience with one company, then move to another company for a raise and a promotion. Your new company might use an entirely different technique for building systems, and many of the skills you learned in those first three years would not transfer to your new job. Besides your frustration and embarrassment, this situation would create very substantial tangible costs for your new employer because the company will have to absorb the costs of training you to use their particular systems techniques. Your productivity with the new company would be very low initially.

Nearly everyone involved in information systems work now agrees that the key to building systems successfully requires a full understanding of the organization's goals and each user's needs before any work is done to actually build (implement) that system. Similarly, it does not make sense to start building a house before completing its blueprints and getting those prints approved by the future owners of the house.

Yet many of the early techniques for systems building allowed exactly that sort of mistake to occur. Analysts would show users early drafts of a system's requirements specification, but often that specification was written or drawn in a way that users could not really understand. Because of frustration, embarrassment, and time pressure, users frequently approved a system specification that they did not fully understand. This is equivalent to your giving a contractor the OK to start building your house, even though you don't know how to read blueprints.

A related problem, which was referred to as "premature coding," is that analysts would often allow programmers to start writing program code for parts of a system while other parts were still in the design stage. The analogy here is starting construction on your garage before the blueprints for the garage were done.

The cost of premature coding was high because the part of the system that was coded early almost always required change later as the full details of the system design were completed. Changing code is much more difficult (and expensive) than changing the specifications for that code, just as it is harder to tear down and rebuild part of a garage than to change a few lines on the garage blueprints.

The accumulated upshot of all these problems was that the information system profession developed the unfortunate reputation of systems delivered late, over budget, and not meeting specifications. Some systems were so bad that they were never used. None of this is surprising given the disorganized and unsystematic way in which analysts and programmers have traditionally approached information system development. This lack of systematic standards created room for each analyst to develop a high degree of personal style in analyzing and designing systems, making systems work more of an art than a standardized, disciplined science.

In a few isolated companies, this approach worked well, but these instances were rare. While systems as art may seem positive from a personal satisfaction and worker motivation point of view, it is generally unsatisfactory, resulting in chaos throughout the industry. Fortunately, a new, standardized, well-ordered scheme for system development has recently emerged. This approach relies on a combination of highly specific and detailed system development methodologies, Computer-Aided Software Engineering (CASE) tools, and a drastic modification of the traditional life cycle.

Now: Structured Methodologies and CASE

A **structured methodology** is really nothing more than a recipe for how to build an information system. But it is a highly detailed, very carefully conceived recipe to ensure that the almost overwhelming number of steps taken to build a system are accomplished in the proper sequence, done in the correct way, and properly documented. No steps should be omitted. Additionally, the methodology ensures user involvement in the design of a system, guarantees the generation of system specifications using standardized pictorial products that users can read and understand, and tries to prevent premature coding.

In its physical form, a structured methodology is simply a big book. One such methodology, Method/1 from Andersen Consulting, resides in a series of 3-ring binders that occupy several feet of shelf space. STRADIS is the system development methodology from McDonnell Douglas Automation (MCAUTO) Inc. It, too, consists of detailed guidance published in several binders. Some methodologies, such as Method/1 and STRADIS, are proprietary and "licensed" from the company that owns the methodology. Other methodologies are created internally by companies for their own use and are generally not available to other organizations. Regardless of their source or ownership, however, most methodologies are quite similar, and lead to standardization and order in system development, rather than the chaos and confusion of the past.

Various software development methodologies have existed since the early 1970s. For nearly 25 years, many people in the information system (IS) profession have recognized the benefits methodologies over the art-and-craft approach to systems work. Structured methodologies were not widely used, though, because of their high "paperwork intensity." Realistically, using the methodologies required that systems analysts fill out a lot of forms—even hundreds of forms. These forms contain the detailed specifications for all the data elements, data record layouts, files, processing logic, screens and reports, and other design details. Additionally, analysts drew numerous specification diagrams by hand. All of this paperwork was so time-consuming that most analysts quickly gave up on formal methodologies and reverted to familiar art-and-craft methods. While they recognized that better systems could result, the paperwork was simply not worth it.

This paperwork drudgery connected with structured methodologies is now replaced by **Computer-Aided Software Engineering (CASE)**. CASE tools are any kind of software that provides automated assistance with any of the activities connected with systems building. CASE tools now provide analysts with the ability to electronically draw and store specification diagrams, create and store data and processing specifications, and quickly lay out and store screen, report, and database designs. Additionally, some CASE tools actually write program code automatically. Like methodologies, many commercially available CASE tools support individual portions of the system life cycle. Specifics aside, however, CASE tools have empowered analysts to actually use methodologies on a practical, day-to-day basis, without the paperwork burden of the past. Analysts who are developing systems in this way are highly successful at overcoming past problems such as non-standardization, lack of documentation, low skill transference, poor system delivery, and underestimation of costs.

Some people view CASE tools as revolutionary, while others see them as evolutionary. This view is quite similar to those of architects in the middle 1980s as Computer-Aided Design (CAD) systems began to impact their businesses. Today you hardly ever enter an architect's office without seeing staff members designing new facilities on a computer that uses CAD software. As we move toward the twenty-first century, CASE software will impact analysts just as CAD impacted architects of the mid-1980s.

The use of CASE and structured methodology offers advantages and disadvantages but, as the following list suggests, the former outweigh the latter. The advantages of CASE and structured methodology include:

- Lower long-term costs
- Improved system reliability, with longer gaps between system failures
- More easily enhanced or improved systems
- More flexible systems, allowing for a wider variety of applications
- Greater user satisfaction as a result of earlier and more constant user involvement
- Lower maintenance costs
- Reduced likelihood of errors in analysis, design, and implementation
- More understandable and accessible programs and systems
- Wider involvement of more people, leading to easier acceptance of the new system

The disadvantages include:

- More time spent analyzing and designing the system, resulting in additional up-front design costs
- Need for retraining of analysts and programmers who have not practiced the new approach

One of the keys to the modern approach is to keep a clear split between the logical and physical phases of design. Remember that the logical specifications of a system are its blueprint. It is not until the system begins to take on physical reality (programs written, databases created and loaded, and screens developed) that has the project enters the physical stage.

There are two advantages to doing a complete logical design before moving on to the physical stage: business focus and ease of modification. When users and analysts work together on developing the logical specifications for a system, they can concentrate solely on the business aspects of that system because—at this stage—there is no need to worry about the technical details. They are concerned with designing the system so it will minimize operating costs, provide faster service to customers, reduce inventory carrying costs, schedule labor more efficiently, or whatever else the system is intended to do, while maintaining a business focus rather than a computer focus.

Logical design also promotes ease of modification because the system is first designed entirely in the abstract, either on paper or electronically with the support of a CASE tool. In this form, the system is easily modified and re-modified until it is satisfactory to the users. This iterative refinement process is often done with the analyst and user seated at a personal computer that is running a CASE software tool on a personal computer.

One of the important activities going on during the development of the logical blueprint for a system is **modeling**. Modeling means creating an abstract or non-real depiction of aspects of the system that will become real when the system is physically built. Two kinds of modeling are done concurrently—process modeling and data modeling—and both result in a set of graphic diagrams that become part of the system's logical specification document.

Process modeling involves figuring out how the data will flow within the system and what steps to perform on the data; in fact, this is the origin of the term *data processing*. Process models are drawn in the form of a set of Data Flow Diagrams (DFDs), and these are later used to guide the creation of the processing logic when the system is physically created.

In data modeling, the types of data that the system produces and uses and the relationships among those types of data are captured in the form of a set of Entity-Relationship Diagrams (ERDs). The ERDs then subsequently guide the creation of the database when the system is physically created. The DFDs and ERDs are central to the logical system specification. When DFDs and ERDs are combined with screen and report layouts, details about the data (the data dictionary), and a few other miscellaneous items, a system's logical specification package is complete. For a small system, this may be only about a 30-page document. For a large one, it could amount to hundreds of pages.

The users and analysts then work with the model, making successive modifications until they are satisfied with the design. Modeling usually results in better quality systems because it is much easier for users to work with a semi-functioning version of a system than with the completely abstract system depicted in logical specification documents.

CASE, prototyping, and modeling software may result in some dramatic changes in software development. Because users can more easily operate these software tools, some end-user development may emerge. While this thought frightens many practicing analysts, the new tools bring users closer to the software development process and will challenge analysts to develop even better systems.

A PERSPECTIVE ON CASE

The use of CASE tools by software developers is expanding at a rapid rate. Five years ago, few CASE tools existed in the commercial marketplace, they were quite expensive, and they were unsophisticated. Now there are many tools, at far lower prices, and they come with graphical interfaces operating on a variety of different computer systems. Analysts of the 1990s must have a familiarity with CASE tools and this part of the chapter allows you to acquaint yourself with a CASE tool. As we move into other chapters in this book, you will see a variety of activities that use CASE tools. Before starting, you need to know some of the more practical aspects of a CASE tool. To prepare for this experience, please answer the following questions (your instructor may need to help you with some of them):

1. What is the name of the CASE tool you will use?
2. Who is the vendor (software developer) of the tool?
3. On what hardware platform does the CASE software system operate?
4. How much does the CASE software cost?
5. CASE tools support all or some of the events in the system life cycle (analysis, design, implementation, and maintenance). Which events does your CASE software support?
6. Most CASE tools advertise that they will allow an analyst to pick from many types of diagrams and methodologies. Which diagramming techniques does your CASE software support? Which methodologies does it cover?

7. Very new CASE software systems advertise that they will allow the analyst to use "Object-Oriented Design." Does your CASE software allow for this newer design discipline?

8. Does your CASE software use the Graphical User Interface (GUI) or does it use a text-based screen?

9. CASE software vendors push their products as easy-to-use, with such features as mouse support, color screens, and pop-up menus. What features does your CASE software have?

10. New to some CASE software systems are real-time analysis and object-oriented analysis and design. Does your CASE system support either of these methodologies?

SUMMARY

Systems surround us: homes have heating, air conditioning, and ventilation systems; our human bodies contain complex muscle, bone, digestive, and circulatory systems; and businesses use information systems to operate.

Most business organizations are now almost totally dependent for their survival on their computer-based information systems. The loss of stored data or a shutdown of their computer systems for more than a few days would create severe economic hardships. The organizations with the greatest computer-dependence are those with the highest degree of information intensity—banks, insurance companies, news services, and companies where information governs the primary product or service that they provide. Beyond the importance of systems to everyday cost-effective operation, organizations have found ways to use information systems to achieve significant advantages over their competition, resulting in market share gains and custom-oriented product differentiation.

Management information system departments are staff functions in most organizations, meaning that they are expected to provide support to the line departments who are responsible for the actual production and delivery of the firm's goods and services. While many MIS departments used to report to the senior financial officer in an organization, they are increasingly becoming autonomous staff organizations that are headed by their own MIS vice president.

The systems analyst plays the role of a problem-solver and works much like an architect. An analyst begins the system life cycle by interviewing the user for a preliminary study. Having defined user needs, the analyst draws up tentative plans for user approval, which—if granted—leads to a detailed study that results in system specifications. The analyst, along with the programmer construct the system; the user learns it and runs it for awhile, evaluating it along the way.

Systems are designed logically first. Once the logical design is complete and is fully approved, the physical design starts. A logical design is like the system blueprint and the physical stage is when the system actually takes on physical reality. Process models in the form of Data Flow Diagrams (DFDs) and data models in the form of Entity-Relationship Diagrams (ERDs) are important components of the logical system specifications. Large and small systems projects adhere to the basic four phases of the system life cycle: analysis, design, implementation, and maintenance. With prototyping software, variations in the life cycle occur because the analyst can implement a small system, collect user comments, then return to the design stage and modify the system quickly. With prototyping, the system life cycle is more circular than linear.

Systems implementation is traditionally a difficult activity with a poor track record in terms of delivering systems on time and within budget. This is due to the lack of a uniform approach to systems work. Analysts were left free to exercise personal preference and personal style when designing systems, and the result was relative chaos in the systems industry. Now, Computer-Aided Software Engineering (CASE) tools—coupled with structured system development methodologies—are making it possible to develop a uniform, disciplined approach to systems building, and to achieve dramatic improvements in on-time, on-budget delivery. These sophisticated technical tools are coupled with a need for analysts to skillfully manage interpersonal relationships with sensitivity. They must communicate clearly when speaking and writing. Analysts should also keep up-to-date on ever-changing technical concepts and tools and understand their impacts on users.

KEY TERMS

System	Systems design
Data	Systems implementation
Information	Systems maintenance
Organization chart	Prototyping software
Systems analyst	Systems specification
Users	Structured methodology
Systems life cycle	Computer-Aided Software Engineering (CASE)
Systems analysis	Modeling

QUESTIONS FOR REVIEW AND DISCUSSION

Questions appear in four categories. Questions in the Review category ask you to study the goals of the chapter. You can find the answers to the Discussion questions in this chapter. The Application questions require you to apply the material presented here, whereas the Research questions necessitate investigation or inquiry.

Review Questions

1. What does the word *system* mean? Give at least three examples of common systems.
2. How do computer-based information systems benefit businesses?
3. What is the relationship between the design of an organization and the design of the business systems that support that organization?
4. What are the job roles of users and analysts as they relate to business systems?
5. Why is it necessary to design a "logical" version of a system before designing the "physical" version?
6. What are the four primary steps in the system life cycle?
7. How have business systems traditionally developed?
8. What are four advantages of the CASE method of software development?

9. Why are modern methods for developing business systems better than the old?

10. What communications skills are important to analysts?

Discussion Questions

1. Some new systems will require new equipment. Does the analyst select new equipment during the analysis, design, or implementation phase?

2. Can an analyst design a system without performing a systems study? Explain.

3. Identify the characteristics of a good analyst.

4. Why would the study of psychology help a systems analyst?

5. Some MIS departments are attached to individual departments. Sketch an organization chart for such a structure. What is the name of such a structure?

6. List three disadvantages of structured methodology.

Application Questions

1. List six systems that currently affect your life.

2. Does an analyst need to be able to program a computer?

3. In a systems design, why does an analyst sketch report formats before determining data-collection requirements?

4. Why are users sometimes reluctant to talk with a systems analyst?

5. Identify whether the *MIS manager* or the *analyst* is responsible for each of the following functions:
   a. Budgeting
   b. Training users
   c. Hiring new staff
   d. Designing files
   e. Supervising operations
   f. Specifying equipment
   g. Dealing with company politics
   h. Managing a systems project
   i. Overseeing a group of systems projects
   j. Developing long-range goals, objectives, and plans

Research Questions

1. Systems designed in the 1960s and early 1970s were mostly batch. Today's systems are on-line. How do these two types of systems differ?

2. What other names can you find for "Management Information Services"?

3. Would you say that the educational requirements for an analyst are greater than for those of a programmer? If they are, why?

4. Find the organization chart for the college you are attending or the firm for which you are working. Is the MIS department line or staff?

5. From a recent issue of *Computerworld* or a similar computer magazine, find recruiting advertisements for analysts. What skills and/or experiences are required?

I n 1992, Cindy and Mark Stensaas decided to start their own business. Their three children had completed college and Cindy and Mark were tired of working for someone else. After many months of examining potential business opportunities, they decided that the videotape rental business would represent an excellent opportunity for the 1990s. One of the hardest parts of their investigation was finding a name for their store. After many hours of talking and research, they settled on "View Video, the Movie People."

Some of their research revealed that video cassette recorders were found in over 50% of all American homes, that people were tiring of network television, that the three main networks (CBS, NBC, and ABC) were facing stiff competition, and that cable television plus satellite antennas were making vast inroads into traditional viewing habits. They felt that many of the stores were missing opportunities for "side" business when they didn't sell popcorn, sodas, blank videotapes, posters, and other video-related items. They knew how frustrating it was to find just the right movie to rent and wanted to provide their customers with some assistance when selecting a video.

With almost $50,000 to invest, the Stensaases decided they couldn't afford a "superstore" with over 5,000 titles. Instead, they wanted to concentrate on having solid titles, fast check-out and return services, and a locator service to find tapes. The Stensaases contacted a videotape distributor to confirm titles, prices, availability, and delivery. They eventually selected over 3,000 titles for their initial stock. At the same time, they remembered the adage that location is important for retailers and successfully sought out just such a site. They rented a 2,500 square foot retail space in a new shopping center with lots of parking and good lighting.

As Cindy and Mark made their business plans, they visited many of their competitors' stores, watching how competing businesses operated and how customers located tapes. They discovered that video rentals in their northern New York community were done manually. Because of their intended focus on quality and efficient customer service, the Stensaases knew that they would need to computerize their operations similar to the video rental stores in some large cities. In their discussions, the Stensaases focused on how to help customers and make customers happy so that they would come back to rent another tape. This customer orientation was reinforced weekly as they read the Tom Peters column, *On Excellence*, in the newspaper and Cindy and Mark agreed with his philosophy that the customer is the key player in successful operations.

The Stensaases found that regardless of geographic location, all stores established certain business policies and routines. They narrowed these practices to three general categories: customer memberships, videotape inventory, and tape rentals/returns.

Many stores required new customers to complete a membership application that collected the customer's personal data as well as some credit history. Customers completed the application, submitted it for approval, and then were eligible to check out tapes. The Stensaases decided that the customer data was a valuable resource, one that could benefit them in the future. They dreamed that the computer could keep track of each customer's viewing preferences and that this data could help them advertise their store and its rental tapes.

Their inventory of videotapes represented Cindy and Mark's biggest financial investment. Since tapes cost between $10 to $40 each, the money spent here was a major financial commitment. With their cash investment, a local bank made them a small business loan to help them finance their original tape purchases.

During their visits to competitors' stores, Mark and Cindy noticed two trends: bar codes on tapes and placing the actual tape in the display cases. Bar codes afforded easy identification on check-out and return since the clerk only had to scan the bar code, minimizing manual work and speeding the customer through the check-out process. A side benefit from this system was reducing the number of clerks who had to work the counter, resulting in a cost savings.

Placing the real tape in the display cases, instead of warehousing them behind a counter, also seemed like a good decision. In her research, Cindy discovered two distinct operational methods for tape checkouts. Some stores placed empty boxes on the shelves. When the customer brought the empty box to the counter, the clerk had to retrieve the real tape from some shelves behind the service area. While retrieving the tape, the customer had to wait for the clerk to return with the actual tape.

The other type of store placed the real tape out in the customer area. When a tape was brought to the counter, the clerk passed it over an electronic device that deactivated a label in the tape's case. This electronic process resembled those used in the local library. When the tape was returned, the clerk passed it over another electronic device that armed the label so that if it left the store while activated, an alarm would sound, telling the clerks that someone was taking a tape without permission.

Using technology was the better solution in Cindy's mind. The customer was never left waiting alone at the counter, only one set of tapes needed filing, shelf space was minimized, and check-out—as well as check-in—would take a shorter amount of time. The tradeoff to this approach was the experience of the electronic label inside of every tape cassette and the turn-on/turn-off device required at the counter.

Cindy and Mark knew that their computerized store could keep track of each tape's rental history and tell them which tapes were renting frequently, which ones were hardly ever rented, the popular times and days that their store was busy or idle, and which customers rented frequently or not at all. They saw all this data as a wealth of information to help them better manage their business. It could also help them win customers away from the store they were now using and into the Stensaases' store.

Working With View Video

The fictional videotape rental business of the Stensaases' is a case study we will follow in all the other chapters in this book. Although imaginary, it will illustrate all the facets of the systems process, starting with an anticipated use of the computer to help an organization solve problems.

Part of the systems process is getting involved and understanding the problem that needs solving. We suggest that you stop by your neighborhood video rental store and take a few minutes to notice how it operates.

Case Study Exercises

1. Does the video rental store use a computer? If so, which one?
2. Are the actual cassettes kept behind the counter or out on shelves in the display area?
3. Is the store customer-oriented or did you feel neglected?
4. Do you have to pay a membership fee to join?
5. Is there an application form to fill out before you can check out a tape? If so, try to get a copy.
6. Do they have a locator service to help you find your favorite Mel Gibson movie?

As we learn more about View Video, we will ask you to return to your neighborhood store and research some common facts. We hope that this active involvement will bring the systems process alive for you and also will help you discover the people-oriented side of systems analysis.

Modeling Tools for the Systems Analyst

After reading this chapter, you should be able to do the following:

- Explain the purpose of modeling
- List the symbols used to draw data flow diagrams
- Read a data flow diagram
- Draw simple data flow diagrams
- Differentiate between a context and a leveled data flow diagram
- State at least five rules for drawing a data flow diagram
- Draw data models using entity-relationship diagrams
- Explain the difference between an entity and an instance type of an entity
- Explain the benefits of modeling in terms of its impact on user involvement and on future system development

PREVIEW

Imagine buying a bicycle so complicated to assemble that only the original designer could put it together or fix it for you. Such a bicycle might represent a unique engineering feat, but it would not provide a very practical form of daily transportation. Yet, until recently, this sort of situation plagued the field of systems analysis and design. With few widely accepted guidelines to follow, each analyst brought his or her own set of unique skills to the development of computer-based systems, and, as a result, many of them bore one analyst's personal stamp. A complex accounting system may contain hundreds of thousands of lines of code in its program, making it difficult for anyone besides the original developer to fix or modify it. Worse, even the person or persons who originally designed the system may not clearly recall all of its intricacies a few months later.

Chapter 1 covered the systems process in general. Before we proceed to a thorough discussion of analysis, we will pause in this chapter to learn how to use two tools that help analysts understand the requirements of a system. The two tools—data model and data flow diagrams—allow the analyst to develop a logical model of the system, which assists in standardizing the systems process.

Just as an architect may use a drafting board, ruler, and drawing instruments to create plans, an analyst can employ certain tools to create systems. The data model and data flow diagram resemble an architect's blueprint. With them, the analyst, users, and management can grasp the overall logic of a system.

THE ROLE OF DATA IN BUSINESS

Data is slippery stuff. You can't weigh it, feel it, touch it, or take its temperature. Yet it is the lifeblood of organizations. The storage and processing of data is the fundamental underpinning of all activity in business organizations.

Data's slippery nature poses an incredibly frustrating problem for the analyst. Somehow, an analyst must come up with a detailed understanding of something intangible and nearly invisible. Every detail of the data storage and processing connected with, say, inventory maintenance, requires thorough understanding, fully depicted on paper, and absolutely 100% accuracy. This is not unlike asking a blindfolded engineer to draw a detailed blueprint for a jet engine. It sounds impossible.

Fortunately, there are techniques and tools associated with structured methodologies that bring the analyst's job into the realm of the possible. **Data modeling**, through the use of **entity-relationship diagrams (ERDs)**, and **process modeling**, through **data flow diagrams (DFDs)**, are techniques that help analysts study the information structure of a business function. Using these techniques, analysts can better understand the business function in detail. They can then capture that detail on paper in a standard graphical "language" that other analysts, users, and programmers can understand. Once the data and process models of a business function are captured on paper, they can guide the design of a new (or modified) computer-based system to support that business function.

To understand why data and process modeling are so central to the task of system-building, you must accept what will at first sound like a pretty far-fetched idea. The data structure associated with a particular business process is like an *act of nature*.

Think, for example, about order processing. To perform this business function, we must store data about

- Products for sale (their names, numbers, size, and weight)
- Prices of the products
- Customers that buy the products (their names, shipping addresses, billing addresses, credit ratings, and so forth)
- Accounts receivable to keep the financial details of the purchase (so that we will not ship a customer an order when they have a large unpaid balance from an earlier order)
- Shippers that will deliver the products (availability of trucks or rail cars, shipping charges for various distances, methods of shipment, and so on)

Processing an order involves checking to see if there is enough warehouse stock to fill the order. It also means figuring out the total price (including shipping), and checking the customer's credit and outstanding unpaid balance. We must still record the receivable so that we may bill the customer, check with the shipper to determine a probable delivery date, and then confirm with the customer that the order is acceptable and will arrive on a certain date.

The reason we say the data structure here is an "act of nature" is that this description of order processing applies just as well to General Electric in Cincinnati as to a small neighborhood bakery in Paris, France. General Electric may process orders for jet engines, while the baker takes a local restaurant's order for the morning's pastries, but they both go through almost the same standard steps—all that differs is the scale.

This universality of data structures applies to a great many organizational processes, regardless of whether they are payroll, production scheduling, fixed assets valuation, human resource management, or any of numerous others. Patient billing in a doctor's office is not very different for a small clinic in the Australian outback than for Massachusetts General Hospital. The scale and depth of detail differ, but the fundamentals are the same.

So what? This leads us to a fundamental principle of systems analysis: our job is to design computerized systems to fit with, and provide support for, these naturally occurring data structures—not the other way around. The data structures existed well ahead of the computer and the analyst, and will still exist after we are gone. While this horse-before-the-cart principle may seem overwhelmingly obvious, much of past systems practice has ignored this principle and instituted systems that either disregarded or tried to change these fundamental data structures.

Not surprisingly, many of these systems projects failed, caused dissatisfaction among users, and disrupted the normal operation of businesses. Now, modern methodologies and tools lead the analyst through the task of discovering these underlying data structures and modeling them on paper. The analyst then designs systems to support and complement, rather than conflict with, these business structures.

MODELING: A PORTRAIT OF THE BUSINESS

At first glance, the analyst's job seems simple: discover the underlying data and processes in the business function in question, capture them on paper in the form of models, then provide automation whenever it could cut costs or speed things up. The problem is that the data and processes are not at all visible or obvious. Thus, the analyst takes on the role of an archaeologist, carefully sifting through evidence and digging down through the layers to find treasures.

In some cases, the data and process models may lie obscured by layers of old, poorly designed information systems, confusing organizational and departmental structures, users who do not fully understand or can't explain their jobs, political hostilities that make them unwilling to talk fully or honestly, and a whole host of other contaminating influences. Through patience, personality, and skill, analysts can uncover the data and process structures that underlie a business function, and capture them on paper in the form of models. While this may seem like a bleak picture, it is realistic portrayal of what we need to do. Some analysts even relish working on these unconventional discovery tasks.

Although there is no prescribed sequence, many analysts prefer to perform data modeling before process modeling for two reasons. First, data structures are considered more static—less changeable over time—than process structures. Even though the finer details and rules of the process for determining a commercial customer's credit limit may change over time, the underlying data structure connected with credit evaluation will change only minimally over a period of years. Remember that the fundamental data connected with order processing has not changed for centuries. Hence, data models provide a firmer foundation for the systems analysis activity than process models.

The second reason for starting with data modeling is that it is a bit easier than process modeling. Users may not know the step-by-step details of how they perform a business function, but they usually can identify the data they store and maintain. In reality, the data and process modeling tasks are usually interleaved.

MODELING WITH DATA FLOW DIAGRAMS

Data flow diagrams (DFD) depict how data interact with a system. Some professionals call them data flow graphs, bubble charts, or Petri networks. Whatever they're called, data flow diagrams are extremely useful in modeling many aspects of a business function because they systematically subdivide a task into its basic parts, helping the analyst understand the system that they are trying to model.

What Are Accounts Receivable?

Accounts receivable, or AR as it is often known, tracks money owed to an organization for goods it has sold or services it has rendered, but for which it has not yet received payment. During any business day, an organization makes sales to their customers. Some sales may come in the form of cash, meaning the customer pays for the product or service at the time of purchase. Today, many sales are credit sales, meaning that the customer will pay for the sale at some future date.

When the organization sells the customer a product or service, the customer signs a receipt and the organization issues an invoice to record the transaction. The invoice identifies who made the purchase, a description of the items or services, the cost of the items or services, and the date payment is due. Invoices remain "unpaid" until the customer makes a payment.

At the end of the month (or on some other time schedule), the organization sends customers a statement (or bill) that lists each unpaid invoice, payments received, and the total balance due for the month. As American consumers, we see all types of bills every month, but the AR system is the same in countries all over the world.

To cancel their debt to the organization, the customer can make a payment. In America, most payments are made by check. In European countries, payments are often made to the post office or some other collection agency that forwards the money and data about the payment to the company.

To encourage their customers to pay early, an organization can offer a discount for early payment. If the customer does not pay on time, the organization can assess a finance charge based on a percentage of the past due amount. Finance charges are also shown on the statement.

AR provides one of the most important sources of financial data to the organization. A company that does not keep a close watch on the money that its customers owe and how much it expects to collect will not stay in business long.

To see an AR system in action, consider how a transaction flows through Mountain Motors. Customer Katie Baker purchases a pair of floor mats for $45.60. Instead of paying by check, cash, or credit card, Katie charges her purchase on her Mountain Motors charge account by signing an order form. Since the order form contains all data relevant to the sale, it becomes the invoice and can serve as the primary source of data for Mountain's accounts receivable.

At the end of the month, Mountain Motors sends Katie a statement or bill, showing the month's purchases, payments received during the month, any unpaid balances from previous months, finance charges, and the total balance due. When Katie sends a check for $45.60 to Mountain, the company reduces her balance by an identical amount.

Most AR systems track a variety of information, such as:

- Amount the customer has purchased so far this month, this year, and each of the previous 12 months, and the customer's highest balance
- Amount the customer currently owes
- Due date for each unpaid purchase
- Discount date and amount for each unpaid purchase

All of this information is available in a variety of reports that the owners/managers can use to follow the business transactions between the organization and their customers. Each report helps management better monitor the organization's operation and profitability.

Drawing the Context DFD

To see how we construct data flow diagrams, let's consider Mountain Motor's accounts receivable system. Our review of the AR system has given us a good understanding of the data and the processes involved, which is the right way to start modeling efforts.

Data flow diagrams use a variety of symbols to represent processes, data, and those who provide or receive data (Figure 2.1). A square represents a provider of data (**data source**) or receiver of data (**data sink**), such as a customer or management. The parallel lines depict a file of data or data store; for example, a file holding data about purchases made by various customers. Arrows show the flow of data between the other symbols. Circles stand for a process that converts or transforms data. We'll cover these symbols in more detail later in the chapter, but this should give you the basic idea.

This DFD notation is often referred to as "Yourdon/DeMarco" notation. Another widely used DFD notation is called "Gane/Sarson." The primary difference between Yourdon/DeMarco and Gane/Sarson DFDs is that the latter uses rectangles with rounded

Figure 2.1 A data flow diagram shows the flow of data through a system. It does not specify how we convert data from one form to another. The diagram uses only four symbols.

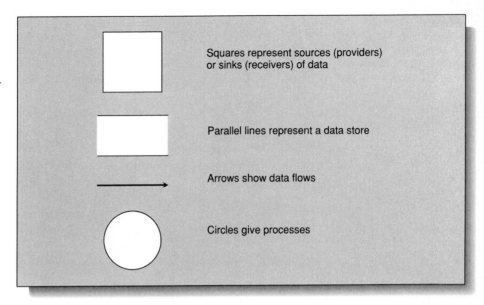

Squares represent sources (providers) or sinks (receivers) of data

Parallel lines represent a data store

Arrows show data flows

Circles give processes

corners for processes, rather than circles. Yourdon/DeMarco uses parallel lines for a data store while Gane/Sarson uses an open-ended rectangle.

Squares, parallel lines, and circles receive names. Arrows also receive names, except when they point to or from a data store file. Some data flow diagram authors also name the data flows to or from a data store, especially when the data flow uses only part of the data held by the data store.

We start all data flow diagrams by drawing the **context diagram**. This diagram displays "a big picture" of the system under scrutiny: this is the system, these are the players, this is how the players interact with the system. It shows the system at an overview level.

Understanding the system makes drawing the context diagram easier. First, we place a circle in the middle of the page, labeling it Accounts Receivable, as in Figure 2.2. Surrounding the AR circle, we use squares to depict the "players" in the AR system: customers and management. The squares either provide data (act as data sources) or receive data (act as data sinks) and are outside the system in question. We link both sources and sinks (squares) to the AR system circle with arrows that reveal the flow of data.

The customer square is involved with three data flows: customers sign sales receipts, receive bills, and pay their bills with checks. The management square is involved with two data flows: management receives data via the aging report and sends a request for billing back to the AR system via delinquent customers.

As Figure 2.2 illustrates, the context diagram defines the boundary of the system: It identifies the participants. The context diagram is a statement that says that this is all we are concerned with—nothing more and nothing less. This is a very important step in the concept of modeling because it specifies what the system under study is and what is its interface with other systems outside the model.

Notice the absence of files or data stores in the context diagram. At this point, we do not want to cloud our view of the general nature of the system with details. A correct context

Figure 2.2 A context data flow diagram. Most of us interact with an accounts receivable (AR) system on a regular basis. We receive bills from all types of businesses: credit card organizations, utility companies, gasoline companies, cable television companies, and so on.

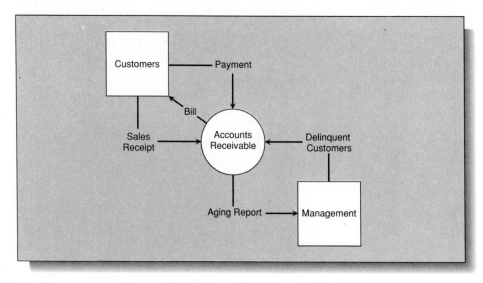

diagram is so simple that even people unfamiliar with computers or the technical uses of data flow diagrams can easily understand it. Thus, it can serve a management purpose— as well as a technical purpose—an especially valuable characteristic in a business setting.

Leveling DFDs

Once we have drawn our context diagram, we can begin breaking it down into its detailed parts. We call this **leveling**, which refers to dividing the high process diagram into many "levels" of finer detail. From Figure 2.2, we know that we must allow for five specific systems functions:

- Record a sales receipt
- Print a statement or bill
- Apply a customer payment check
- Print the aging report
- Identify delinquent customers

Each of these activities represents pieces of the big picture and a greater level of detail.

Faithful to the top-down approach of the structured methodology, we identify these activities with their own process circles, each with a single purpose. We can now draw a more detailed data flow diagram showing all five activities, as in Figure 2.3. At the center of it will sit two files, CUSTOMER-MASTER and PURCHASES-PAYMENTS. The first file keeps customer account numbers, addresses, telephone numbers, account balances, and other pertinent facts. The second file stores all the data about a purchase or payment made by a customer, the customer account number, date, amount, and so on.

The five activities we have defined will interact with the CUSTOMER-MASTER and the PURCHASES-PAYMENTS files. For example, when Mountain Motors receives a

Figure 2.3 The level 1 data flow diagram of accounts receivable expands into 5 subtasks. Each task performs one event in the accounts receivable billing system.

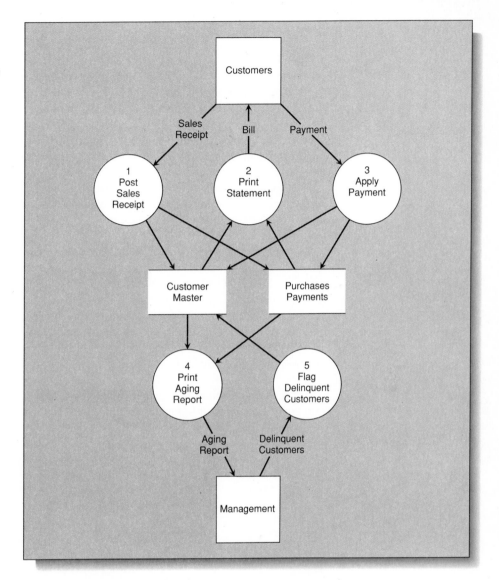

payment, the company will apply it to the customer's account and reduce the balance appropriately. The details of the payment are stored in PURCHASES-PAYMENTS so that we have these details when we print the statement. Similarly, printing the accounts receivable aging report requires looking up the amount each customer owes before printing this amount on paper. The system must adjust the customer's account balance when a debt is incurred or payed off, and it must contain the customer's name, address, and account balance in order to print a bill or record a check.

Level Numbering

Now we have gone beyond the context diagram, sometimes called the level 0 diagram in (as in Figure 2.2), to our first detailed breakdown of the system, the level 1 diagram. Each circle in the level 1 diagram receives a numeric label to help us keep track of its relationship to other tasks when we divide the system into further levels of detail.

In some less complex situations, such as printing the aging report in our example, we need not go beyond a level 1 diagram. Figure 2.4 shows that activity standing alone. Pulling it out and letting it stand on its own sometimes makes an activity a little easier to visualize and understand.

The AR aging report involves a simple list, as shown in Figure 2.5, which lists each customer number, name, telephone number, and debt. The last item is divided into the following categories: current, overdue 31–60 days, overdue 61–90 days, overdue over 90 days, and total owed.

Management receives the report for review and to identify delinquent customers. Look at customer number 000003 (Webb, S.), who made a purchase this month of $948.23, and owes Mountain $500.00 from over 90 days ago. Management might worry about this customer's payment record and take action to collect. On the other hand, customer 000234 (Lee, J.) has paid his bill regularly and requires no special action. Customer 000512 (Jacobs, T.), who owes $25.00 from last month, falls somewhere in between.

The grand total line at the bottom of the report tells management how well the company collects moneys owed to it. If too much debt appears in the overdue 90 days column, management must take steps to improve its cash flow by demanding payment, turning some accounts over to a collection agency, or cutting off further purchases. Obviously, this report provides essential information for running the business profitably.

Figure 2.4 The "Print Aging Report" activity is straightforward. We need to take data from the CUSTOMER-MASTER file and print them on paper.

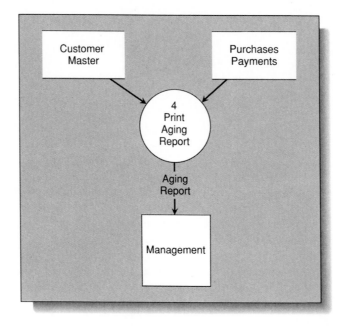

```
Dec. 14, 1993                          Mountain Motors                      Page:  1
8:05 A.M.                           Accounts Receivable
                                        Aging Report


=========    ==========    ===========    ========    ========    ========   ========    ========
Customer     Contact       Telephone       Current     31-60       61-90     Over-90       Total
Number       Person        Number                        days        days       days        owed
=========    ==========    ===========    ========    ========    ========   ========    ========
000003       Webb, S.      (213)665-3456    948.23       0.00        0.00     500.00     1,448.23
000234       Lee, J.       (213)789-2341     23.89       0.00        0.00       0.00        23.89
000512       Jacobs, T.    (213)678-9000    125.00      25.00        0.00       0.00       150.00
   *            *               *              *           *           *          *            *
   *            *               *              *           *           *          *            *
   *            *               *              *           *           *          *            *
   *            *               *              *           *           *          *            *
Grand Totals --------------------------->  8,907.23     456.00      245.00   1,200.89    10,809.12
```

Figure 2.5 The aging report tells management who owes how much money and how old the debts are.

While the "Print the Aging Report" activity is a relatively simple task, others are not, especially "Recording a Payment." When Mountain applies a payment to an account, it retires the oldest debts first. If customer 000003 (Webb) owes $500 for an over-90 day debt and sends Mountain a check for $1,100, the company applies $500 to the over-90 days debt and $600 to the current debt of $948.23. This reduces the current debt to $348.23. Webb now owes Mountain $348.23, which will appear as a 30-day old debt on the next statement. Upon examining this activity, we would want to break it down into a level 2 diagram, showing all the activities required to reduce the debt according to the amount of the payment.

When drawing a data flow diagram, we often discover later that we overlooked something. Data from the CUSTOMER-MASTER file can provide quite a range of different reports to aid management's decision-making. For example, management may want other reports besides the aging report, possibly one listing customers by number or name, one summarizing finance charges on unpaid amounts, or one providing sales histories. If so, we can modify the original context diagram, as shown in Figure 2.6, by adding an arrow to depict generation of the new reports.

The addition of the new data flow also requires modification of the level 1 diagram, to which we add a new circle to show the printing of management reports, as Figure 2.7 illustrates.

Since the new arrow represents many activities, we must now draw a level 2 diagram, as shown in Figure 2.8 (page 44). Each circle in this diagram represents a report that the system will produce. Notice the numbering process.

The new circle in Figure 2.7 received a label of 6. Subsequent circles in this figure will follow a similar pattern: 6.1, 6.2, 6.3, and 6.4. This numbering system shows the relationships that exist between parent circles and child circles. If we wished to level activity 6.2 one more time, we would label the new processes 6.2.1, 6.2.2, 6.2.3, and so on.

This handy labeling convention can go on indefinitely without losing track of ever smaller details and their relationships to the original picture. Remember that the context diagram starts at level number 0 and is the "parent" over all successive diagrams.

Figure 2.6 The new context account receivable diagram shows that management will receive a series of reports.

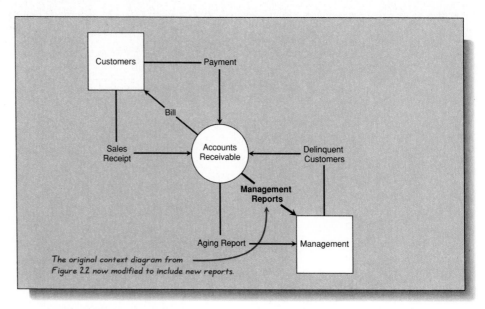

DFD Symbols and Rules

As you can see in our series of accounts receivable data flow diagrams, we draw them using four symbols. These four symbols ease the learning process for those unfamiliar with data flow diagrams and allow the analyst to hand draw the picture without any special templates.

We can use data flow diagrams to represent either an existing system or a proposed one. Representing the existing system with a DFD helps the analyst to understand it and even explain it to others. Recently, an analyst drew one for a small savings and loan deposit system. After seeing it, the vice president remarked that now he understood how their system really worked. No one could explain it to him until the analyst drew the DFD.

In a similar vein, we draw DFDs to represent what we want to have a new system logically do. Architects draw blueprints of their new designs, whereas analysts draw DFDs.

Our accounts receivable system had two squares: customers and management. In all DFDs, squares represent a person or persons, department, division, office, or agency associated with, but outside of the application. People can come from within the organization (management) or from outside the organization (customers).

Some applications require an interaction with another system outside the scope of the application under study. For example, a payroll system must send earnings and withholdings data to federal and state income tax agencies. What the tax agencies do with the data we send them is not a part of our system, but a part of theirs. This is not to say that we do not care what they do, just that it is out of the interest of this application.

The arrows we draw are **data flows**. An arrow represents a collection of data values passing between two data stores, processes, sources, or providers.

We have some very specific rules on data flow arrows. All data flows have a name, except those that point to or out of a data store (file). These type of arrows are the only ones that are not usually labeled since it is understood that we are writing data to the data store or reading data *from* the data store.

Figure 2.7 The new level 1 data flow diagram for the accounts receivable system.

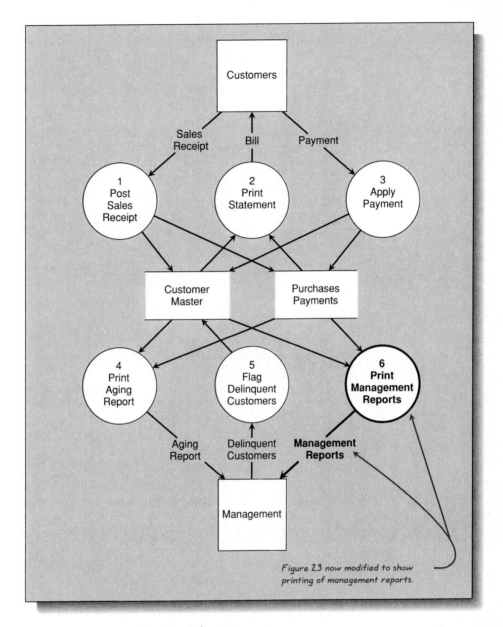

Figure 2.3 now modified to show printing of management reports.

All data flows have arrowheads telling which way the data are traveling. In this sense, the arrowhead indicates whether the data are input, output, or both.

We do not draw a data flow between providers or receivers of data (squares). This would indicate that we are sending the data from a provider to a receiver and that the data does not undergo any change; the receiver is simply getting a copy of the data. Data must pass between a process symbol. An arrow between two squares is totally external to our system and is of no interest.

Figure 2.8 A level 2 data flow diagram. The activity "Print Management Reports" is subdivided into four subactivities.

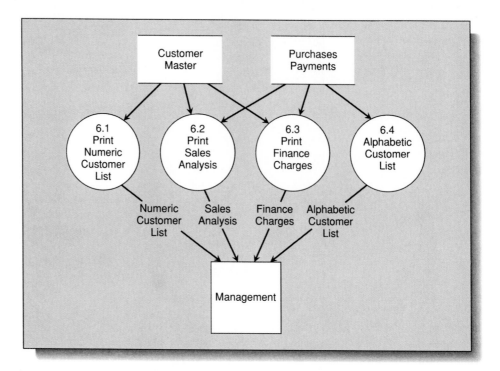

Data flow arrows are nouns (a person, place, thing, or quality). We are naming the data flow, not telling what we want to have happen to it (which is an action, or verb). Names should describe the flow of data in a few words, three or four at a maximum. Pick meaningful names (such as payment, bill, or paycheck) so that the reader can visualize in his or her mind what is traveling along the arrow.

We never connect two data stores (parallel lines) with an arrow. This would imply that data are copied between data stores and that is incorrect. It takes a process to make this happen.

The process symbol (circle) depicts the conversion of incoming data to outgoing data. Each process symbol is given a number, usually placed in the top center. Process names start with verbs that depict the action we want performed on the data. Pick strong verbs such as print, read, sort, compute, balance, post, and summarize (see Figure 2.7 for examples). Avoid weak verbs such as process, repeat, and perform.

A process at this stage in the systems analysis represents a "black box" that magically converts the data. We make no attempt to provide any details about how the data are converted, only that they somehow are changed. During the systems design stage, we determine how to make the conversion actually happen.

In most cases, data enter a process symbol from a data source or file and leave it for a data receiver (square) or file. It is legitimate to have a data flow connect two process symbols. These situations mean that the data are temporary and that one process is simply passing the data onto another process symbol.

A process symbol can have multiple "in" arrows and multiple "out" arrows. It is not correct to have a process symbol without an in arrow or an out arrow. This would imply that the process never receives any data or that it never produces any data. Processes convert data by definition.

Data stores represent collections of data. We usually think of these collections as files. Every data store has a name. Data flows pointing to and from the data store are not usually named. It is understood that the data store is the repository of data. If the arrow is named, the implication is that a selected part of the data is passing between the two data stores.

No two data stores in a data flow diagram can have the same name. Violation of this rule implies that two different data stores have the same data and one would overwrite the other.

Most data flow diagrams have data stores in the center, with process symbols surrounding them. On the outside, we have sources and sinks of data. Imagine a target with three circles. In the bull's eye (center) are the data stores. The next ring out contains the process symbols, and the outermost circle has sources and sinks. Data flows from sources and sinks through processes to stores. We get information back to users in the opposite way: data stores through processes to sources and sinks.

DRAWING DFDs WITH CASE

Have you visited an architect's office lately? If you have, you are aware of the revolution this profession has undergone in the last four years. If you haven't, you are in for a surprise. Gone are the drawing boards, T-squares, and plastic templates to draw special symbols. Computers are now the architect's most widely used tool. The software running on nearly every machine is CAD, which stands for computer-aided design. With this match of hardware and software, architects can draw buildings and other structures, plus lay out plumbing, electrical, heating, ventilating, and air conditioning faster than ever before. The CAD systems use full color, show the designs in 3-dimensional perspectives, collect specifications (such as window sizes and types), check drawings for correctness, and a host of other similar functions.

The CASE Revolution

This same revolution has taken place in the offices of America and the rest of the world. Computers using word processing software have replaced typewriters in many offices. Why? Increased productivity, higher quality documents, greater flexibility, and more enjoyable work are some of the reasons given.

What CAD is to architects, CASE is to software developers. The hardware/software match allows analysts to automate the development process with the same benefits that CAD provides to architects.

CASE software comes from many different vendors, operating on a variety of hardware platforms, and with various levels of functionality. Some CASE systems are best suited for analysis and design, called **upper CASE**, while other systems are tailored for implementation and maintenance, called **lower CASE** (Figure 2.9). Other CASE software systems try to cover all four phases of the systems process. Still others, such as Hewlett Packard's

Figure 2.9 The traditional structured systems life cycle as viewed from a CASE perspective.

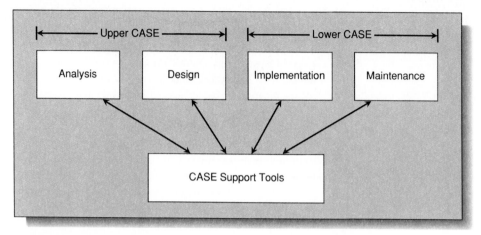

CASEdge, integrate other tools with a common interface, allowing data to pass between them with apparent ease. HP advertises that CASEdge helps software developers to understand and measure the development process better.

CASE Drawings

CASE software systems come in many "flavors," just like ice cream, but there are some consistencies among all of them. Every CASE software tool allows you to draw diagrams.

It is during the analysis phase of the Systems Life Cycle that the analyst uses a DFD and an ERD to capture and understand the objectives and requirements of the system. Designers level context diagrams to capture more details and technical aspects of any size system. Remember how we started with the idea of the need of an accounts receivable system and drew a picture of it. Next, we drew a leveled data flow diagram that showed subsystems for posting sales receipts, printing statements, applying payments, printing reports, and flagging delinquent customers. And for each of these subsystems, we can draw diagrams that give more details about them, such as showing primary inputs and outputs.

As you learn more about the CASE software at your disposal, you will find that it probably supports other types of diagrams (such as state transition or decision tables) and various styles within each diagram (such as Yourdon/DeMarco or Gane/Sarson for data flow diagrams). Most CASE systems support entity relationship diagrams.

Regardless of the type of diagram, your first step in learning your CASE software is to draw a diagram. In making your drawing, you will learn how the CASE software operates and how you perform some ancillary functions such as saving your diagram to a diskette, recalling your diagram from the diskette, and printing your diagram. You will also learn how to start up the software and how to exit it.

Graphical User Interfaces

Most CASE software has a graphical user interface with a mouse and pull-down or pop-up menus similar to the Apple Macintosh or Microsoft Windows. You will have a "toolbox" or "palette" from which you can pick and create symbols. Either during or after creation of your drawing, you can resize the diagram objects, drag them to new positions, change

Like most things in life, using technology for the first time may intimidate you. However, as your skills develop, it gets a lot easier. Here are ten rules to guide you in drawing any DFD, whether it's your first or fiftieth:

1. Erase when you make a mistake.
2. Throw away your first data flow diagram and draw a fresh one. Your second is better than the first.
3. Arrows should not cross one another.
4. Squares, circles, and files must bear names.
5. No two data flows, squares, or circles can have the same name.
6. All providers and receivers of data flows (sources and sinks) should appear around the outside of the diagram.
7. Assign meaningful names to data flows, processes, and data stores.
8. Use strong verbs followed by nouns (such as "print statement," "post charge," or "apply payment").
9. Omit irrelevant control information such as validation requirements.
10. If too many events seem to occur at a given point, level that process.

the names you have assigned to objects, delete an object if you make an error, declare an object primitive, or define the child diagram for the object.

When writing a term paper for a class, most people stare at the empty paper for a long time before they write any words; getting started is hard. The same is probably true about your first CASE data flow diagram. To break the ice and get you started, use your CASE software to re-create the context accounts receivable data flow diagram from Figure 2.2. Once you have drawn it, level it to show the data stores and processes for each activity, as done in Figures 2.3 and 2.4. Assign level numbers to processes that match Figures 2.7 and 2.8.

The best part of the CASE software is that if you make a mistake, you can fix it with some function such as delete or prune. The CASE software operates remarkably similar to a word processor. You can add new symbols, delete them, move them around, and edit them if you want to change the words that you have written.

Once your first diagram is drawn, the second is even easier; in fact, you may even enjoy it. Work with a few of your classmates and have them each draw a part of the leveled diagram and then you can combine them into one large diagram. Working together as a team is a challenging, yet rewarding experience; it is how most organizations develop high-quality software.

MODELING WITH ENTITY-RELATIONSHIP DIAGRAMS

The analyst's task in data modeling is to figure out what types of data underlie the business function under study, and to capture a replica of that structure. Let's say that you have spent some time interviewing users and understanding the placing of an order for our accounts receivable application. The following aspects of order entry have emerged as things about which data are stored: products, customers, accounts receivable, shippers, and orders. A preliminary sketch of these might look like Figure 2.10.

Figure 2.10 A simple entity-relationship diagram for order entry.

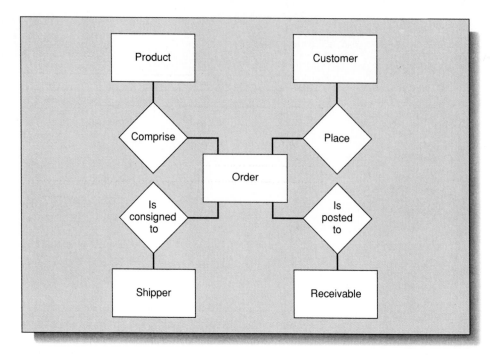

That's all there is to it—the diagram in the figure is a data model diagram. Soon we'll refine the notation and add a few more details, but this diagram is the basic idea.

Entity Types and Instances

A number of subtle but important concepts and terms are connected with data modeling. First, the "things" in the figure are called **entities.** The term *entity* is very broadly defined: it is an object or event that is of interest to the business function that we are trying to model. It is real, and has data content associated with it. In the order entry example of Figure 2.10, CUSTOMER, PRODUCT, ORDER, SHIPPER, and RECEIVABLE all qualify as entities. Other examples of entities are FLIGHT (airline), PATIENT (hospital), DONOR (blood bank), VENDOR (accounts payable), CLASS (College), and PALLET (in a warehouse).

Another important concept is that of an "instance" of an entity. In our example, CUSTOMER is an entity, but Westinghouse is not considered an entity. Westinghouse is one **instance** or occurrence of the entity CUSTOMER; it is analogous to a record in a file. Some data modelers use the term "entity type" in place of just "entity." Thus, CUSTOMER, PRODUCT, SHIPPER are types of entities, and Westinghouse, Square D, General Electric, WalMart, Shell, and Allen Bradley are instances of the CUSTOMER entity type. Similarly, United Parcel Service, American Airlines, Delta Airlines, Federal Express, Union Pacific, and PIE Nationwide are instances of the SHIPPER entity. It takes a little practice, but most analysts can quickly learn to identify the things that are and are not entities in a business situation.

Attributes of Entities

Taken alone, an entity itself (CUSTOMER, for example) does not convey or contain any data. The data resides in the **attributes** of the entity. Think of attributes as characteristics or properties that describe an entity or describe what we want to store for this entity. An attribute corresponds to a field in a record. Figure 2.11 shows attributes that could typically describe the entity type CUSTOMER.

Notice that the attributes of CUSTOMER in the figure are technically attribute types, and are the names of things that we would use to describe the entity CUSTOMER. When we give actual data values to these attribute types, we have a set of attribute types that describe a specific entity instance of CUSTOMER. Figure 2.12 shows one instance of CUSTOMER as described by a set of instances of attributes.

Figure 2.11 *(right)* Sample attributes for the entity CUSTOMER.

Figure 2.12 *(below)* Sample instance of the entity CUSTOMER.

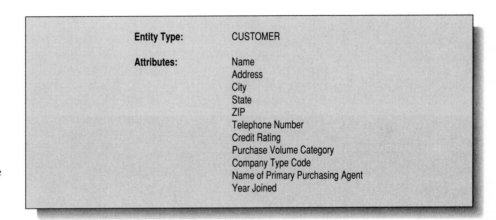

Entity Type:	CUSTOMER
Attributes:	Name
	Address
	City
	State
	ZIP
	Telephone Number
	Credit Rating
	Purchase Volume Category
	Company Type Code
	Name of Primary Purchasing Agent
	Year Joined

Entity Type:	CUSTOMER	Instance of CUSTOMER:	Westinghouse Electric Corp.
Attribute Type:	Name	Instances of Attribute Types:	Westinghouse El. Corp.
	Address		1 City Center One
	City		Pittsburgh
	State		PA
	ZIP		15212
	Telephone Number		(412) 432-7892
	Credit Rating		AAA
	Purchase Volume Category		2
	Company Type Code		D
	Name of Primary Purchasing Agent		Wilma D. Stephanic
	Year Joined		1989

This figure also shows that we have inserted or "plugged in" values for the instance. Now we can begin to see how the data model provides guidance for the business function.

It is very important to realize that the data model we constructed here is purely logical—it is part of the "blueprint" of the underlying information structure of order entry. Nothing in the data model or in the attribute description says anything about what kind of computer contains this data; whether the files are sequential, indexed, or direct; where the stored information physically resides; or whether we have a centralized or distributed database. At this point, the model is a logical—not a physical—one.

This is what a model represents. Remember that one of the primary advantages of logical modeling is that it separates physical and technical computer system design issues from business issues. It allows us to focus on the business in the all-important early stages of system design. And it is a business problem that we are trying to solve.

Relationships

Turning back to the original sketch of the accounts receivable order entry data model (Figure 2.10), we see that the entities are connected by **relationships**. CUSTOMERS place ORDERS. "Place" is the relationship. We could just as well state this relationship in reverse: ORDER "is placed by" CUSTOMER. ORDER "is consigned to" a SHIPPER. RECEIVABLE "is posted to" ORDER. One or more PRODUCTS "comprise" an ORDER. Thus, the data model diagram can conveniently depict information not just about the names of entities, but the much more useful information about how those relationships interrelate in a particular business setting. This is why we call data models entity-relationship diagrams (ERDs).

Cardinality

Going one step further, data models can also show the **cardinality** of relationships—the number of instances allowed between two relationships. One CUSTOMER might place one ORDER or many ORDERS. An ORDER is consigned to one SHIPPER, many SHIPPERS (in the case of a large multi-part order), or even to zero SHIPPERS (if the order is picked up at the factory directly by the customer). This cardinality information allows the data model to contain deep and detailed information about how the business functions.

Suppose, for example, that a company's policy absolutely forbids customers to pick up orders at the plant. The cardinality of the ORDER "is consigned to" SHIPPER relationship becomes a one-to-one or one-to-many. This cardinality is sometimes written as 1:1 or 1:N, in which N means many. It is impossible in our example to have a 1:0 relationship between ORDER and SHIPPER because that is forbidden by company policy.

As another example, the company might allow several customers to place joint orders. If so, the ORDER "is placed by" CUSTOMER relationship would have a 1:N cardinality. But if the company policy forbids joint orders, then ORDER "is placed by" CUSTOMER has a cardinality of 1:1. This feature enables the analyst to capture many subtle but important nuances of how the business functions, and to conveniently and compactly show those details on the ERD. There are many different notations for depicting the cardinality of relationships on data models, but Figure 2.13 shows the most common notations.

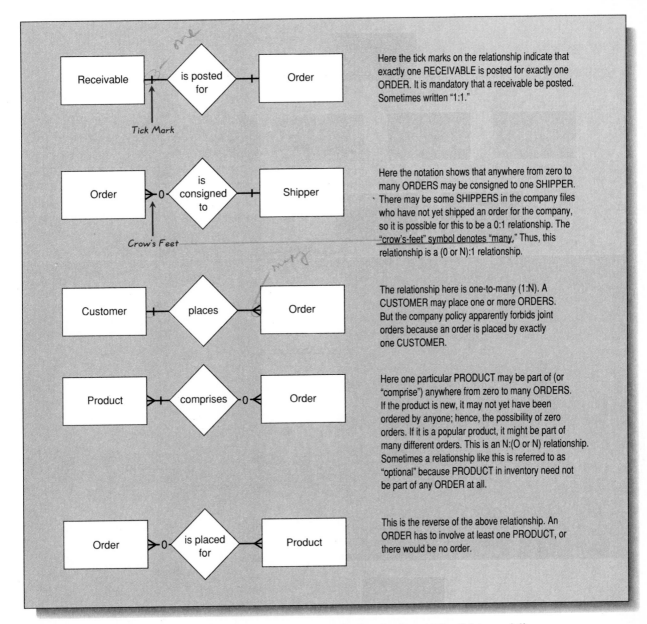

The diagram shows common ER notations with the following explanations:

Receivable — is posted for — Order
(Tick Mark labeled on the relationship line)
Here the tick marks on the relationship indicate that exactly one RECEIVABLE is posted for exactly one ORDER. It is mandatory that a receivable be posted. Sometimes written "1:1."

Order — is consigned to — Shipper
(Crow's Feet labeled on the relationship line)
Here the notation shows that anywhere from zero to many ORDERS may be consigned to one SHIPPER. There may be some SHIPPERS in the company files who have not yet shipped an order for the company, so it is possible for this to be a 0:1 relationship. The "crow's-feet" symbol denotes "many." Thus, this relationship is a (0 or N):1 relationship.

Customer — places — Order
The relationship here is one-to-many (1:N). A CUSTOMER may place one or more ORDERS. But the company policy apparently forbids joint orders because an order is placed by exactly one CUSTOMER.

Product — comprises — Order
Here one particular PRODUCT may be part of (or "comprise") anywhere from zero to many ORDERS. If the product is new, it may not yet have been ordered by anyone; hence, the possibility of zero orders. If it is a popular product, it might be part of many different orders. This is an N:(O or N) relationship. Sometimes a relationship like this is referred to as "optional" because PRODUCT in inventory need not be part of any ORDER at all.

Order — is placed for — Product
This is the reverse of the above relationship. An ORDER has to involve at least one PRODUCT, or there would be no order.

Figure 2.13 Common notations for showing the cardinality of relationships on ERDs.

At this point, we have seen the key fundamentals of data modeling:

- Entity types as things in the business world about which we store data
- Entity instances as specific occurrences of particular entity types
- Attribute types to describe the properties of an entity type
- Attribute instances that contain specific values for data
- Relationships and their cardinality

There are many variations and extensions to this basic data modeling idea. These variations are associated with different system development methodologies, different CASE tools, and different company standards and procedures.

ERD Notations and Symbols

Figure 2.14 shows one such notational variation. This is the Chen notation, named for Dr. Peter Chen, the originator of ERDs. Entities appear as rectangles, relationships as diamonds, and the cardinality is written twice alongside the relationship lines—once for each direction of the relationship.

Another variation in ERDs, known as Bachman diagrams, is shown in Figure 2.15. This technique was devised by Charles Bachman, one of the originators of the concept of database management systems. Bachman notation lists entities as rectangles with specific attributes listed within the rectangle. Cardinality is shown by lines linking the relationships to one another.

Regardless of which technique you use, this data model allows for very accurate estimating of the logical data requirements of either an existing or a proposed system. They are very valuable business analysis tools.

Figure 2.14 Chen notation for entity relationship diagrams.

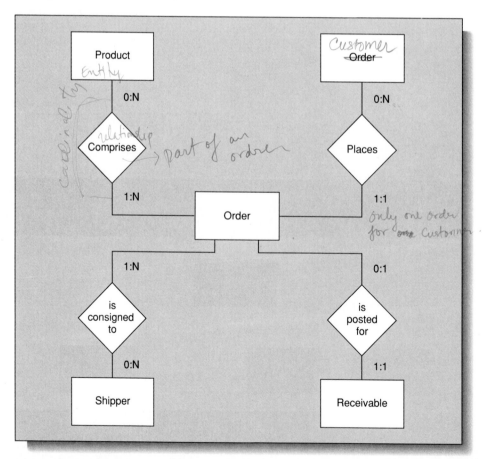

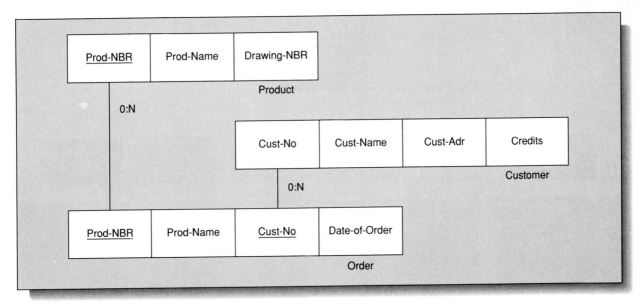

Figure 2.15 Bachman notation for ERDs.

How to Model Data

Data modeling is not as easy as it looks. Often, it is hard to tell whether one should consider an item as an entity or an attribute. In fact, it is sometimes even difficult to distinguish an entity from a relationship. Figure 2.16 shows two possible ways to model a portion of the order entry function.

Is ORDER a relationship or an entity? Since most order entry situations record information about the order that is independent of both PRODUCT and CUSTOMER, it is probably more correct to show ORDER as an entity. The bottom half of this figure represents this model.

You will probably encounter other situations where such a decision is not as easy as this one. The final decision will come from you and the users after careful study, negotiation, and debate.

As another example, suppose you were trying to separate entities from attributes. Is CREDIT-RATING an entity on its own or an attribute of CUSTOMER? Probably the latter, because if one particular instance of CUSTOMER were deleted from the database, we would not have to keep that customer's CREDIT-RATING. When there is no need for an item to exist independently of its entity, that item is probably an attribute.

In a tougher example of this same problem, consider DRAWING-NBR, the company's internal identification number for the blueprint of a particular PRODUCT. You could argue that DRAWING-NBR is an attribute of PRODUCT, because if a product is deleted, so is its drawing. On the other hand, another information system elsewhere in the company uses DRAWING-NBR for controlling engineering changes. In this situation, DRAWING-NBR is very much an entity in its own right.

Thus, there are many gray areas in data modeling that require joint resolution by users and analysts. This resolution process is usually a healthy one because it often leads to clarification of the way the business operates and reveals places where systems analysts can

Figure 2.16 Two ways to model order entry data.

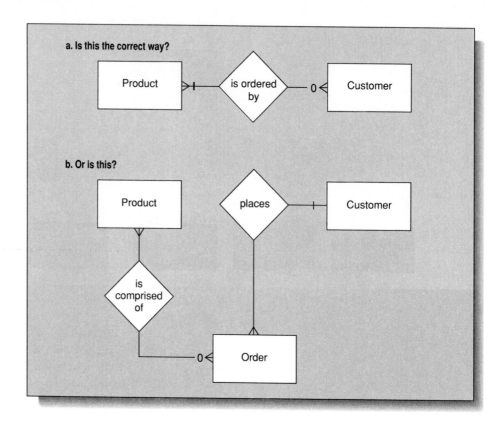

make improvements. The model also provides a firm foundation for design of the present system and provides a lasting blueprint that greatly simplifies the task of designing future systems. Fortunately, there are some specific guidelines that remove some of the ambiguity and confusion from the data modeling process.

Recognizing Entities

Recall that your task as an analyst, like the archaeologist, requires you to uncover hidden things. Entities are often partially hidden. The overall process of finding entities reduces to three key steps:

- User interviewing sessions
- Separating the entities from the attributes and relationships
- Feedback and refinement in user/analyst sessions

User interviewing sessions work best when several users and analysts are involved. Ideas are generated more quickly in a group, the meeting is usually more relaxed, and several users are likely to have a broader grasp of the business function that the model will represent. The fundamental purpose of the sessions is to produce a list of things, objects, or other nouns that apply to the users' business environment. This interview is an open-ended brainstorming process, with no effort made to distinguish entities from attributes and relationships. The result of this session or series of sessions is a "laundry list" of nouns (PACKING SLIP, INVOICE, BILL OF LADING, SHIPPER ID, BACKORDER CATEGORY, BILL-TO ADDRESS, UNIT OF MEASURE, DISCOUNT CATEGORY, ELECTRONIC DEBIT, PRESHIP LEAD TIME, and so on.) that apply to the business function. For a large application, the list could have over 100 nouns.

The second step in finding entities is identifying nouns on the laundry list that truly are entities. Important to the second step is the idea of a **primitive entity**. In data modeling, primitive entities are fundamental or basic. It cannot be further decomposed or broken down; it is fundamental to the business function. For example, in a payroll system, TAX DEDUCTIONS is not primitive because we can further decompose it into other entities: FEDERAL INCOME TAX WITHHELD, FICA TAX WITHHELD, and STATE INCOME TAX WITHHELD. The entity STATE INCOME TAX WITHHELD is a primitive entity. If it were broken down into pieces, such as TOTAL AMOUNT WITHHELD, SOCIAL SECURITY NUMBER, and so on, these pieces could not stand on their own as entities.

The concept of a **derived entity** is also useful to the analyst. As the term suggests, a derived entity is composed of or derived from primitive entities. PROJECT COST (the total dollar cost of building an addition to an electric generating plant) might seem like an entity. But since cost is calculated from other entities that contain information about hours worked, labor rates, materials used, and costs of materials, PROJECT COST is considered a derived (calculatable) entity. It is not primitive. Only primitive entities belong in the ERD.

Another useful trick in finding primitive entities and distinguishing them from attributes is the **existence test**. The existence test says that if an entity disappears, all of its attributes will disappear also. Anything left over is an entity. Consider ORDER-NUMBER. If a particular instance of the entity ORDER suddenly disappeared, then the ORDER-NUMBER for that order would also cease to exist, indicating that ORDER-NUMBER is an attribute. In another situation, you might think that CUSTOMER is an attribute of the entity ORDER because there is no point in talking about customers unless there are orders. But CUSTOMER passes the existence test—if an ORDER disappears, the CUSTOMER will still exist. We would want to keep the customer's name, phone, and other information on file because they've ordered from us in the past and/or we expect them to order in the future.

Thus, the second step of finding entities—separating out the entities from the relationships and attributes—requires these checks:

- Is the item primitive? If not, subdivide it down to the primitive entities that it contains.

- Is the item derived? If so, find out from where it originated. Continue this process until you can go on further. These constituents are primitive entities.

- Is the object an attribute of an entity? If it fails the existence test, then it is an attribute.

By systematically mixing the above checks with careful judgment and study, you should reduce the list of noun items generated in the user interview sessions down to a list containing only primitive entities. A first draft of the ERD comes from this list of primitives.

The third step in finding entities—feedback and refinement—is simple but time-consuming. Here, successive drafts of the ERD are shown to the users for critique and comment. This iterative process continues until both the users and the analyst(s) agree that the ERD is an accurate reflection of the information structure that underlies the business function.

Beyond these important fundamentals, several universal situations can further complicate the data modeler's job. One of these is intersection entities. These arise when a data type is a product of two entities, such as:

CUSTOMER + ACCOUNT —> TRANSACTION
STUDENT + COURSE —> GRADE
PRODUCT + CUSTOMER —> ORDER

Intersection data are not difficult to understand as long as the analyst realizes that their occurrences are normal and they are legitimate types of entities. Although intersection entities can't exist independent of the entities that forced them, we must show them in ERDs. Finally, note that intersection entities and derived entities are different. Intersection entities represent something new that is formed from other entities. Derived entities do not contain any new data, and thus do not belong on ERDs.

Another oddity is the recursive entity. To the data modeler, "recursion" refers to the definition or restructuring of data in terms of itself (see Figure 2.17).

As with intersection entities, recursive entities do not require any special treatment in ERDs. Analysts must acknowledge them, know that they are legitimate components of data models, and that they will occur occasionally in the practice of data modeling.

Other minor peculiarities are connected with drawing ERDs. These variations are further evidence that data modeling is not a hard-and-fast, rule-based scientific activity. Rather, it is a mixture of art, science, intuition, and common sense.

Levels of Modeling

Data modeling with ERDs and process modeling with DFDs are universal tools. We can apply them to modeling problems that have widely varying degrees of scope as well as levels of detail. First, let's consider the issue of scope.

Scope refers to the breadth, or wideness of a modeling effort. Some organizations have attempted data modeling projects that encompass essentially the entire organization. These expensive, time-consuming initiatives are very broad in nature and are often referred to as enterprise modeling projects. They involve the entire corporation, organization, or enterprise.

Other organizations model with intermediate scope. DuPont, for example, successfully constructed several models for systems at a textile plant in Waynesboro, Virginia. This intermediate scope of modeling, which is less costly than enterprise-wide modeling, is usually referred to as a business unit in scope. Divisions, large profit centers, and geographically distinct operations are examples of the business units use of immediate scope modeling projects. In these cases, only part of the enterprise participates in the modeling effort.

The narrowest modeling scope is for an individual application system. When an analyst prepares data and process models for a labor scheduling system, the modeling task is said to have only an application-wide scope. The modeling does not extend beyond the individual system project.

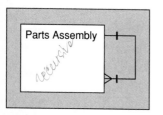

Figure 2.17 Example of recursive entities. A parts assembly could be made up of another parts assembly.

The decision about whether to undertake a modeling project with enterprise, business unit, or individual application scope depends mainly on available resources. In general, the scope decision is a trade-off between the benefits of broad scope modeling and its high costs. The important point is that the modeling process and the tools used are exactly the same, regardless of the scope at which the modeling process is undertaken. The only difference is in the size of the effort. Thus, the principles and techniques are applicable in a wide variety of situations.

The concept of levels of models differs from the idea of scope. Level has more to do with the detail of the modeling than with the size of the modeling. Both data and process modeling are done at high, middle, and low levels. These levels of modeling are applicable to all three scopes, as indicated in Figure 2.18.

High-level models are somewhat abstract, summary-level diagrams that show the "big picture" of the data or process structure. High-level data models are constructed using ERDs. High-level process models are usually called context diagrams and/or system DFDs.

Mid-level data models are needed because of the inability of ERDs to show all of the details of the data structure underlying a business function. For example, a systems analyst might identify an entity called CUSTOMER on an ERD, but several types of customers are possible, such as JOBBER, OEM, BROKER, RETAILER, and LESSOR. The mid-level data model diagram can show these different types of the entity CUSTOMER. Mid-level modeling supports entities for storing many records. For example, many logical transaction records result in connection with each instance of an entity called ACCOUNT. Mid-level diagrams show detail, rather than the big picture of the ERD. Notation, techniques, and symbols for mid-level modeling differ greatly among methodologies and CASE tools. Mid-level process models are more straightforward than mid-level data models, and are nothing more than successively more detailed "explosions" of DFDs.

Low-level models contain the greatest detail, and often some of the physical design details of systems begin to appear at this level. Here, the boundary between logical and physical modeling blurs. Physical data models vary in format, depending upon the system development methodology in use. They almost always show actual data field lengths, and data types (character, decimal, packed, and so on). Low-level process models are also somewhat physical in character in that they contain detailed descriptions of how to do the

Figure 2.18 Modeling may be done at three levels and with three degrees of scope.

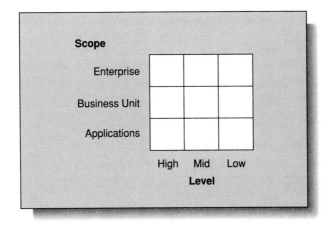

actual processing. These descriptions take the form of pseudocode, decision trees, decision tables, and other methods of notation.

Taken together, a full set of data and process models for all three levels constitutes a very important part of the logical specification package for a system.

BENEFITS OF MODELING

Experienced systems analysts know that the key to building successful systems is to keep the future users of a system highly involved in its design and construction. Analysts that involve users have a good chance of completing a system that possesses the features that users need to accomplish their business functions.

Unfortunately, it is very difficult to get and keep users involved with system specification documents because of the complicated, technical, computer-oriented symbology and terminology. Often, analysts push these system documents across the table for users to review, and the users push them right back, complaining that they are "over my head" or that "I can't understand these—I don't know anything about computers." The users give up, saying, "I trust you to do it right. Don't bother me with these details—just call me when the system is ready." This abdication of responsibility by the users often leads to dissatisfaction when the system finally emerges because the analysts had to guess at system specifications and features.

User Involvement

When analysts push DFDs/ERDs across the table for users to review, they don't automatically push them back. Most users can read DFDs/ERDs and understand how they represent the business function in question. Now they will comment on errors and omissions since the symbols used in DFDs/ERDs are simple, self-explanatory, and show business-centered facts rather than computerese.

The easy comprehensibility of the diagrams—plus their ability to draw users into the system design process at a high level of detail—yield tremendous benefits in the quality of systems. Users become partners with the analysts. Users work closely with analysts in studying, refining, criticizing, and improving the diagrams until they are accurate replicas of the business function under study and modeling.

Furthermore, the modeling process causes users to study the fundamentals of how the business works on a day-to-day basis. It is fairly common for users to find tasks they are doing that don't make sense, are awkward, inefficient, or otherwise need improving. Users are often so busy with the daily pressures of the business that they don't stop to consider whether their routine methods are efficient until they are forced to do so by a systems modeling project.

A Blueprint of the Business

For all its benefits, modeling also carries substantial costs and is not without its opponents. Most of the opponents are users and nontechnical senior managers who do not understand the long-term benefits that a model can create. Their frustration and opposition stem from

Helen Langley had decided to send her key analyst, Kurt Wurst, to a database management systems training seminar in Baltimore. Helen hoped that the knowledge Kurt picked up would help her staff switch from a record file system to a database environment.

Kurt looked forward to the seminar and carefully read the manual that the seminar leader, a software vendor, sent after Kurt enrolled in the program. As part of the seminar, each participant would actually analyze and design a small system. Since he had designed many such systems in the past, Kurt felt confident about doing so again and hoped that the new software would add an exciting new element to the task.

By the end of the first day of the seminar, Kurt felt that he had learned a lot. He eagerly took the class assignment back to his hotel room, where he spent three hours drawing and refining his flowcharts. When Kurt returned the next morning, he noticed that the person sitting next to him had not used flowcharts at all, but an amazing array of tools with which Kurt was only superficially acquainted: ERDs and DFDs, a HIPO, a decision table, and a Warnier-Orr diagram. It all looked so complicated that Kurt assumed his old method would out-shine any "new fad."

As the second day's sessions unfolded, the seminar leader distributed sample data to the participants, asking them to test their solutions to the problem with it. Kurt's 15-page solution failed, but his neighbor's 4-page solution passed. Kurt took little solace from the fact that most of the other participants also failed. He felt confused and embarrassed. When questioned by the seminar leader, most of the "failures" revealed that they had stayed with their flowcharts, whereas Kurt's successful neighbor had employed modern modeling techniques, coupled with modular problem-solving and top-down development.

As the seminar leader explained the advantages of the so-called structured methodology, Kurt's feelings of embarrassment slowly changed to appreciation. He vowed that he would master the new technique, so he would really have something to show his colleagues at Bay Area Telephone.

Noticing how the seminar leader carefully avoided poking fun at flowcharts and simply offered new tools as useful alternatives, Kurt decided to use the same nonthreatening approach with his colleagues. He knew that they, like himself, would learn much more rapidly without the anxiety or guilt that accompanies the fear of being left behind.

the fact that modeling takes a long time to create, is expensive, and does not appear to create immediate tangible payback. These are fair objections. The benefits of modeling are somewhat intangible and long-term. But when the benefits finally are realized, they are large and very tangible.

Consider a new payroll system for a large corporation. Payroll systems appear simple, but turn out to be one of the most complex of all systems. In our example, the payroll must:

- Track 100,000 employees
- Allow pay frequencies as a mix of weekly, bi-weekly, or monthly
- Permit exempt salaried, non-exempt salaried, and hourly employees
- Provide for employees working in 50 different plant and office locations in the U.S., Canada, and France
- Allow for correct currency exchange rates
- Withhold taxes according to the rules of 25 different states, thousands of local governments, and at least four countries
- Withhold Social Security taxes
- Keep union dues and overtime rates for over 100 contracts with union locals for various crafts and trades

This list of system requirements goes on and on, but you've heard enough to appreciate the system's complexity.

In the pre-methodology and pre-modeling days, the payroll system analyst would have conducted extensive interviews and an exhaustive system study, and eventually would have come up with a system design. The design might consist of a mixture of prose descriptions and other types of nonstandard drawings. It would probably have less user involvement than the analyst would have liked. The odds were high that the payroll system would not work exactly as required.

Using the modern approach, the analysts would work closely with users to develop the logical data model and process model diagrams. Their joint objective during this effort (like the archaeologist) is to uncover the fundamental underlying data and processes that drive the payroll system. The bad news is that this takes tremendous time and effort, both by the users and by the analysts. In all likelihood, it would take longer and cost more to build the payroll system by using this approach than by using the old methods. The modeling part of the project causes very high "up-front" costs. These costs occur early or near the front of the system life cycle, not spread out through the entire project. Up-front costs are high because this payroll system requires data modeling.

The good news is that the models, once they're finally done, will provide a lasting logical blueprint of the entire payroll function, and the next time any systems work is done in the payroll area, it becomes easier, cheaper, and faster. Future changes or additions to the payroll system will require very little study or effort because the data and process models will already exist to provide the foundation for the additional work.

Many companies—such as the America International Group, Amoco, DuPont, John Deere, Touche Ross, and others—have recognized the usefulness of modeling. They have decided to invest in the up-front costs in exchange for the long-term benefits. Properly maintained, these models provide a valuable, lasting blueprint for the business, and make future systems work much easier and more effective.

SUMMARY

Increased productivity is a key to success in most organizations, but that only happens when an organization improves its efficiency. To boost the efficiency (and hence the productivity) of analysts and programmers, a variety of modeling techniques are available. These techniques form the backbone of the structured methodology of system analysis, design, implementation, and maintenance.

The goal of modeling is to uncover the data and process structures that occur naturally as part of the business function, and to commit them to paper using standard, easily read diagrams. These techniques have proven successful in increasing user involvement in systems work, and they aid future projects because they provide a lasting blueprint of the business. Two tools that help analysts achieve these goals are entity relationship diagrams (ERDs) and data flow diagrams (DFDs).

Data flow diagrams are pictorial representations of the system by using four symbols: square, arrow, circle, and parallel lines. Squares represent providers (sources) or receivers (sinks) of data. Arrows name the data that providers or receivers are giving to or getting from the system. Circles or processes convert incoming data to outgoing data. The parallel lines represent a data file or storage. Data stores hold data for later processing. Data flow diagrams help the analyst organize and communicate the system's contents. They also lead to software that is more reliable, less expensive, easier to use, less difficult to change, and more timely.

Entity relationship diagrams and data model diagrams are a pictorial representation of the information needed by the business function. Several different sets of symbols are in use, but they all show the same concepts: entities and relationships.

KEY TERMS

Data modeling	Upper CASE
Entity-relationship diagram (ERD)	Lower CASE
Process modeling	Entity
Data flow diagram (DFD)	Instance
Data source	Attributes
Data sink	Relationship
Context diagram	Cardinality
Leveling	Primitive entity
Data flow	Derived entity
Data store	Existence test

QUESTIONS FOR REVIEW AND DISCUSSION

Review Questions

1. What is the purpose of a model?
2. Draw a data model of any system of your choice, using an entity-relationship diagram.
3. What are the benefits of modeling in terms of its impact on user involvement and on future system development?
4. Is there any difference between an entity type and an instance of an entity type?
5. How do you read a data flow diagram for View Video?
6. What would a simple data flow diagram for a payroll system look like?
7. What is the difference between a context and a leveled data flow diagram?
8. What are the symbols used to draw data flow diagrams?
9. What are some rules for drawing a data flow diagram?

Discussion Questions

1. Write a one-paragraph description of how to apply the existence test.
2. Why is it necessary to supplement ERDs with mid-level data models?
3. Explain how the data modeling process can capture certain aspects of company business policy in ERDs.
4. What is the name of the ERD notational system in which relationships are shown in diamonds?
5. What level number and name do we assign to the original data flow diagram?
6. What data are shown on an aging report?

Application Questions

1. Draw an ERD for this scenario: An organization purchases items from a number of suppliers. It keeps an inventory of each item type purchased from each supplier.

2. Draw an ERD for this scenario: Drivers take out vehicles to make deliveries. A vehicle may be taken out of any depot. It is possible for any vehicle to be taken out more than once in a given day. Any vehicle can be taken out by any driver any number of times a day.

3. If a process is numbered 1.5.3.1, what does this mean?

4. Match each of the following:

1. Circle	a. Source of data
2. Square	b. File or data store
3. Arrow	c. Conversion process
4. Parallel lines	d. Data flow

Research Questions

1. Find a friend who works in a small business such as a video store, florist shop, or restaurant. Interview him or her and then draw an ERD for one of its major business functions, such as inventory management.

2. Repeat the previous question, except this time draw a process model for the business function. Construct the context diagram and a second diagram that is an explosion of the context diagram.

3. Interview a systems analyst or application programmer and find out his or her ideas on the usefulness of logical modeling, structured methodologies, and CASE. If the interviewee is opposed to these techniques, find out why.

4. Many other tools are available to a systems analyst. Learn about two of them and write a comparison of them.

5. Often, another symbol is used for the process symbol. What does this symbol look like?

6. Draw a context data flow diagram of a payroll system.

7. Draw an ERD for a university registrar's office. Concentrate your modeling effort mainly on the academic registration activity.

Completing their business plan was just the start of Mark and Cindy Stensaases' search for software to help their new business. They spent some time thinking of their next step and decided to hire a consultant. Mark and Cindy both felt that the software was too important a part of their operation for them to try to resolve themselves; they needed an expert. Cindy had a friend, Lawrence Lee, who managed the information systems department of a very large international data communications company. Cindy asked Lawrence for the name of a consultant. Her friend suggested two people, and the Stensaases contacted both of them, asked for résumés, held a two-hour interview with each, contacted their references, and chose Frank Pisciotta.

When initially assigned to look into View Video's new tape check-out and check-in rental system, Frank Pisciotta first thought about the overall goal of linking it to the BusinessWorks PC accounting system the Stensaases had purchased for accounting purposes. Besides tracking the tapes customers have rented, the system must collect data about the amount of money the customer owes or pays, and pass this data to the BusinessWorks PC accounts receivable system. A good system will make sure a customer pays the proper amounts and it will help View Video maximize its income.

Frank decide to visit the computer retailer that the Stensaases recommended and saw the demonstration of BusinessWorks PC. While there, he was also impressed with the integrated accounting system and called its developers, Manzanita Software. Their chief software developer, Mark Havener, replied that they had an existing interface with many popular cash registers and that he foresaw that linking between the two systems was quite easy, as long as the link came via the cash register.

With a background in process and data modeling, Frank now decides to sketch his view of the proposed system as a data flow diagram. He meets with the Stensaases and, working in a conference room with a blackboard, Frank writes down the names of the ERD entities as they come up in the conversation. After about half an hour, the following items appear on the board as entities: customers, movie distributors, management, and accounts receivable. The group then starts to draw the first draft of the DFD, as shown in Figure 2.19. Customers rent videotapes, receive the newsletter, return tapes, complete an application form, and receive overdue notices. Film distributors receive a movie order from View Video and fill the order. Management receives a variety of reports (e.g., a list of movies for rent) and makes adjustments to the system (such as price increases). Accounts receivable collects financial payment data for posting to customer account balances via BusinessWorks PC.

Frank takes the hand-drawn diagram home and enters it into his CASE software. A few days later, he brings back the diagram to Mark and Cindy for their approval. He explains the symbols that the CASE software tool uses and reviews his understanding of the business operation. He knows from his 14 years of experience with a variety of individuals and organizations that employees, managers, and owners can read and understand ERDs and DFDs quite easily. They are useful tools in analyzing and designing the system.

Frank points out that the DFD does not require them to decide how they will rent tapes or return them. These are activities they will decide later during design.

Frank begins the task of leveling by reviewing his context diagram, his logical model for the tape rental system. Because of the number of data flows, he decides to draw two level 1

Figure 2.19 The context diagram for View Video's tape rental system shows the system as an overview.

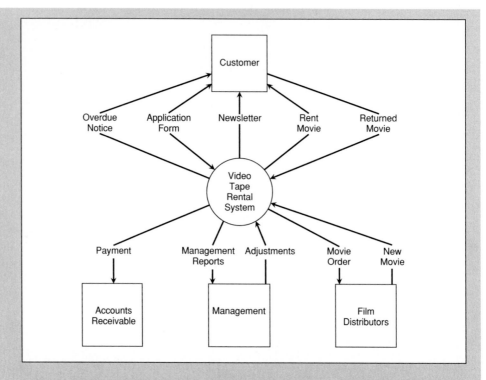

diagrams: one for customers and the other for movie distributors, management, and accounts receivable.

Starting with the customer portion of the context diagram, Frank recognizes that each data flow will have at least one processing activity, as Figure 2.20 illustrates. Franks sees the need for three data stores (files): CUSTOMERS, MOVIES, and RENTALS. CUSTOMERS will hold all the specifics about each customer that rents tapes. MOVIES keeps the specifics on each tape View Video owns. RENTALS records all the data about a specific movie rented by a particular customer.

When Frank decomposes the context diagram (level 0) into its various components, each new circle receives its own number (1, 2, 3, and so on). The numbering lends emphasis to the dependent relationships between various processes. With this first of three level 1 diagrams completed, Frank decides which processes require further leveling. At this phase in his modeling, he feels that checking out and checking in are complex processes (4 and 5) and require further work. Processes (1, 2, and 3) seem more straightforward and may not require further leveling.

Next, Frank takes on the second DFD (Figure 2.21). He finds five data flows and makes five processes, one for each data flow. He examines each and picks the files that it will have to use to produce the necessary results. For example, stocking a new movie requires the MOVIES file and has no need for data in CUSTOMERS or RENTALS (this is a new movie and no one can rent it until it is placed on the shelves).

From his interviews with Mark and Cindy, Frank knows that the system must produce a variety of management reports. Again, Frank divides his data flow diagram into finer levels

Figure 2.20 The level 1 data flow diagram of View Video's tape rental system shows more details of how the customer portion works.

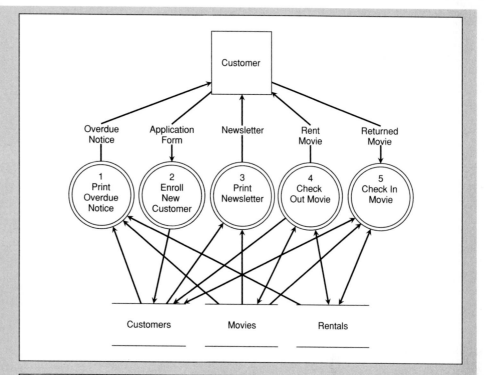

Figure 2.21 The second of Frank's level 1 data flow diagrams shows how the movie distributor, management, and accounts receivable portions work.

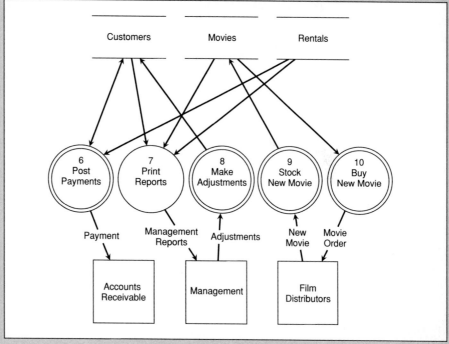

of detail, leveling process 7 (Print Reports) into a new data flow diagram that contains 7.1, 7.2, 7.3, and 7.4 (Figure 2.22). The customer list includes names, addresses, and telephone numbers. The movie analysis shows each videotape, the number of times it has been rented, date of last rental, income earned, and date purchased. The overdue tapes report shows which tapes are abnormally late and are not returned from customers. The cash generated report tells the income generated by day of week. These reports help View Video management better monitor customer renting habits.

Some processes, such as printing the newsletter, do not require leveling. However, the process of checking out and checking in tapes will eventually require further refinement, as will ordering and receiving tapes from film distributors.

To help Mark and Cindy understand the process that he will eventually perform, Frank decides to define the attributes of CUSTOMERS, as done in Figure 2.23. His visits to a number of video stores reveal that a customer's telephone number is the unique identifier for each customer. Besides this attribute, Frank lists others and asks the Stensaases if they can think of any they'd like to add. Mark and Cindy ask for an example of a customer and Frank makes a sample instance (Figure 2.24). All three agree that this is an accurate description for now, but realize that they may have to alter it later in the analysis and design phases.

The resulting data model and data flow diagrams form part of the documentation that Frank will need in later stages of analysis. He makes sure that Mark and Cindy have copies and places a spare copy in a folder—as well as the one kept by his CASE software.

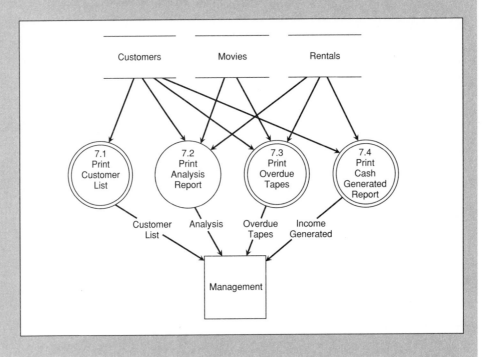

Figure 2.22 Frank breaks down circle 1.7 (Print Reports) from Figure 2.21 into more elementary components.

Figure 2.23 Attributes for the entity CUSTOMER in the View Video tape rental system.

View Video	Entity and Attribute Definition
System	Video Tape Rental
Analyst	Frank Pisciotta
Date	10/93

Entity Type:	CUSTOMER
Attributes:	Telephone Number
	First Name
	Last Name
	Middle Initial
	Street Address
	City
	State
	ZIP
	Income Earned
	Date Joined
	Date Last Rented Tape
	Number of Tapes Rented
	Credit Card Number
	Credit Card Type
	Credit Card Exp. Date
	Sex Code
	Date of Birth
	Profession

Figure 2.24 Sample instance of the entity CUSTOMER for View Video.

View Video	Entity and Attribute Definition
System	Video Tape Rental
Analyst	Frank Pisciotta
Date	10/93

Entity Type:	CUSTOMER
Attribute	Instance of Attribute
Telephone Number	8780666
First Name	Moore
Last Name	Sylvia
Middle Initial	W.
Street Address	467 Horseshoe Bar Rd.
City	South Bend
State	IN
ZIP	46603
Income Earned	22,900
Date Joined	11/23/88
Date Last Rented Tape	10/05/93
Number of Tapes Rented	57
Credit Card Number	4075512909512415
Credit Card Type	VISA
Credit Card Exp. Date	08/95
Sex Code	F
Date of Birth	1960
Profession	Substitute Teacher

Working With View Video

During this episode of our View Video case study, Frank Pisciotta began to plan the logical system to check out and check in videotapes. He drew data flow diagrams and wrote some definitions, but they are far from complete. In later chapters, when we examine the analysis and design phases of the system life cycle, we will expand these first drawings and add much more to the software that will eventually help run Mark and Cindy's store.

Case Study Exercises

1. Draw an entity-relationship diagram for these three files: CUSTOMERS, RENTALS, and MOVIES.
2. What is the cardinality among them?
3. List the attributes of MOVIES.
4. Make up an instance for MOVIES.
5. Are there any other entities that you can foresee for View Video?
6. List two other adjustments that management may want to make to the videotape rental system.
7. What other management reports would you want if you were the owners of View Video?

Structured Methodologies

GOALS

After reading this chapter, you should be able to do the following:

- Explain the principle of functional decomposition
- Define a module and list the rules for composing one
- Explain the purpose of CASE software products
- Describe the relationship between data dictionaries and CASE repositories
- List the symbols used to write a data dictionary
- Write a data dictionary definition of a file
- Describe other tools available for the systems analyst
- Read the pseudocode for a problem
- Describe the purpose of a structured walkthrough

PREVIEW

Earlier, we saw data model and data flow diagrams, the visual tools that many analysts use to depict a system. In this chapter, you will learn more about structured methodologies and their central importance to building systems. Also, you will learn more about CASE software and its role in making structured methodologies practical, and about other tools that analysts use during the analysis, design, implementation, and maintenance of systems. Like data model and data flow diagrams, some of these tools are graphic (the structure chart), whereas others are composed of words (pseudocode). All of these tools help analysts achieve the goal of providing users with the simple, flexible, and efficient computerized systems they need. Finally, we will cover structured walkthroughs and explain how they help users and analysts produce quality systems.

In our look at these new tools, we will compare them with two systems: assembling a bicycle and printing a check.

WHAT ARE ACCOUNTS PAYABLE?

To illustrate all these tools, we will examine a chain of retail shoe stores, Fleet Feet. Specializing in athletic shoes (running, walking, jogging, and tennis), Fleet Feet has a need for an **accounts payable** system to track the money that Fleet Feet owes its creditors. The chain has over 40 stores across the United States, with corporate offices in Sacramento, California. Each store is locally owned and operates semiautonomously, keeping local in-store records on sales and payroll. However, all stores must buy through the corporate offices in order to benefit from volume purchases. Stores are linked to a central computer in Sacramento, where daily sales data are downloaded. Fleet Feet's analyst is Peggy Adams-Russell.

Shoes and other goods are shipped to local stores by vendors; payment is made by the Sacramento office. Reviewing the context diagram for Fleet Feet's accounts payable system, Figure 3.1, we see that data enter the system from vendors in the form of sales invoices, vendor data, and a packing slip. The system produces reports for management and the all-important vendor check, the single most significant function of the system (Figure 3.2). In fact, the vendor check is the reason that the accounts payable system exists at all.

Figure 3.1 The context diagram for Fleet Feet's accounts payable system shows the system in an overview.

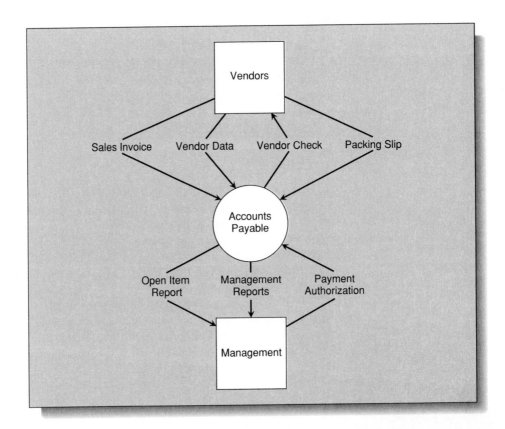

Figure 3.2 An accounts payable check pays a vendor for goods or services purchased at an earlier date. This check has two parts: the remittance advice and the check stub.

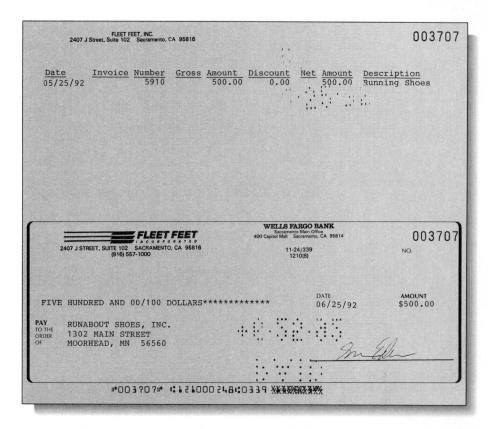

Fundamental to structured methodology is dividing a system into tasks. This process is often called **functional decomposition** or leveling because the system is divided into tasks that each perform a particular function. In the case of printing the accounts payable check, we can divide it into subtasks: extract flagged vouchers, sort by vendor number, and print vendor check. Eventually, the dividing stops when each task becomes single purpose, with a single entry and exit point. For simple systems, we may stop after a single dividing; complex systems may take five or more levels of dividing before we stop.

As you see, the structured methodology divides tasks into finer details. DFDs and ERDs show us overall data and processes, but they do not allow us to write the detailed descriptions of what takes place inside a data flow diagram's process circle symbol. Until now, we have ignored how a process actually converts incoming data to outgoing. Now we will concentrate on the processes.

THE NEED FOR A STRUCTURED METHODOLOGY

During the 1970s and 1980s, many systems were not what the user wanted, but what the analyst thought the user wanted. Communication between analysts and users was minimal, due to the lack of tools available to express a system's objectives.

Use of CASE and the structured methodology creates an engineering or disciplined approach to the software life cycle. It improves communication among analysts and between analysts and users. More than a programming technique that eliminates GO TO statements, more than a simple set of hard and fast rules, and more than a series of popular buzz words, it is a philosophy aimed at delivering high-quality, easy-to-maintain systems within budget, with promised features, and within time constraints. When properly applied, the use of CASE and structured methodology leads to systems with minimal errors and it produces clear, concise, and accessible documentation. In the end, it results in systems that effectively solve users' problems.

→ breaks tasks into constituent subpsies.

The Top-Down Approach: Functional Decomposition

Crucial to the philosophy of CASE and the structured methodology is the **top-down** concept, which establishes a hierarchy of components within a system. When an analyst studies a system, he or she wants to understand the priorities and relationships among the components of that system. Earlier, we used the top-down approach to break down the overall systems process into individual components—analysis, design, implementation, and maintenance—and we further divided each of these components into even smaller units, as shown in Figure 3.3. Analysis divides into two units, preliminary and detailed; preliminary analysis further divides into three subunits; detailed analysis into four.

Imagine a landscape painter working on a picture of a forest. First, the artist sketches the general idea of the painting—three large boulders in the foreground, a towering peak rising above the forest. Having established the sizes and spatial relationships of the larger components, the painter then begins to add details—veins to the granite, branches to the trees, and snow to the mountaintop. Eventually, of course, the artist will use very fine brushes to produce minute details: pine needles, a squirrel perched on a branch, a hawk circling high overhead.

Figure 3.3 A top-down view of the systems process. We will add components under design, implementation, and maintenance and in later chapters.

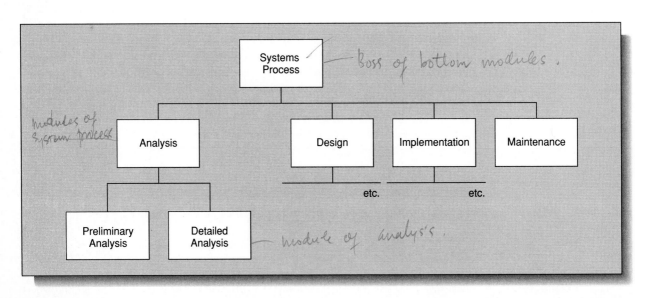

Notice how the artist follows a decreasing set of priorities: from general to more detailed concerns. In much the same way, a systems analyst begins with the general idea of a system (the problem that the user wants to solve), then proceeds to ever smaller details (exactly how the system will solve the problem).

Suppose a user wants a payroll system. Initial study reveals that the system must allow for hourly, salaried, and commissioned employees, then more detailed study discloses that hourly employees fall into five shift subgroups: day, evening, graveyard, weekend, and split shifts. The payroll system must work with the unique pay rates of each shift.

The top-down approach, then, subdivides components into units, and then further divides the units into subunits. This process is called functional decomposition because it breaks tasks into constituent subtasks. The task listed at the top encompasses all of those listed below, and each subtask forms an umbrella over even smaller ones. Tasks at the top "boss" those at the bottom. Thus, the task with the highest authority in Figure 3.3 is systems process, followed by four subordinate tasks of equal authority: analysis, design, implementation, and maintenance.

Modules — unit w/a single specific function of a larger system.

We use functional decomposition to divide a system into its parts or **modules**. A module is a unit with a single specific function of a larger system. Figure 3.3 shows the systems process decomposed into four modules: analysis, design, implementation, and maintenance. Analysis itself has two modules: preliminary analysis and detailed analysis.

We can functionally decompose preliminary and detailed analysis into further modules (Figure 3.4). Preliminary analysis has three modules: user request, analysis of request, and management action. Detailed analysis shows four modules: detail study, fact finding, develop alternatives, and presentation to management.

Some modules relate to one another as "children" of "parents." Analysis, design, implementation, and maintenance are "children" of the "parent" systems process. Similarly, preliminary and detailed analysis are "children" of the "parent" analysis. In this way, analysts have borrowed the modular concept from programming, where each module performs a single routine or task.

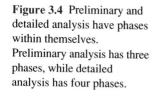

Figure 3.4 Preliminary and detailed analysis have phases within themselves. Preliminary analysis has three phases, while detailed analysis has four phases.

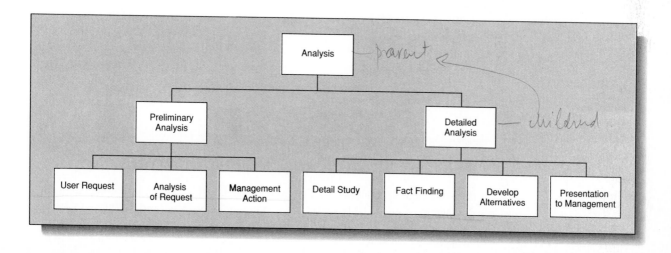

CASE AS AN ENABLING TECHNOLOGY

Managers are relying more and more on computers to provide them with information that will help keep their organization running smoothly, profitably, and more productively.

CASE is considered an "enabling technology" because it makes it relatively easy to actually use the structured methodology. Even though structured methodologies have existed since the 1970s, they generally received only lip service because they require large amounts of paperwork. Analysts had to fill out forms for all data elements, draw modeling diagrams and then redraw them by hand, and sketch screen and report layouts using pencil and eraser. The whole array of paperwork was overwhelming.

With CASE, the picture is very different. Now, almost all of the paperwork is automated. CASE software replaces pencils, erasers, paper, and drawing templates with easy-to-use computerized techniques. CASE tools enable analysts to take advantage of using structured methodologies without the paperwork overhead of the past.

A top-selling CASE software systems is Intersolve (Excelerator) from Index Technology (Cambridge, MA). Excelerator uses a personal computer as the analyst's design workstation and helps analysts by:

- Standardizing analysis and design specifications
- Generating high-quality graphics, such as data flow diagrams, ERDs, and structure charts
- Integrating the design diagrams with the data dictionary
- Prototyping data collection screens, as well as printed reports
- Maintaining analysis and design documentation
- Ensuring the completeness, consistency, and accuracy of the design

In 1992, over 100 other CASE software systems were available from software suppliers. The list grows yearly as new CASE tools come into the market.

CASE software is such a powerful new tool for analysts that some view it as revolutionary. These zealots want analysts to abandon all the old ways of software development and take up the CASE flag. Instead, we view CASE as evolutionary, the first step in using the computer to help develop systems for the computer. We see CASE as another tool in the analyst's toolbox: one that can help develop the software that the user wants. In a later chapter, we will walk you through a design and implementation of an actual system. Then, you can see exactly how an analyst uses a personal computer to prototype and automate the phases of the systems life cycle.

CASE REPOSITORIES AND DATA DICTIONARIES

A **data dictionary** (abbreviated DD) defines each data flow or data store found in the data flow diagram. A good DD provides a warehouse of facts about the data that can speed application development. A DD's definition breaks the data store or data flow down into data elements. It names each data element, specifies how much space the data element requires, and provides the format for such things as dates, telephone numbers, and cash amounts. Properly developed, a data dictionary creates a framework for organizing and controlling a system's data. In many of the newer CASE products, the concept of a data dictionary takes the form of a **CASE repository**. The repository is the central storehouse of information for the CASE tool, containing not just a data dictionary, but a complete catalog of a system's specifications.

A Catalog of the System's Specifications

If an analyst uses a CASE tool when beginning work on a new system project, the repository for that project starts out empty. As he or she gets an idea of the capabilities and features required in the new system, they start entering information about these requirements into the CASE software tool, and thus into the repository contained within that tool. The repository begins to fill up with specifications for the system.

These specifications include information about data, processes, graphs, and rules. Details about data stores in the repository include data element names and specifications, details about data records, specifications, and the details of each entity and relationship on all the data model entity-relationship diagrams. Process information in the repository includes the names and numbers of each process, how the levels (or explosions) of the processes interrelate, and detailed logic descriptions of the processing associated with the basic (primitive) processes.

The CASE repository also contains information about each drawing created by the analyst. This means that there is an entry in the repository for every data flow diagram, entity-relationship diagram, structure chart, screen layout, report layout, and data model diagram for the entire system.

Finally, the repository stores information about the rules of the business function connected with the system. Insurance underwriting, for example, involves a complex set of rules about what types of risks are acceptable to the company, the extent of coverage the company is willing to sell in certain situations, and the premium rates that they should charge. A system designed to automate underwriting would need to contain precise specifications of these rules, and the CASE repository is the place where these rules are stored. By the time the analyst completes a system design, the CASE repository within the CASE tool holds a large amount of very detailed information on that system. Every piece of the system specification package is kept electronically in the CASE repository. At this stage, the repository represents the electronic equivalent of hundreds of printed pages of system specification documentation.

The repository serves many purposes. The most obvious is that it provides a sort of central coordination facility for the CASE software tool itself. The repository is at the center of the architecture of nearly all CASE tools. For example, when an analyst draws a new data flow arrow on a DFD, he or she must enter the detailed specifications of the data in that flow into the CASE tool. The next time that same flow is drawn, the analyst can simply call up the data description for that flow and the CASE tool will automatically enter the details of the data. These data details came from the repository. In another example, an analyst preparing screen layouts with a CASE tool can retrieve most of the data element specifications for that screen from the repository because they were inserted into the repository during the drawing of the DFDs. The repository replaces the need to key them in a second time for the screen design. The repository thus integrates all the capabilities of a CASE tool.

There are other less obvious but equally important purposes for the repository, the most important of which are specification integrity, documentation support, and automatic system generation. Specification integrity means that the information in the repository is available to the CASE tool to help ensure that the system's design is as free of errors and inconsistencies as possible. Examples of errors include data elements that are defined but never used, leveled DFDs in which the inputs/outputs do not match those of the parent process, conflicting names for data items, and output fields that are produced but never used.

These kinds of errors are easy to spot and correct for very small systems, but can easily creep into the design for larger systems, which can have hundreds of processes and data elements.

Without CASE, these errors aren't detected until *after* the system development, which is a very expensive stage for correction. With CASE, analysts can find the errors at the logical (blueprint) stage and correct them easily and cheaply. It is like building a house and then finding out that the garage is not going to fit. You would rather have found that error by analyzing the blueprints prior to construction. CASE tools do that sort of checking, resulting in much higher specification integrity for system designs.

CASE repositories also support automatic generation of system documentation. Since all the specifications for the system are in electronic form in the CASE repository, it is easy for the CASE software to extract automatically complete, organized, accurate, and well-formatted system documentation. The documentation support that results is consistent for different systems. This standardization will make it much easier for analysts in future years to make modifications to systems. All the documentation books for all the systems appear the same as long as the company continues to use the same CASE tool.

Automatic system generation is an exciting aspect of CASE. The most powerful (and expensive) CASE tools are actually able to write COBOL or C programs, compose database definitions, write job control language statements, create load libraries, and code screen layouts. This results in automation of extremely difficult technical tasks that require enormous amounts of time and effort by very highly trained people. None of this could occur without the CASE repository. Automatic system generation requires the definition of every detail of a system's logical specifications before the automatic routines can function. When it is time to invoke the automatic features, most CASE tools transmit the information in the repository from the PC on which the system was specified to a mainframe. The portion of the CASE tool that resides on the mainframe then takes over and performs the system-generation tasks. In reality, no CASE tool automates 100% of these system-building activities, but many claim that they are in the 80–90% realm, with improvements taking place all the time.

All CASE tools have a repository of sorts, but the sophistication of the repository varies with the complexity and sophistication of the CASE tool itself. Nearly all CASE tools, however, contain a data dictionary. In the high-end CASE tools, the dictionary is just one part of the repository. In the low-end tools, the dictionary and the repository are often the same thing.

A Closer Look at Data Dictionaries

Because of the universal importance of data dictionaries, let's re-examine our accounts payable system (refer back to Figure 3.1). An analyst would first take a hard look at the data flows in the context DFD and finds data flows for sales invoice, packing slip, vendor check, vendor data, open item report, payment authorization, and management reports.

After identifying the basic data flows for the context diagram, the analyst looks at level 1 and 2 diagrams, identifying any other data flows, as well as data stores. Spotting data stores helps the analyst to design the files that the new system will utilize.

Data dictionaries and repositories describe files, printed reports, screen designs, or data flows. For example, consider how an analyst might define the data elements in the check data flow:

→who you going to write check to,

Check = Vendor-number +
               Amount-of-check +
               Date-check-written +
               Check-number

This explicit definition becomes a part of the system's overall data dictionary that will ultimately contain all key terms related to various data flows, reports, and files in the system.

Data dictionaries generally use the following symbols:

= Equivalent to
+ And
[] Either/or
() Optional entry

And four rules govern the construction of data dictionary entries:

1. Pick words to stand for what they mean: use VENDOR-NUMBER, not XY\PQR or DATA13. Capitalization of words helps them to stand out.
2. Employ unique words.
3. Use aliases or synonyms such as VENDOR-NUMBER, VENDOR-NO, or VENDOR-NUM when two or more entries show the same meaning. However, aliases should be used only when absolutely necessary.
4. Redefine complex words; do not redefine self-defining words.

The backbone of the accounts payable system is the VENDOR-MASTER file, which tracks all the data about vendors, the organizations from whom we buy goods and services. Using the data dictionary format, the analyst could list the following data elements for VENDOR-MASTER file:

VENDOR-MASTER = Vendor-number +
                    Vendor-name +
                    Vendor-address +
                    Telephone-number +
                    Vendor-type +
                    Discount-type +
                    Purchases-YTD

The analyst could further decompose the dictionary definition for the Vendor-Address:

Vendor-Address =   Street +
(Apartment-number) +
City +
State-abbreviation +
ZIP-code

part of Vendor - Master.

Self-defining or obvious words and terms, such as a ZIP code, state abbreviation, or middle initial, are included in the data dictionary. This is essential to the proper functioning of the specification integrity and automatic system generation capabilities of CASE tools. The complete data dictionary for VENDOR-MASTER includes the key field of the file, its order, length, media, and security, as Figure 3.5 depicts. This definition specifies completely what the analyst means with the term VENDOR-MASTER.

We can use a data dictionary to define the accounts payable numeric vendor list, as shown in Figures 3.6 and 3.7. In the case of the accounts payable system, the vendor list must regularly go to management. Among other things, management uses the report to monitor purchases this year to date, check discounts taken, and examine discounts available

Figure 3.5 File or entity data dictionary for VENDOR-MASTER. This file chronicles all the data about a vendor.

Database.

```
System:                    Accounts Payable
File Name:                 Vendor-Master
Analyst:                   Peggy Adams-Russell
Date:                      10/03/93


Element Name          Length      Data Type


Vendor number            8        Alphanumeric
Vendor name             30        Alphanumeric
Vendor-street           27        Alphanumeric
Vendor-city             12        Alphanumeric
Vendor-state             2        Alphabetic
Vendor-ZIP               9        Numeric
Telephone-number        10        Numeric
Vendor-type              2        Alphanumeric
Discount-type            2        Numeric
Purchases-YTD            9        Numeric
Purchases-last           9        Numeric
Discounts-YTD            9        Numeric
Discounts-last           9        Numeric

Key field:                 Vendor-number.
Order of file:             Indexed by Vendor-number.
Length:                    Approximately 40,000 records.
Media:                     Disk.
Security:                  Internal use only.
                           Telephone numbers and
                           financial accounting data
                           present in file.
```

```
Oct. 24, 1993                    FLEET FEET                       Page:  1
                              Numeric Vendor List

=========   ==============================  =======  ============  =======  ============
Vendor      Name                            Type     Terms         Disc     Purchases
Number      Address                                                         YTD
=========   ==============================  =======  ============  =======  ============
0-000000-1  Acme Office Supply              Cap      2% 10  Net 30  2.00       1,594.00
            123 Elm Street
            Scarborough, NY 10510

0-123456-7  Star Electronics                Mtl      Net 30         0.00       8,212.90
            423 Bancroft
            Berkeley, CA 94709

4-223344-5  Utah Telephone                  Adv      5% 5 Net 10    5.00       9,895.00
            490 Broadway
            Orem, UT 84057
```

Figure 3.6 The numeric vendor list contains vital data about purchases, including each vendor's terms for payment.

Figure 3.7 Report data dictionary for the numeric vendor list.

```
System:              Accounts Payable
Report Name:         Numeric Vendor List
Analyst:             Peggy Adams-Russell
Date:                10/26/93

========          =======   ============   ============
Element Name      Length    Data Type      Format
========          =======   ============   ============
Vendor-number        8      Alphanumeric   X-XXXXXX-X
Vendor-name         30      Alphanumeric   none
Vendor-address      50      Alphanumeric   none
Vendor-type          2      Alphanumeric   none
Due-days             2      Numeric
Discount-days        2      Numeric
Discount-percent     2      Numeric
Purchases-YTD        9      Numeric        money

Order of report:     Ascending by Vendor number.
Subtotals:           None.
Final totals:        Purchases-YTD.
Counts:              Number of vendors appearing in report.
Frequency:           Monthly before writing checks.
Length:              Approximately 100 pages.
Type of Paper:       8 1/2 by 11 white with perforations.
Distribution:        To accounts payable clerk.
Security:            Internal use only.
                     Financial data on report.
```

to the organization. This type of information helps management know more about the operation of the business so that they can make better decisions.

The DD for the numeric vendor list includes a list of every field in the report, how the data are ordered (ascending by Vendor-number), totals and counts, frequency, length, who receives the report, and what confidential data it contains. The definition does *not* include a description of such format requirements as titles, column headings, or dates.

For each data element, data store (file), data flow, or report in the system, the analyst could complete a form with all the necessary details. For example, note the definitions of the data element "Vendor-number" in Figure 3.8, and of the "Purchases-YTD" in Figure 3.9. The analyst starts with a list of any aliases or synonyms, such as vendor-no. Then the analyst goes on to describe the term verbally; to specify its length, data type, and format (for output purposes); to stipulate allowed values and heading text (caption); and to list the data stores and reports where the data element appears. Some data elements may have no aliases, may appear in many files, or may have a limit on the values they can hold.

Data dictionaries allow analysts and users to define precisely what they mean by a particular data store, data flow, or report. In that way, they improve communication and reduce misunderstanding between analyst and user.

Some commercial software packages, usually called Data Dictionary Systems (or DDS), help maintain dictionaries with the help of the computer. One of these systems, Cognos' Powerhouse QDDR, keeps track of each term, its definition, which systems or programs use the term, aliases, the number of times a particular term is used, and the size (in bytes of memory) of the term. It also provides a link to commercial database or more

Figure 3.8 Data element dictionary for Vendor-number. The term is defined, given a verbal meaning, length, format, and so on.

```
System:              Accounts Payable
Element Name:        Vendor-number
Analyst:             Peggy Adams-Russell
Date:                10/26/93
_____

ELEMENT NAME:        Vendor-number
ALIASES :            Vendor-no, Vendor-num, Vendor-id
DESCRIPTION:         Unique identifier for each Vendor
TYPE:                Alphanumeric, 8 characters
OUTPUT FORMAT:       X-XXXXXX-X
ALLOWED VALUES:      Not zero or spaces
HEADING TEXT:        Vendor
                     Number
PROMPT:              Vendor Number
HELP MESSAGE:        Enter the Vendor's number
DATA STORE:          Vendor Master
DATA FLOWS:          Any with Vendor data
REPORTS:             Aging report
                     Numeric Vendor List
                     Alphabetic Vendor List
                     Open Item Report
                     Check Register
                     Vendor Analysis
                     Distribution to General Ledger
```

Figure 3.9 Data element dictionary for Purchases-YTD.

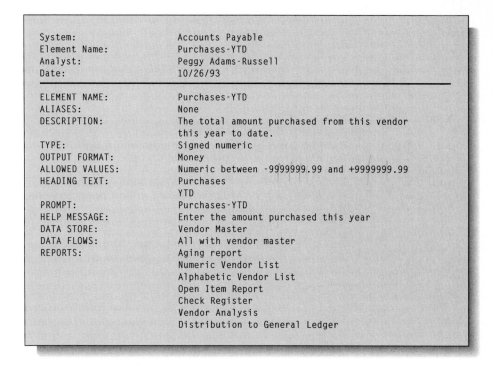

```
System:              Accounts Payable
Element Name:        Purchases-YTD
Analyst:             Peggy Adams-Russell
Date:                10/26/93

ELEMENT NAME:        Purchases-YTD
ALIASES:             None
DESCRIPTION:         The total amount purchased from this vendor
                     this year to date.
TYPE:                Signed numeric
OUTPUT FORMAT:       Money
ALLOWED VALUES:      Numeric between -9999999.99 and +9999999.99
HEADING TEXT:        Purchases
                     YTD
PROMPT:              Purchases-YTD
HELP MESSAGE:        Enter the amount purchased this year
DATA STORE:          Vendor Master
DATA FLOWS:          All with vendor master
REPORTS:             Aging report
                     Numeric Vendor List
                     Alphabetic Vendor List
                     Open Item Report
                     Check Register
                     Vendor Analysis
                     Distribution to General Ledger
```

traditional file managers. The data dictionary can even reside on a personal computer, which can then upload, or transmit, the definitions to a mainframe computer. Thus, automated data dictionary capability is available as part of a CASE tool or as a separate piece of software. In the latter case, none of the system integrity or automated system-building capabilities is available.

ADVANTAGES AND DISADVANTAGES OF MODELING AND DATA DICTIONARIES

Devotees of data models, data flow diagrams, and data dictionaries insist that no other analyst's tool expresses a system's underlying information structure so well. However, like most new tools, modeling diagrams and DDs present both advantages and disadvantages.

The advantages include a proper focus on the structure and flow of data through a system, the ease with which nontechnical people, especially management, can understand them, and a **balancing** feature that builds error detection into a system. Balancing helps maintain proper relationships between parent and child modules. For example, if a parent has three inputs and two outputs, the leveled child diagrams must reflect that fact. If an imbalance occurs between the two, an error exists in their relationship.

On the other hand, some professionals feel that modeling diagrams and data dictionaries are disadvantageous because they divert attention from the details of processing. They feel that modeling diagrams show only data structures and the data inputs and outputs, omitting processing descriptions. Detractors also suggest that users and analysts find it difficult to convert to DFDs and DDs. Despite the time and trouble it might take to master them, data models, data flow diagrams, and data dictionaries provide useful tools to the contemporary computing environment. They are essential parts of structured methodologies.

OTHER SPECIFICATION TOOLS

At this point, we've covered the important tools and techniques that support most system design activity. Data and process models, together with CASE repositories/dictionaries, are the mainstays of modern systems work. But the evolution of methods for systems work was not particularly orderly. As a result, there is quite a collection of other techniques that are in use besides these primary ones. Some companies use these other techniques out of habit because they have not yet picked up on the new methods. Others use these tools on the advice of consultants. Regardless of the reason for their adoption, it is likely that you will encounter one or more of these other specification tools in actual practice.

In the past, software developers relied heavily on the flowchart, which, like the data flow diagram, graphically describes a system. In the last few years, the advent of newer tools have minimized the importance of flowcharts. While many are still around, their use is declining rapidly and we will only mention them as a tool of the past, preferring to focus on the tools for the 1990s.

Structure Charts

The diagrams we drew to depict the systems process in Figures 3.3 and 3.4 are really structure charts. A **structure chart** is a graphical tool that allows the analyst to break a system's processes down into finer components. Unlike the data flow diagram, which shows what data passes between processes, the structure chart focuses on the processes themselves.

Following the structured approach, which begins with generalities and descends to details, structure charts break a system into its components. Therefore, the name of the system appears at the top of the chart, while the names of the even smaller subfunctions or tasks lie on the second, third, and succeeding levels.

Structure charts arrange the modules in order of priority. We read them from the top to the bottom and from left to right. Each module appears as a rectangle surrounding a brief description of the module's purpose, two to four words, beginning with a verb followed by an object (for example, "compute net pay").

The structure chart to assemble a bicycle might include five major modules: open carton, remove parts, group parts, assemble wheels, and finish assembly, as depicted in Figure 3.10. Notice that the order of the modules is listed in the structure chart in the order specified, from left to right.

Each module or task may involve several subtasks. For example, "assemble wheels" involves the separate subtasks of front wheel and rear wheel assembly. When we read a subtask,

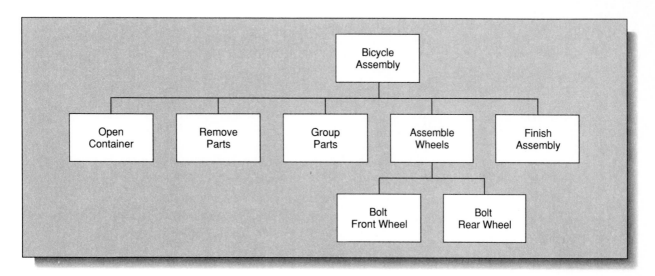

Figure 3.10 Structure chart
for the assembly of a bicycle.

we must complete all the subtasks underneath the parent task before we can slide to the right and start the next task or subtask.

We can use a structure chart to show the details for each subtask, as Figure 3.11 illustrates. "Finish assembly" requires bolting the wheel assemblies to the frame, then attaching handlebars, chain, and seat.

Switching to another example, suppose you wish to prepare a structure chart for an accounts payable system. At the top of the chart, you would draw a box and write "Accounts Payable." Among the subtasks, you would find print check, post invoice, and print reports (Figure 3.12). Remember that you should name tasks according to their function (with the verb-object format) so that anyone looking at the structure chart could identify the purpose of each task or subtask.

After naming tasks or modules, you assign a number to each, just like a data flow diagram. For example, you could label the overall system 0, the print check module 3, post invoice 5, and print reports 6. Such a number system clarifies the relationships among tasks or modules and shows the hierarchy among modules that bear corresponding numbers.

Figure 3.11 Structure chart
for finish assembly.

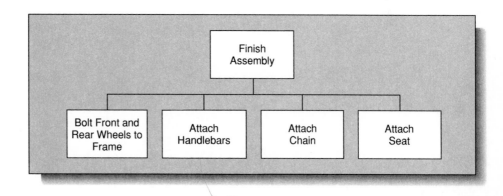

HOW IMPORTANT IS KEEPING CURRENT WITH THE STATE-OF-THE-ART?

"We had a big shoot-out at the office last week," confided Bob Handcock, a 52-year-old analyst, to carpooling friend Sandra. "The new kids wrote Don Price a memo telling him how outdated I am. They said I used obsolete techniques. I've been in this field since its beginning, but sometimes I feel as though I'll never catch up. It changes too fast! I have a feeling that my job is in jeopardy. What should I do?"

Sandra pondered the question. She was an ambitious young accountant who had passed her CPA exams, but had not yet accumulated the 4,000 hours of work experience for the coveted CPA. She responded to Bob's worry by agreeing that computers and electronics are two of the most rapidly changing fields in U.S. enterprise.

"How could *anyone* keep up with it?" she asked.

"Well," replied Bob, "the young kids seem to have less trouble than I do."

Sandra laughed, "That's because they're fresh out of school. But no matter how current they are, textbooks can't replace your 15 years of experience. Maybe you just need a little more book learning."

A few days later, Don Price stopped Bob in the hall and asked him to come by his office in a few minutes. Don greeted Bob with a smile. "I guess you know why I asked you to stop by."

"I suppose it's the memo," Bob replied.

"You're really upset by it, aren't you?" Don queried.

"Well, wouldn't you be if you were in my shoes?"

"I suppose so," responded Don. "Let's talk it out. Where do you feel your strengths and weaknesses lie, Bob?"

Bob thought for a minute. "My strengths are in my abilities to communicate with people and in my writing skills. I seem to have good rapport with my users. They've told me that my training manuals are written with the right amount of detail, while eliminating most of the computer jargon. As for my weaknesses, I would have to say that I don't have the most up-to-date technical knowledge."

"I agree. And what are the strengths and weaknesses of the memo writers?"

"Just the opposite of mine," Bob said. He frowned. "They mean well, I think. They're really bright and eager. But some of them are in for a rude awakening when they try to use some of their theories on real systems."

"Why not offer to trade some of your strong practical experience for a little training in new techniques?"

Bob mulled this over. "That's a good point. I've been using the same methods for quite awhile now. Maybe it's time to learn some new tricks."

Bob left Don's office feeling much better. Don was right. Technical skills are important, but other skills can also make or break a practicing analyst, and some of them don't come from books or school.

Figure 3.12 The structure chart for an accounts payable system.

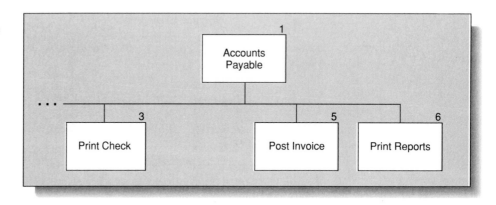

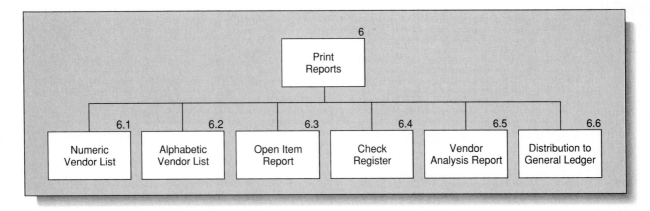

Figure 3.13 The structure chart for the Print Reports module with level numbers.

After assigning level numbers to each module, you can consider whether or not you need to decompose that module further, as done in Figure 3.13. Print Reports (module 6) has six subtasks, each assigned numbers that range from 6.1 to 6.6.

Figure 3.14 highlights Figure 3.12's Print Check module, number 3, because it represents two tasks (remember, a module must have a single purpose). In Figure 3.14, we decompose it to include each of the two parts, or stubs, of the check. The upper stub contains the remittance advice (the date, number, discount, balance, and total of each invoice covered by the check). The lower stub contains the check itself, complete with check number, dollar amount, vendor name, and number.

Applying the top-down concept to the Print Check module, we can add another level of modules: one for printing the remittance advice and one for printing the check itself (Figure 3.15). We assign a third set of numbers (3.1 and 3.2) to this third level of modules. Decomposition ends at any level in which all modules are single purpose.

From the analyst's point of view, decomposition of the check module need not progress beyond the third level of decomposition. However, the programmer who eventually receives the structure chart for this module would probably decompose the module to even lower levels, thus establishing a new series of numbers. The programmer decomposing

Figure 3.14 This structure chart reveals the two components modules within the Print Check module.

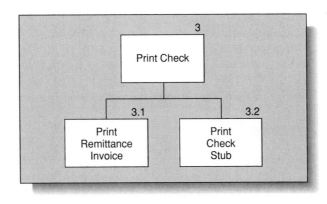

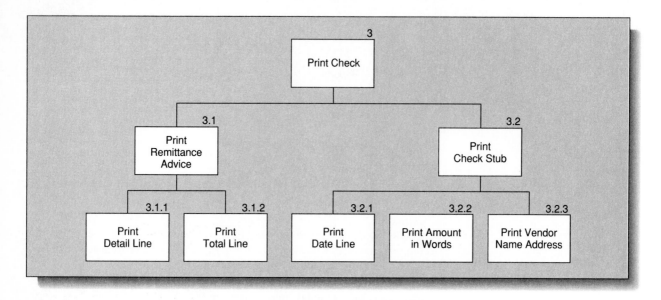

Figure 3.15 Module development of the AP check system from the programmer's perspective.

this module would represent the individual tasks he or she would have to write in their program.

When the programmer finally begins programming the accounts payable Print Check module, he or she would begin at the top left, programming each level before moving to the right and then moving them down. Thus, the programmer first would code Print Remittance Advice (3.1.1), followed by Print Detail Line (3.1.1), and then Print Totals Line (3.1.2). Having coded all of module 3.1, the programmer would next tackle module 3.2, beginning with Print Date Line (3.2.1), moving on to Print Amount in Words (3.2.2), and finishing with Print Vendor Name Address (3.2.3).

Figure 3.15 shows three levels of modules, but complex systems may require many more. Regardless of their number, modules should receive unique and brief names that contain just enough detail for readers to understand their purposes.

The structure chart helps establish valuable system documentation, and it aids as the system eventually enters the design and implementation phases of the systems life cycle. These charts offer several advantages. First, they take little time to draw or modify and many CASE tools allow the analyst to draw them. Second, they graphically convey the system to noncomputer people in an understandable form. Third, by employing standardized symbols, they enable some future analyst to grasp the system quickly. Finally, analysts can draw structure charts of each module, outlining the tasks it must do and making it easier to estimate the time it will take to program a module.

Pseudocode

Pseudocode provides a concise, step-by-step, English-like description of what users want the computer to do. Also called structured English, pseudocode uses simple imperative sentences; it omits most punctuation and all adjectives and adverbs.

The pseudocode for "finish assembly" of a bicycle might read:

1. Bolt front-wheel assembly to frame.
2. Bolt rear-wheel assembly to frame.
3. Attach handlebars.
4. Attach chain.
5. Attach seat.

Pseudocode produces a detailed description of a module, using only three types of sentences: sequence, selection, and iteration (repetition). We call such sentences **control structures** because they dictate the flow of the module's details. Control structures are the patterns for building the logic of a computer program.

Sequence control structure shows an event or action and subsequent events or actions without interruption between events. Figure 3.16a illustrates sequence control structure using pseudocode.

Figure 3.16 Pseudocode sentences fall into one of three control structures: sequence, selection, and iteration. Each has the single entry and exit points required by the structured methodology.

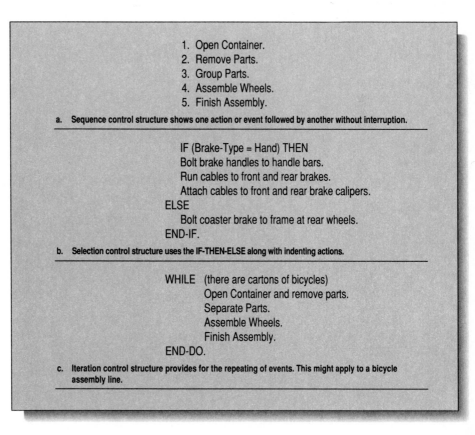

1. Open Container.
2. Remove Parts.
3. Group Parts.
4. Assemble Wheels.
5. Finish Assembly.

a. Sequence control structure shows one action or event followed by another without interruption.

IF (Brake-Type = Hand) THEN
    Bolt brake handles to handle bars.
    Run cables to front and rear brakes.
    Attach cables to front and rear brake calipers.
ELSE
    Bolt coaster brake to frame at rear wheels.
END-IF.

b. Selection control structure uses the IF-THEN-ELSE along with indenting actions.

WHILE   (there are cartons of bicycles)
        Open Container and remove parts.
        Separate Parts.
        Assemble Wheels.
        Finish Assembly.
END-DO.

c. Iteration control structure provides for the repeating of events. This might apply to a bicycle assembly line.

Selection structure provides a means for testing a certain condition to determine which of two possible events or actions should take place next (Figure 3.16b). This structure also goes by the name of IF-THEN-ELSE, with IF representing the test, THEN indicating event(s) that should occur if the test proves true, and ELSE indicating event(s) that should take place if the test proves false. A selection ends with the words "END-IF."

The final control structure, **iteration** (Figure 3.16c) sometimes goes by the name DO/WHILE. This control structure calls for the repetition of all subservient instructions as long as a certain condition holds true.

Figure 3.17 shows what happens when we apply pseudocode and the three control structures to the accounts payable check print logic that we already diagrammed with a structure chart. As the figure shows, the rules of pseudocode include indention of sentences that are subordinate to an iteration sentence. Thus, the four sentences in the middle of Figure 3.17 (IF, Add, Print, and READ) depend on the "WHILE (this Vendor)" iteration sentence. Similarly, the entire center of the pseudocode depends on the "WHILE (not end of file)" sentence. Unfortunately, no hard and fast rules govern the construction of proper pseudocode because each unique problem requires its own unique pseudocode solution.

Both computer and noncomputer people can easily learn to read and/or write pseudocode. Analysts who currently employ older flowcharts to depict program logic can quickly translate them into pseudocode, which more closely follows the rules of structured methodology. As a system evolves, so can the corresponding pseudocode.

Figure 3.17 Process data dictionary showing the pseudocode for the AP Check module. It contains all three control structures.

```
System:            Accounts Payable
Process Name:      Print Check
Analyst:           Peggy Adams-Russell
Date:              10/26/93

    READ invoice record
    READ vendor master record
    WHILE (not end of file) DO
        Initialize total of amounts to zero
        WHILE (this Vendor) DO
            IF (discount available) THEN
                Calculate discount amount
                Subtract discount amount from invoice amount
            ELSE
                Set discount amount to zero
            END-IF
            Add amount to the total
            Print data about this invoice
            READ invoice record
        END-WHILE.
        Print the date line
        Print the amount line
        Print the vendor name and address lines
        Update vendor master record
        READ vendor master record
    END-WHILE.
```

The terseness of pseudocode poses its major disadvantage. Its lack of adv jectives and limited number of verbs can make it sound quite cryptic. Some (grammers feel that pseudocode wastes time because it so closely resembles the COBOL language itself. They prefer to sidestep pseudocode altogether and start by writing their COBOL programs, a practice frowned upon by analysts.

Guidelines to Writing Pseudocode

Many analysts find these general guidelines helpful to writing useful pseudocode:

1. Each sequence control must contain at least one sentence.
2. Selection control structures may have IF-THEN, without ELSE. This occurs when a condition leads to only one possible activity. If the condition is false, then nothing should happen. In such cases, you may omit the ELSE in the pseudocode or write a phrase indicating that nothing should happen.
3. An iteration can specify a critical condition (repeat five times) or the instruction to repeat an activity until a file is complete (for example, "WHILE not end of file").
4. Subordinate sentences in an iteration do not necessarily have to be performed (for example, if we have no invoices for a customer, we do not print a check to that customer).
5. Indent subordinate sentences.
6. Show the end of a selection or iteration structure with an END-IF, END-DO, or END-WHILE.
7. If the pseudocode for a module is longer than one page or CRT screen, divide it into two or more subservient modules.
8. Place parentheses around conditions in an IF or WHILE statement.
9. Avoid weak verbs. Instead, pick strong action verbs such as combine, read, write, or calculate.
10. Consider using an editor or computer word processing program for writing and maintaining the pseudocode.

STRUCTURED WALKTHROUGHS

Analysts and programmers almost always work alone. Now, however, there is considerable teamwork involved in system-building, and this new cooperative spirit is no more apparent than in the growing industry-wide use of structured walkthroughs. A **structured walkthrough** is a meeting of peers in which one of the peers explains a program, data flow diagram, or some other product that they have created. The peers in the room suggest ways to improve it.

The philosophy in the walkthrough is that all of the programmers and analysts will work together to help each other produce better systems. Structured walkthroughs give the

WORKING WITH TECHNOLOGY
TEN GUIDELINES FOR A SUCCESSFUL WALKTHROUGH

Here are ten rules that will help turn any walkthrough into an enjoyable learning experience:

1. A walkthrough is not a formal meeting, but simply a group evaluation by one's peers.

2. Participants should state criticisms without offering value judgments. They are reviewers, not rivals.

3. Participants should adopt an impersonal manner, reviewing the *system*—rather than the *person*—who analyzed, designed, or developed it.

4. The analyst should listen to all comments and suggestions with an open mind.

5. Everyone should remember that each problem has more than one solution. In matters of taste, the benefit of the doubt should go to the system's creator.

6. A walkthrough should be delayed if the material is not ready, if it turns into a witch hunt or inquisition, or when reviewers are not prepared.

7. If participants detect serious errors, the analyst should delay the walkthrough until he or she has had time to review them.

8. A set of attainable objectives should guide the walkthrough.

9. The ultimate number of walkthroughs should be determined when the project is assigned to the analyst or programmer.

10. Materials should go to participants at least two days in advance of any walkthrough.

However, no set of guidelines can ensure successful walkthroughs. Usually, good ones depend on the right attitude, one that says, "Hey, we're trying to solve a problem here. Let's help each other solve it in the best way possible."

analyst or programmer an honest and objective appraisal of the system. Management does not attend these fairly technical reviews and should never use them to evaluate the staff. Otherwise, the participants might not offer candid criticism and advice.

Participants in a walkthrough receive copies of data flow diagrams, programs, data dictionaries, and other pertinent material prior to the review. Note that the walkthrough focuses on error detection rather than error correction, which will come later. Walkthroughs often include personnel at all levels: key users, junior programmers and programmer-analysts, and data entry personnel, as well as senior analysts.

Conducted during all phases of the systems life cycle, these reviews strive to

- Uncover errors, omissions, and misunderstandings in specifications, terminology, or coding
- Familiarize the staff with the system
- Serve as a teaching device for the staff
- Motivate the analyst and programmer to produce higher quality systems
- Promote uniformity across departments and systems
- Serve as a project management tool
- Seek expert advice
- Involve others in the systems process

The focus of a walkthrough depends on the systems phase of which it occurs sis **walkthrough** centers on the analyst's understanding of the problem, the correctness and completeness of the data flow diagram and data dictionary depicting the system under study, and the proposed solution. Participants at this stage may include users, the analyst, and other members of the computer services staff.

A **design walkthrough** concentrates on the completeness and correctness of the design. Do all of the components mesh as planned? Reports must give users the information they need, the database must store all required data elements, data collection must capture data (and check for errors as data flow into the system), programs and modules must get the job done efficiently, new equipment must achieve the system's goals, and costs must remain under control. Attendees at this stage may include people from the computer services staff (analysts or programmers), but not users because of the technical nature of the undertaking.

Implementation walkthroughs ensure that the completed system actually solves the original problem. This walkthrough occurs just before the system goes into use, and it should include careful review of all manuals, training materials, and system documentation. Again, users, the analyst, and members of the computer services staff (manager, other analysts, programmers, and operators) may attend this meeting.

Maintenance walkthroughs certify that the operational system is still performing as planned. Walkthroughs at this stage in the system's life cycle are opportunities for users to critique the system and point out errors, omissions, or enhancements that they need or would like to see.

During the walkthrough itself, the project's analyst, designer, or developer gives everyone a brief overview or tutorial of the material. The analyst then controls and monitors ensuing discussions, recording errors, omissions, or suggested improvements. Prerequisites to a successful session include limiting the number of participants (three to six participants are ideal), distributing materials at least a day in advance, limiting the session to about an hour, emphasizing the error-detection priority of the session, and providing a climate of cooperation in which people feel free to give their honest opinions.

The leader of the session must watch out for trivial discussions, gratuitous criticisms that may simply reflect a matter of taste, lack of preparation by the participants, and having as aggressive superprogrammer try to take over the walkthrough. If the meeting bogs down over deficiencies in the materials, the leader should terminate the session and reschedule it after the analyst repairs deficiencies.

Analysts unaccustomed to having their work examined by their peers may feel threatened by a walkthrough, but the results are valuable and cannot be obtained any other way. Some inexperienced analysts fear that the walkthrough can turn into a competitive witch hunt or that detected errors will damage a reputation or relationship. However, if conducted in an objective and nonjudgmental way, walkthroughs can be enjoyable experiences for the true professional who wants to learn and will not suffer ego damage over constructive criticism. A walkthrough is a peer review and provides an opportunity, of course, to show off impressive work.

SUMMARY

Structured methodologies are needed because they enable analysts to take an organized, standardized approach to the complex task of designing systems. They help ensure that no important steps are left out of the process and that the steps are taken in the proper order. Furthermore, structured methodologies make it easier for analysts to understand each other's work because they will produce approximately the same kinds of specification documents, in roughly the same way, with about the same formats.

One of the basic principles of structured analysis is functional decomposition. This "divide and conquer" philosophy enables analysts to break down complicated large tasks into manageably small subtasks. Also, it leads to a logically organized system structure because each module of the system is designed for a particular, clear purpose.

CASE enables analysts to take advantage of all the benefits of structured methodologies without all of the related paperwork and overhead that accompanied them in the pre-CASE era. As a result, structured approaches are now not only practical, but quite easy to implement.

Data dictionaries were first used to provide a place to store and organize "data about the data." As such, the emphasis was on data elements and data records, and all of the specifications related to them. The data dictionary provided a central reference that reduced naming errors and definition conflicts for data. The idea of the data dictionary is now incorporated in the CASE repository.

The repository is at the center of the architecture of most CASE products. It includes not only data about the data, but much more. Every screen, every modeling diagram, and, in fact, every symbol on every modeling diagram in a system is included in the repository. The repository allows for sophisticated integration of all of the capabilities of a CASE package, supports checking of system specifications for correctness and integrity, and supports code generation and other automatic systems-building capabilities of advanced CASE products.

Modeling, dictionaries, and the general structured approach to systems building are not perfect. They require a long time to learn and use up extra system resources. Most information system professionals agree, though, that the extra effort is justified because of the long-term payback of systems that are easier to maintain and modify.

Several other adjunct tools are useful to the main data modeling and process modeling methods: the backbone of the structured approach. The structure chart uses rectangles to produce an overall picture of a system. Structure charts look like organizational charts, with boxes representing modules rather than employees.

Pseudocode uses simple imperative sentences without adjectives, adverbs, and punctuation to describe the activities of a module, but not the data inputs or outputs.

Increased productivity is the key to success in most organizations, but that only happens when an organization improves efficiency. The following chart compares the major tools that all analysts should have in their toolbox.

Category	Structure Chart	Pseudo-code	DFD	DD	ERD	Walk-through
Analysis	No	No	Yes	Yes	Yes	Yes
Design	Yes	Yes	Yes	Yes	Yes	Yes
Implementation	Yes	Yes	Yes	Yes	Yes	Yes
Maintenance	Yes	Yes	Yes	Yes	Yes	Yes
Graphic	Yes	No	Yes	No	Yes	No
Verbal	No	Yes	No	Yes	No	Yes
Modularity	Yes	Yes	Yes	NA	NA	Yes
Top-Down	Yes	Yes	Yes	NA	NA	NA
Decade of Origin	1960s	1970s	1970s	1970s	1980s	1970s

Most of the tools and techniques are particularly useful in specifying the processing details for primitive processes—those at the bottom of a leveled set of DFDs. The trick is to pick the right tool at the right time. Some tools work best for design activities; others, for analysis or implementation. A few of the tools can cross among two or more activities. With practice, you will learn which tools will work most effectively for each particular problem you encounter.

KEY TERMS

Accounts payable
Functional decomposition
Top-down
Modules
Data dictionary
CASE repository
Balancing
Structure chart
Pseudocode

Control structure
Sequence
Selection
Iteration
Structured walkthrough
Analysis walkthrough
Design walkthrough
Implementation walkthrough
Maintenance walkthrough

QUESTIONS FOR REVIEW AND DISCUSSION

Review Questions

1. What is the principle of functional decomposition?
2. What are the rules for composing a module?
3. Why do we have CASE software products?
4. What are the symbols used to write a data dictionary?
5. Describe the relationship between data dictionaries and CASE repositories.
6. What makes up a data dictionary definition of a file?
7. How do we read the pseudocode for a problem?
8. What is the purpose of a structured walkthrough?

Discussion Questions

1. What are three goals of a structured methodology?
2. What constitutes a poor module?
3. Describe the tools that an analyst would use for design.
4. List six rules to follow when writing pseudocode.
5. Name three rules for conducting a structured walkthrough.
6. When do we hold walkthroughs?

Application Questions

1. Who writes the data dictionary?
2. Write a data dictionary definition of vendor-address.
3. Write a data dictionary definition of vendor-type.

Research Questions

1. Who are the developers of structured methodology?
2. When was the structured methodology first stated?
3. Write the data dictionary definition of the VENDORS-PURCHASES file.
4. Write the pseudocode that compares the date of a transaction with today's date and determine whether payment is over 30, 60, or 90 days past due.

Data Dictionary and Pseudocode for View Video's Customer List

O ne of the primary reports of View Video's video cassette tape rental system is the "Customer List," as seen in Figure 3.18. This report lists each customer (in order by customer number, which is their telephone number), the customer's name, balance, number of tapes rented, date joined, date of last rental, and income earned.

The report is useful in that it tells View Video's management how much business each customer provides, as well as their activity level. From the report, we can see many different types of customers. Hakimoto owes a significant amount of money. Another customer, Lanning, joined, rented two tapes in a month, and has not come back to rent a tape in six months. Macauley has rented many tapes (142) and earned View Video a substantial amount of money. Data are valuable and tell us information about our customers that we would not know unless we had access to the facts and figures.

With the report designed, Frank writes the data dictionary definition for the report (Figure 3.19). For each entity, he lists the entity's name, length (in character count), data type (alphanumeric or numeric), and editing requirements. The editing tells how the data will appear on the report and he uses familiar words to describe them, such as date, money, phone number, or none.

Frank's dictionary definition also tells all the physical facts about the report: expected length, type of paper, who receives the report, frequency of printing the report, and security. The physical factors are important because they help the maintenance phase of the systems life cycle.

Figure 3.18 View Video's Customer List.

```
Date:    9/20/93              View Video, The Movie People              Page:  1
                               C U S T O M E R   L I S T

                                                                    > not personal
                                                                    ============ Income
=================   ===============   ==========  ==========  =========  =========  ==========
Customer            Customer          Customer    Tapes       Date       Last       Income
Number              Name              Balance     Rented      Joined     Rental     Earned
=================   ===============   ==========  ==========  =========  =========  ==========
(916)122-5463       Macauley, Tammi       5.00       142      01/23/91   11/29/92      388.00
(916)122-6782       Nancebo, David                    56      01/12/92   12/12/92      150.00
(916)333-4182       Laki, Michelle       25.00        14      11/21/92   12/19/92       28.00
(916)333-8111       Hakimoto, George     50.00        35      03/05/91   12/23/91       80.00
(916)652-1140       Lanning, Floyd                     2      06/19/92   06/23/92        5.00
       *                   *              *            *          *          *            *
       *                   *              *            *          *          *            *
       *                   *              *            *          *          *            *
                                       ==========  ==========                       ==========
Grand Totals                            1,230.00     9,267                            15,789.00
```

Figure 3.19 Report data dictionary of the Customer List.

```
System:                   Video Cassette Tape Rental
Name of Report:           Customer List
Analyst:                  Frank Pisciotta
Date:                     10/03/93

=====================     =======   ================   ===============
Element Name              Length    Data Type          Format
=====================     =======   ================   ===============

Customer number             8       Alphanumeric       phone number
Customer name              16       Alphanumeric       none
Customer balance            8       Numeric            money
Tapes rented                6       Numeric            9,999
Date joined                 6       Numeric            date
Last rental                 6       Numeric            date
Income earned               7       Numeric            money


Order of report:          Ascending by Vendor number
Order of report:          Ascending by customer number
Subtotals:                None
Final totals:             Balance, tapes rented, income earned
Counts:                   None
Frequency:                On demand
Length:                   One page for every 50 customers
Type of Paper:            8 1/2 by 11 white with perforations
Distribution:             To owners and store manager
Security:                 Internal use only; telephone numbers
                          and financial data present on report
```

Frank Pisciotta chooses pseudocode for the report because he can write it with his word processor and make changes as he develops it (Figure 3.20). He is careful in his pseudocode to follow the structured methodology of sequence, selection, and iteration. His logic reads a record before printing headings in case the file is empty and reads another record at the end of the WHILE iteration structure. He counts lines on each page, subtracting 1 from lines remaining, until finally there is no more room on the current page; in other words, lines remaining works its way down to 0. At this point, he calls the printing of a new page, along with headings.

This is just one report that Frank will have to design. The others will wait until the system moves into the design phase. The early development of a report helps the users (Mark and Cindy Stensaas) get a grasp on what the system will produce. It also helps boost their confidence that the system is what they want.

Figure 3.20 Pseudocode for the Customer List report.

```
System:              Video Cassette Tape Rental
Process Name:        Customer List
Analyst:             Frank Pisciotta
Date:                10/03/93

Initialize totals to zero.
Initialize the page number to 1.
Get the report date from the operating system.
Set lines remaining to 0.
READ a record from the Customers Master File.
WHILE (not end of Customers Master File)
    IF (lines printed < = 0) THEN
        Advance paper to top of page.
        PRINT heading lines.
        ADD 1 to page number.
        Set lines remaining to 50.
    END-IF.
    PRINT detail line.
    ADD balance, tapes rented, income earned to grand totals.
    SUBTRACT 1 from lines remaining.
    READ a record from the Vendor Master File.
END-WHILE.
PRINT grand total lines three lines after last detail line.
```

Working With View Video

This chapter and the associated View Video case study illustrates many tools that analysts have at their disposal. It takes practice to gain skill and confidence with any tool. The more you use it, the better you'll get.

The same is true for the data flow diagram, data dictionary, or pseudocode. You must practice them to get better.

Case Study Exercises

1. Write an entity data dictionary for View Video's MOVIES file. See the example for VENDOR-MASTER (Figure 3.5).
2. Design a report that lists the elements in MOVIES similar to the one Frank did for the Customer List (Figure 3.18).
3. Write the pseudocode for the report you designed. See the example in Figure 3.20 for the View Video Customer List.
4. Write a report data dictionary for the MOVIES report. See the example in Figure 3.19.

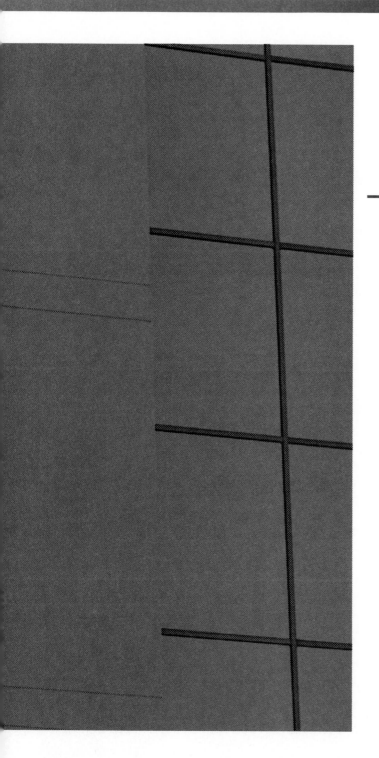

PART II

Systems Analysis and Modeling

CHAPTER 4
Preliminary Systems Analysis

CHAPTER 5
Detailed Analysis

CHAPTER 6
Prototyping and Fourth-Generation Languages

Preliminary Systems Analysis

GOALS

After reading this chapter, you should be able to do the following:

- Describe preliminary analysis and the analyst's main function during this phase of the systems process
- Cite the five stages of an interview
- List four steps for setting up a meeting
- List the topics that a preliminary report to management should cover
- Outline at least four factors that management will usually take into account before allowing an analyst to proceed with detailed analysis

PREVIEW

Suppose you develop an idea for a new digital watch. Before you spend a lot of time actually making your watch, you might make a sketch of it and write up a short "sales pitch" to stimulate an investor into providing venture capital. After receiving some start-up funds, you could then buy the supplies necessary to draw more detailed diagrams. Detailed plans would enable you to determine accurate costs for manufacturing your watches. You might think of such an initial sketch as a preliminary analysis and your later, more detailed, plans as a detailed analysis. Detailed analysis, of course, could lead you to design a manufacturing facility and to actually start producing your new watches. Detailed analysis could also lead you to conclude that the idea has no advantages over existing watches and you should drop it now, not after building a multi-million dollar factory.

In Chapter 1, we saw a broad overview of the systems life cycle and all the steps that an analyst takes to develop a system. This chapter and the next will look at the first phase in the systems life cycle: analysis. Later chapters will deal with systems design, implementation, and maintenance.

AN OVERVIEW OF ANALYSIS

Analysis begins when a "user" such as an accountant or sales manager recognizes a problem and requests a solution to it, such as a new report or an improvement to an old report. We will call the initiator of the systems process the "user" even though the person may not function as a true user of the system. In some situations, the initiator can take the form of a new law or regulation imposed by some branch of the government. In other cases, a customer, manager, vendor, or supplier may start the systems process.

The structure chart in Figure 4.1 will help you visualize the entire analysis phase. As you can see, systems analysis has two subphases: preliminary and detailed.

During **preliminary analysis**, the analyst reviews the original request from the "user." The analyst then returns to the user, asking for a more complete description of his or her needs, in an effort to understand the full details and implications of the request. All along the analysis phase of the systems process, management should provide expertise and sufficient funds to staff the analysis study.

Detailed analysis, as its name implies, is a more in-depth review of the user's request and results in a **feasibility study**. This report offers a complete description of the system's objectives, including alternative solutions, costs, revenues if there are any, and benefits. If management personnel approve the recommendations made in the feasibility study, they will authorize the analyst to proceed to the design phase of the systems process.

Before formally responding to the user's initial request, the analyst should develop an overall perspective of an organization's information needs, its structure, and all relevant

Figure 4.1 A structure chart for analysis. This chart levels analysis to show more detail, breaking it into two subphases: preliminary and detailed analysis.

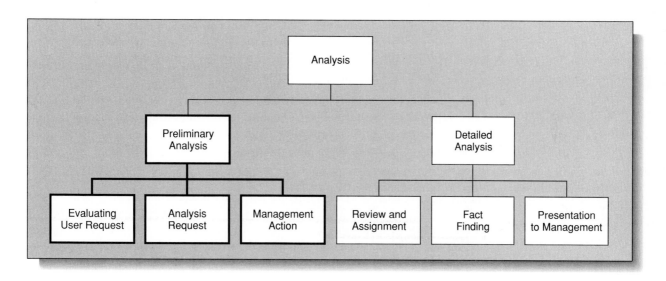

personnel. The analyst can best accomplish this by reviewing the company's organization chart and relating it to currently operating systems. For instance, the analyst involved with Fleet Feet's accounts payable system will want to determine who manually tracks purchases since that person will probably have a direct involvement in decisions about the proposed system.

Then, the analyst can construct a context data flow diagram to illustrate the existing logical system, if there is one. A data flow diagram of the current system will help the analyst understand the problem that the user wants to solve. Furthermore, the diagram enables both analyst and user to visualize how the current system functions, and it helps assure the user that the analyst understands the current system and how it relates to the business.

Preliminary analysis is a fairly general investigation of the request. Assuming management approval of preliminary recommendations, the analyst would then proceed with a detailed study of the current system and the ways it might be improved. In a later chapter, we will explore the subject of detailed analysis more thoroughly.

THE PROBLEM-SOLVING ATTITUDE

Since the user's request initiates the first phase of the systems process, an analyst must treat it seriously. Even if the request seems ill-informed or unreasonable to someone familiar with what computers can and cannot do, a technical specialist should always keep in mind that the user does have a genuine problem. The user may state the request poorly or vaguely and the user may even offer unworkable suggestions based on a lack of computer knowledge. If so, the analyst must not forget to maintain a problem-solving attitude and treat the user with respect, sensitivity, and tact. Otherwise, an adversarial or competitive relationship may obscure the need to solve a problem and create an atmosphere in which the user fights against the analyst's recommendations.

Remember, the analyst may understand technical issues better than users, but he or she seldom understands the business or its organization as well as someone who has worked in it for years. Bear in mind, too, that the user may fear or dislike computers. If so, a smug, flippant, or condescending attitude will do little more than reinforce that focus. Good analysts form partnerships with users, basing their relationships with less computer-oriented personnel on trust and open, sympathetic communication.

Education of users can also strengthen the analyst's role in problem-solving. Some users do not fear or dislike the computer as much as they simply misunderstand its capabilities and limitations. To them, the computer takes on almost magical qualities. When that happens, the analyst can demystify the machine and reduce it to its proper place in the organization. A good analyst knows that the more knowledgeable users become, the easier his or her tasks can become.

The extent to which the analyst clearly understands the user's problem will determine the success or failure of the system that the analyst proposes. Good analysts have an uncanny ability to identify the right problem and to then stay focused on that problem until they find the best solution.

Figure 4.2 Memorandum from a user requesting a systems study.

FLEET FEET
INCORPORATED

MEMORANDUM

DATE: July 20, 1993

TO: Peggy Adams-Russell, Systems Analyst

FROM: Sally Edwards, President, Fleet Feet   *SE*
Sid Moore, Vice President, Franchise Sales   *SM*

SUBJECT: Accounts Payable

For the past few months, I've watched our accounts payable system and have noticed that we're not taking full advantage of discounts offered by our suppliers. I've talked with our book-keeper, Debbie Delfer, and she assures me that we have sufficient cash to make these payments, leading me to feel that this isn't a cash flow problem. I think we could save a substantial amount of money by taking advantage of more, if not all, of these discounts. My calculations show that we could have saved over $1259 in just March and April.

Will you perform a study of our accounts payable system to discover how we can take advantage of these "lost" discounts?

PRELIMINARY ANALYSIS

The structure chart in Figure 4.1 levels preliminary analysis into three activities: evaluating the user's request, analyzing the request, and management's action on a preliminary report. These activities reflect a series of logical problem-solving steps, and they apply to all situations, from something as simple as preparing a new report to the ambitious and complex task of reorganizing an entire operation.

Evaluating the User's Request

A user's request often takes the form of a memorandum, as shown in Figure 4.2. If the organization hires an outside consultant, that person usually receives a formal letter outlining the problem and asking for fee and schedule estimates. In a company with its own computer services or management information systems department, an interdepartmental memo may briefly describe a problem or request a certain improvement. Some organizations require users to complete a special form for any request. This form should include spaces for all relevant facts, such as:

- The date
- The user's name and telephone number
- The subject of the request, in title form
- A description of the problem or situation
- Comments by the analyst

As shown in Figure 4.3, the user fills out the top portion of the form and sends it to the computer services department. In this case, Sally Edwards (Fleet Feet's president) and Sid Moore (vice president of Franchise Sales) request an analysis of the accounts payable system. They want to determine the feasibility of establishing a procedure that would allow the company to take advantage of vendors' discounts for prompt payments. Copies of the form will go to the user, to the personnel assigned to the project, and to the analyst's files.

In evaluating Sally and Sid's request, systems analyst Peggy Adams-Russell considers several factors:

- Could this enhancement enable the organization to achieve its goals more effectively?
- Does the request attack a critical or pressing problem?
- Does the computer services staff possess the expertise necessary to solve the problem?
- How will the request influence existing software and/or hardware?
- Will the system attain any savings or cost avoidances?
- Can the system be developed in an acceptable amount of time?
- Given the backlog of work, what priority should this request require?

Using rough estimates, experience, intuition, and perhaps some quick research, the information services manager should answer these types of questions promptly. If the answers seem mostly positive, the manager will authorize the second stage of preliminary analysis; if they seem mostly negative, the study might stop at this stage. In this case, the manager may decide that perhaps the solution to the problem does not require the application of a computer at all.

Most of the questions in this particular situation attempt to answer an obvious question: "What is the problem?" However, the answers are not so simple because they imply a rather vigorous consideration of the problem. For example, savings may be extremely difficult to project, especially if they accumulate or accelerate over a long period of time.

In another case, a large West Coast manufacturer received a request to buy a faster photocopy machine. The problem sprung from the fact that the current copy machine made 7,000 copies per day, but was intended to produce only 2,500. As a result, people were spending an inordinate amount of time in line as they waited for their turn at the machine! The solution to the problem might seem obvious: Buy a faster copy machine. However, the analyst performing the preliminary analysis asked the right question: "What are all these copies for?" Further investigation revealed that two users were monopolizing the

Figure 4.3 A completed user's request form. The user completes the request portion and sends it to the computer services department. The manager evaluates the request and then makes the assignments, as shown here.

REQUEST FOR SYSTEM SERVICES

Request Date: July 20, 1993

Requester: Sally Edwards, President
Sid Moore, Vice President

Telephone Number: 371-8665

Subject of Request: Accounts Payable System

Description (Use additional pages if necessary):

For the past few months, I've watched our accounts payable system and have noticed that we're not taking full advantage of discounts offered by our suppliers. I've talked with our bookkeeper, Debbie Delfer, and she assures me that we have sufficient cash to make these payments, leading me to feel that this isn't a cash flow problem. I think we could save a substantial amount of money by taking advantage of more, if not all, of these discounts. My calculations show that we could have saved over $1259 in just March and April.

Will you perform a study of our accounts payable system to discover how we can take advantage of these "lost" discounts?

This portion of the form is for MIS staff use:

Date Received: July 24, 1993   Assigned To: Peggy Adams-Russell

Action Taken: Request approved. To begin by August 1 and be completed by August 15.

File Number: 84-023   Approved By: Peggy Adams-Russell

Form No. IS-36-1993

machine. In the end, the analyst approached those two users about the real problem, helped them solve it by limiting their copying to certain hours, and avoided the cost of a new high-speed copy machine.

Assuming an overall positive reaction to the request, however, and given a clear and concise definition of the problem, the computer services department manager assigns an analyst to the task. To confirm this decision and alert staff to it, management distributes a second memo to all concerned (Figure 4.4). In addition to informing everyone about the impending study, the memo solicits their cooperation and helps to eliminate rumors and misconceptions that might impede the project. By now, the request has progressed from the idea stage to the point where preliminary analysis has actually gotten underway.

Analyzing the User's Request

Preliminary analysis continues with two steps of thorough research. First, the analyst must determine whether the user has identified the real problem or need. For example, Sid Moore has requested that Peggy Adams-Russell undertake an analysis into ways the computer can help her more easily identify invoices that offer discounts for prompt payment. Peggy discovers that since its inception, Fleet Feet has paid all bills on the 24th of each month. After examining Fleet Feet's monthly checking account records, she determines that the company enjoys sufficient cash resources to permit bill payment at any time during the month. She concludes that Sid's request could indeed save the company money: The real problem stems simply from tradition, not from a possible cash flow crunch.

Figure 4.4 Letter of authorization sent to the personnel involved with the system under study. This memo confirms management's approval to proceed with preliminary analysis.

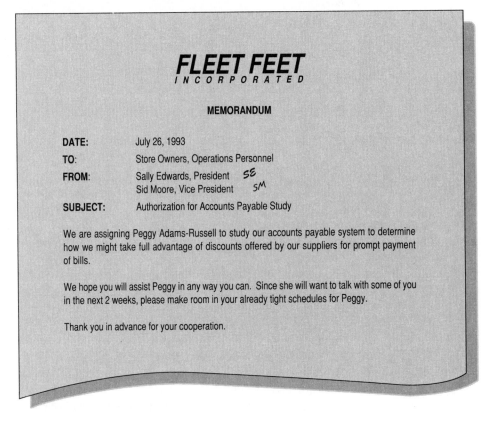

FLEET FEET
I N C O R P O R A T E D

MEMORANDUM

DATE: July 26, 1993

TO: Store Owners, Operations Personnel

FROM: Sally Edwards, President   *SE*
Sid Moore, Vice President   *SM*

SUBJECT: Authorization for Accounts Payable Study

We are assigning Peggy Adams-Russell to study our accounts payable system to determine how we might take full advantage of discounts offered by our suppliers for prompt payment of bills.

We hope you will assist Peggy in any way you can. Since she will want to talk with some of you in the next 2 weeks, please make room in your already tight schedules for Peggy.

Thank you in advance for your cooperation.

Open-ended question.
ex. How do you do your job?
Tell me about yourself.

Second, the analyst must estimate the costs of a detailed study. For Fleet Feet's accounts payable request, this requires a review of past invoices, profit and loss statements, balance sheets, bank statements, stockholders' reports, and other historical documents. A quick review should roughly determine whether a new, or revamped, system could profitably solve the problem. Of course, not all problems require a computer solution. For example, when Sally requested a computerized fixed-asset system for Fleet Feet, she soon learned that it would not really benefit a small firm with only 30 pieces of equipment. Manual calculation of depreciation records, it turned out, still made the most sense.

FACT FINDING AND THE INTERVIEW

closed-end

People usually provide one of the best sources of information for an analyst's research. Therefore, an analyst examines organization charts and job descriptions (Figure 4.5) in order to zero in on the right subjects for **interviews**. An interview may take place in a formal meeting or in an informal setting, but in any case it will provide face-to-face interaction between the analyst and the people associated with the system. For example, Fleet Feet's organization chart reveals that James Taylor, the finance manager, and his staff share full responsibility for the accounts payable system. Thus, Peggy Adams-Russell, the analyst assigned to conduct the preliminary analysis, decides to interview the key people in Taylor's department: Carol Lindsay, Shelby Katada, and Debbie Delfer.

Before scheduling interviews, analysts should follow organizational protocol, first contacting managers or supervisors to ask for their cooperation. Therefore, Peggy would certainly talk to Taylor before approaching his staff.

Preparing for the Interview

In some cases, selecting the correct people for interviews can be tricky because an organization chart reveals how a firm is supposed to run, not necessarily how it actually works on a day-to-day basis. If Peggy Adams-Russell suspected that this is the case with Fleet Feet's chart, she could follow an invoice from its receipt all the way to payment, noting along the way all those who handled it. A list of these people would confirm the actual operation of the system and indicate the right people to interview.

It helps to plan interviews carefully, making an appointment for a specific time and place and providing in advance a written summary of questions that the analyst wishes to ask (Figure 4.6). This memo allows interviewees to gather relevant information and to consider their responses before the interview takes place. As with most aspects of systems analysis, careful planning helps eliminate inaccurate guesses and saves valuable time. One of the most common complaints within organizations is that too much time is wasted in meetings, so analysts should try to make their meetings as efficient as possible.

With the list of questions well-formulated beforehand, the interview can focus on collecting pertinent answers. Most effective interviews include three steps: establishing rapport, questioning, and summarizing the interview.

Figure 4.5 A sample job description that the analyst might find in Fleet Feet's personnel file.

FLEET FEET
INCORPORATED

TITLE:	Bookkeeper
DATE APPROVED:	March 18, 1992
REPORTS TO:	Finance Manager

DUTIES AND RESPONSIBILITIES:

1. Post accounting transactions to receivables, payables, payroll, and general ledger.
2. Prepare a list of overdue accounts needing follow-up contact.
3. Prepare a list of vendors to be paid to take advantage of discounts.
4. Train new employees in the department.
5. Lead an office staff of two clerks.
6. Perform related work as assigned.

PRIOR EXPERIENCE:
At least 3 to 4 years of progressively responsible experience keeping books for a private business. Ability to post a set of books with transactions of all types, analyze data, draw conclusions, speak and write effectively, work well with others, and conduct oneself in a professional manner. Some background in computing is mandatory. Must be able to operate a 10-key adding machine and a personal computer/terminal. Must be bondable.

EDUCATION:
Two-year community college degree in accounting or similar area. Prior experience can be substituted for education. High school diploma mandatory.

SALARY RANGE:
Professional level 5.

good for Resume

Establishing Rapport

During this part of the interview, the analyst seeks to establish rapport with the interviewee, striving to create a relaxing atmosphere and making the interviewee feel like a partner in the process. Topics for opening the discussion can range from the weather to the outcome of a ballgame or some other informal topic. If the analyst has dealt with the interviewee in the past, reminiscences can help establish the needed rapport.

Analysts conduct interviews with a wide range of people who display varying educational and cultural backgrounds: corporate officers, middle managers, clerks, and assembly-line workers. A sensitive and friendly warm-up can establish the analyst as a concerned problem-solver and help open up the channels of communication.

Figure 4.6 To prepare people for interviews, an analyst should send a memorandum sufficiently in advance to allow them to gather relevant information and thoroughly consider their responses.

FLEET FEET
I N C O R P O R A T E D

MEMORANDUM

DATE: August 3, 1993
TO: James Taylor, Finance Manager
FROM: Peggy Adams-Russell, Systems Analyst *PAR*
SUBJECT: Accounts Payable Systems Study

This confirms our meeting for August 8 at 3:00 p.m. in your conference room. As you are aware, I am analyzing our AP system to determine how we can take advantage of supplier discounts. I want to cover the following topics.

1. On the average, how many invoices do you pay each month?

2. What is the average balance in Fleet Feet's checking account?

3. Are there any time periods during which we need extra cash to meet specific needs?

4. How long does it take an invoice to get from the warehouse to your office?

5. Once an invoice is in your possession, how long does processing take?

6. Are invoices complete when you receive them or do you have to contact the warehouse for missing data? Please give me some typical examples of missing data.

7. Since all checks must be signed by the vice president, how long does it take for them to be returned to you for mailing?

8. After the signed check comes back to you, how long before you mail it?

9. Do you have all the necessary equipment and staff to perform the duties assigned to your office?

10. Are you aware of any other problems in the AP system?

If you need any clarification or further information, please feel free to contact me.

Questioning

Once the people are relaxed, the analyst sets the scene for **questioning**, beginning with a review of the reason for the interview, citing management's authorization, and stating the problem. Although questions should generally follow the outline prepared beforehand, an effective communicator knows how to adapt the lines of questioning based on the answers given by the interviewee.

During questioning, the analyst listens carefully, asks relevant questions, and tries to elicit complete answers. Good listeners allow sufficient time for full responses. Try not to interrupt unnecessarily, avoid provoking expected responses, and resist answering for the interviewee.

If an interviewer attacks an answer or telegraphs a value judgment, the interviewee may feel threatened, and the initial rapport will evaporate. Most importantly, the interviewer should minimize computer jargon and buzzwords, relying instead on common language people use in their daily work. Remember that the interviewee, not the analyst, is the expert on the subject under discussion. Facial expressions and body language should not reveal the interviewer's reactions, attitudes, or feelings about answers. If clarification of a response seems necessary, the interviewer may ask, "Just to make sure I understand your answer, do you mean…?" rather than assuming that he or she knows what the other person really means to say.

If the analyst's list of questions is long, is technically complex, or requires detailed answers, the interviewer should take brief notes summarizing the responses, but should not need a verbatim transcript. Some analysts make video- or audio-tape recordings of interviews to make sure nothing escapes their attention. Because many people feel uncomfortable with a microphone or camera present, the analyst should obtain their permission before using such equipment.

Summarizing the Interview

After all the planned questions are answered by the interviewee, the interviewer terminates the discussion, perhaps by asking whether the interviewee has any questions. The interviewer then thanks the other person and indicates whether or not a second interview might follow. Sometimes it makes sense to conclude with a little idle conversation, a sort of "cooling down" counterpart of the establishing rapport stage.

Following Up After the Interview

Shortly after the interview, the analyst should summarize the findings in writing, forwarding a copy to the person interviewed, as seen in Figure 4.7. A good summary:

- Produces a record of information gained from the interview (which can also help protect the analyst if, later on, the interviewee remembers the conversation differently than the analyst does)
- Allows those interviewed to have the opportunity to check for accuracy
- Permits the analyst to express formal appreciation, which helps maintain rapport between the parties

Interviews afford the analyst an opportunity to personalize the fact-finding process, helping everyone relate to each other as people, rather than as just names on a chart, voices over

Figure 4.7 An example of a memorandum summarizing an interview.

FLEET FEET
I N C O R P O R A T E D

MEMORANDUM

DATE: August 9, 1993

TO: James Taylor, Finance Manager

FROM: Peggy Adams-Russell, Systems Analyst *PAR*

SUBJECT: Summary of August 8 interview

I greatly appreciate your time yesterday, since I know how busy you are. Your advice was valuable and confirmed some of my feelings about our AP system. My notes show that we covered the following:

1. We are paying the following number of invoices:

Month	Number of Invoices Paid	Average Value
April	78	$455.66
May	92	$603.90
June	84	$505.23

2. Fleet Feet's average checking account balances for April, May, and June were: $20,331.56, $22,099.51, and $23,347.96.

3. Fleet Feet needs extra cash ($3,656.55) the first week of each month to pay utilities and the mortgage on the building and parking lot.

4. It takes 6 days for an invoice to get to your office from the warehouse.

5. Once an invoice is in your possession, it takes an average of 7 days to process it.

6. All necessary information appears on the invoice when you receive it from the warehouse. Occasionally, invoices do arrive stained as a result of careless handling.

7. The process of securing signatures on checks takes 2 days.

8. Once the check comes back signed, you mail it within 24 hours.

9. You feel that you need an additional staff person and a typewriter and calculator to keep up with the workload.

10. You do not see other problems in the AP system.

If you would like to add or change any information in this memo, please let me know. Thanks again for your input.

a telephone, or faceless authors of impersonal memoranda. Among the disadvantages of the interview approach are the time that it takes to schedule, conduct, and follow-up on interviews, and the danger that personal biases on the part of either participant may creep into the process.

CONDUCTING A MEETING

Muttering under her breath, Heidi Kantola walked back to the office that she shared with Douglas Manufacturing's other systems analyst, Mike Lee. After working on a preliminary report for a new system for almost three months, Heidi had finally finished the task and scheduled a meeting to present her recommendations to Douglas's department managers. Although she began the meeting with great enthusiasm, it turned into a disaster. When people entered the meeting room, Heidi stood at the door, handing out a three-page detailed agenda and a 20-page report that she had spent days preparing.

Hoping to impress everyone, she invited assistant managers as well as department heads, and she soon ran out of reports. The small room did not contain enough chairs for all present, so before Heidi could call the meeting to order, she wasted time finding more chairs. Scuffling feet, rattling chairs, and the shuffling of papers got the meeting off to a shaky start, but when it finally got underway, Heidi took a deep breath and asked if anyone had questions about her agenda and report.

Of course, no one had had time to read through all those documents. The head of the advertising department asked a question about word processors, an item not on the agenda. Heidi spent 15 minutes providing a detailed answer while others in the room began to chat among themselves about the upcoming company picnic. When the national sales manager excused himself to tend to important business, five other people sneaked out of the room. Heidi struggled to establish control, but finally adjourned the meeting when the conversation in the room drifted to the weekly football office pool and to the point spreads among the teams playing that week. No one had discussed the key items on the agenda, much less her recommendations.

Fearing that she had made a fool of herself, Heidi approached Joe Sandvan, a retired Douglas management information systems manager with a reputation for handling people skillfully. Joe had hired Heidi, so she felt that he might agree to help her recover from her setback.

Joe listened as Heidi described her meeting and all the problems that she had encountered. He then offered several suggestions. First, he pointed out that Heidi had not planned the meeting well. Should she have arranged for a larger room with sufficient seating, or should she have restricted attendance to department heads? Heidi

THE PRELIMINARY REPORT FOR MANAGEMENT

After researching the request and estimating the cost of a detailed analysis, the analyst presents findings to management. Incorporating information gleaned from job descriptions, organization charts, files, and interviews, the analyst submits a written **preliminary report**, which usually contains four sections:

1. Problem review
2. Findings
3. Cost and schedule estimates
4. Recommendations

Whenever possible, analysts should pose alternatives so that management can weigh the costs and benefits of more than one solution to a given problem. Since most business decisions

wished that she had done the latter. Why hadn't Heidi distributed the agenda a few days earlier, giving those attending a chance to prepare? Heidi hadn't thought of that. Joe went on to point out that a meeting's leader should avoid losing the audience's attention and should deflect irrelevant questions or those of interest to only one or two people.

Meetings cost time and money. A one-hour meeting for 16 people, with an average cost per person of $50 per hour in lost productivity, amounts to $800.

Considering the expenses involved and the reluctance of most people to attend meetings, Joe recommended the following "rules" for more effective meetings:

1. Avoid calling a meeting if a memo and a series of telephone calls would serve your purpose.

2. Refrain from calling a meeting to decide something you could or should decide yourself.

3. Invite only personnel vital to the discussion.

4. Insist on punctuality. If you're 2 minutes late for a meeting with 20 people, you've wasted 40 minutes.

5. Keep the purpose of the meeting in mind.

6. Draft an agenda. Even a lengthy agenda, if well constructed, could mean a short meeting.

7. Circulate the agenda sufficiently ahead of time so that those involved can prepare for the meeting.

8. Set time limits for each item on the agenda. Discussion, like work, expands to fill the time available.

9. Limit meetings to about an hour.

10. See that the meeting leader states the issues, sticks to the agenda, lets everyone have a voice in the discussion, restricts the discussion to the subject, and sums up at the end.

11. Check the conference room to make sure it is available, clean, and supplied with paper, pencils, an easel or chalkboard, and other equipment.

While some of these rules may seem like mere common sense, many meetings fall apart because someone breaks one or more of them.

Studies of group dynamics in meetings show that small groups of eight to ten people come to consensus more rapidly than larger groups of 15 or more. Very large groups of 40 or more can take forever and get nowhere.

Joe concluded by saying, "My first meeting was even worse than yours. It took me ten years to learn that the key word for every meeting is 'less': less meetings, less participants per meeting, less time in each meeting."

involve trade-offs between costs and benefits, management has already become accustomed to making decisions this way. The decision may range from halting the process to requesting a detailed analysis.

At this stage in the analysis phase of the systems life cycle, the costs and benefits are not final. They represent the analyst's best estimates at an early stage. During detailed analysis, the analyst will estimate costs and benefits.

Before presenting the report orally to the requesting user and appropriate department managers (Figure 4.8), the analyst circulates it for review. This allows everyone concerned to mull over the analyst's findings, discuss the findings among themselves, and investigate questionable statements. In Fleet Feet's case, Sally Edwards would confer with her staff, discussing the analyst's recommendations, cost estimates, and proposed schedule for detailed analysis.

Although the person or persons who requested the study (Sally Edwards and Sid Moore, in our example) would preside over the oral presentation, the systems analyst (Peggy Adams-Russell) leads the discussion, responding to management's questions. The analyst should pay special attention to communication and try not to condescend to the audience or use unnecessary jargon. If the preliminary report indicates that a new system would not solve the problem or would cost too much, management may halt the process to

Figure 4.8 A preliminary report to which management can react.

FLEET FEET
INCORPORATED

MEMORANDUM

DATE:	August 12, 1993
TO:	Sally Edwards, President
	Sid Moore, Vice President
FROM:	Peggy Adams-Russell, Systems Analyst *PAR*
SUBJECT:	Preliminary Report on Accounts Payable System

Problem Review

We have now completed our preliminary investigation of the accounts payable system that we undertook to ascertain the feasibility of taking advantage of vendors' discounts.

I interviewed the four key individuals involved with the AP system; summaries of these interviews were circulated to everyone involved. We also interviewed two of our largest suppliers, Nike and Converse, to verify potential discounts.

Findings

During the interviews and our search we learned the following:

1. It takes 16 days for the average invoice to be paid.

2. The average value of each invoice during April, May, and June of this year was about $520.

3. For this same 3-month period, an average of 85 invoices were processed each month.

4. The AP system is functioning smoothly. Our credit rating with the two suppliers interviewed is good.

5. The workload of Debbie Delfer's staff is quite heavy, and we have acquired no new equipment in this area for the past 4 1/2 years.

6. Debbie's staff has increased by fifty percent in the last 4 years, while our business volume has expanded five hundred percent.

7. While the majority of our suppliers offer a 2 percent discount for payment within 10 days, at least two offer a 3 percent discount for payments made within 5 days.

(continued)

avoid wasting time and money. If, on the other hand, the report favors a new or improved system, detailed analysis will probably begin.

In some organizations, the manager responsible for the new system makes the decision to start or stop independently; other businesses use consensus decision-making, letting all interested parties contribute. Businesses in the United States tend to use the former method;

Figure 4.8 *(continued)*

Recommendations
On the basis of these preliminary findings, we recommend the following:

1. A detailed analysis should begin immediately to determine the feasibility of computerizing our manual accounts payable system. Based on 85 transactions per month with an average of $520 and a 2 percent discount for payment within 10 days, there is a potential savings of $884 per month ($520 x 85 x 0.02) or $10,608 per year.

2. Debbie Delfer's staff should receive special thanks for their work.

3. Authorize Debbie Delfer to sign checks. This would save 3 days of processing time. Debbie would have to be bonded, at a cost of a few hundred dollars.

Cost and Time Schedule
The computer services staff is ready to perform the detailed analysis and can begin 4 days after management's approval. Peggy Adams-Russell would continue the analysis and expects that it will take her 5 weeks to complete. The cost of a detailed analysis would be:

1. Five weeks (part time) for Peggy Adams-Russell $3,192.00
2. Secretarial aid (20 hours) 210.00
3. Two weeks for other Fleet Feet staff for interviews and discussion 1,300.00
   Total 4,702.00

the Japanese style of management prefers the latter. Using the consensus approach, Fleet Feet accepts Peggy Adams-Russell's recommendations and decides to proceed with a detailed analysis of its accounts payable system.

The first two steps in preliminary analysis (evaluation and analysis of request) provide a cursory look at a problem. The examination remains purposefully superficial, with no attempt to solve the problem; instead, it concentrates on defining the scope and magnitude of the problem. The third step allows management to discontinue any further examination before large amounts of staff time and funds have been spent.

THE ROLE OF CASE IN SYSTEMS ANALYSIS

Regardless of the CASE tool, the first major task you must perform with it is to describe your system graphically by drawing a diagram. Before CASE tools, systems were described with text, which can often mislead or confuse the reader. Even the best writers require their readers to read from beginning to end in order to fully understand the system. A picture, on the other hand, often tells the whole story immediately.

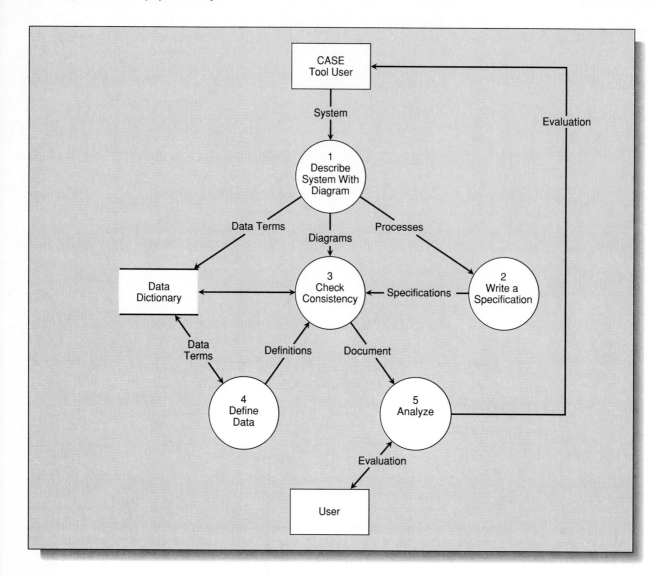

Figure 4.9 A DFD depicting a CASE tool and how it assists in the analysis phase.

Figure 4.9 shows you a picture of CASE software relative to the analysis phase of the systems process. The CASE tool user describes his and her system with a diagram, usually a DFD. The diagram is stored in the data dictionary.

Diagrams are checked for consistency and accuracy. Consistency checks probe deeper into the system design and look for unfinished work or errors in the diagram. Some typical errors on diagrams include:

- A process that has only incoming flows (sometimes called a black hole)
- A process that has only outgoing flows (sometimes called a miracle)

- Unnamed data flows
- Unnamed processes
- Unnamed data stores
- Incomplete or faulty process numbering
- Data flows in which consistency between parent and lower-level diagrams do not match
- Logically impossible data flows between symbols

Consistency checks ensure that the diagram is correctly drawn, although the diagram still may not reflect the system under examination.

For each process in the diagram, the analyst writes a specification, process 2. A specification is a description of the logic of a process. Usually written in pseudocode, the logic uses the familiar IF-THEN-ELSE, WHILE, and REPEAT UNTIL grammar structures. CASE tools often employ a word processor component for the analyst to use in composing the pseudocode.

Once completed, the specification undergoes consistency checks similar to the diagram's (process 3). Some typical errors are

- Use of incorrect key words
- Improper tabbing or indenting of subordinate phrases
- Use of undefined terms

When the specification is correct, each process is defined in logically correct terms.

With diagrams and specifications finished, the CASE user must define data stores and data elements, process 4. The data term definition can include aliases, data type, length, maximum or minimum values, and a remark defining the data term. Individual data elements are grouped into data stores (files), along with an identification of any primary or secondary keys.

With all data stores, processes, diagrams, and data elements completed, the CASE tool can print a document ready for the end-user to review, process 5. The analysis document normally includes:

- A title page
- Context data flow diagram
- Leveled data flow diagrams
- Specifications for each process
- Definitions for all data terms in the dictionary (usually in alphabetical order by each term defined)

Changes in the analysis document go back to the start, with the CASE tool altering the diagram, specification, or data terms. The cycle repeats itself until the user and analyst are sure that they have a proper definition of the system under study.

Most CASE software systems have some restrictions on them, such as:

- Maximum number of data items in the dictionary
- Maximum number of keys per data store
- Maximum number of processes per level
- Maximum number of characters for a data term

Before using any CASE software system, the analyst needs to know the restrictions that the software places on the diagram and on the definitions.

The beauty of the CASE tool is that it automates this important phase of the systems process. CASE tools are indeed the Computer-Aided Design (CAD) system for analysts.

APPLYING PRELIMINARY ANALYSIS: VALLEY BLOOD CENTER

Our first look at preliminary analysis centered on Fleet Feet. In our second, we will look at the Valley Blood Center, which offers another perspective of an analyst's roles and duties as he or she performs a preliminary analysis.

Paul Willey had reached the end of his rope. In charge of recordkeeping for the donation department of the Valley Blood Center, Paul spent most of his time ransacking his filing cabinet. Every donor filled out a "receipt," and Paul had to file each in the proper drawer of one of three filing cabinets that lined his wall. Inside, they looked like the aftermath of an explosion: Scraps of paper overflowed from hundreds of manila folders stuffed inside hundreds of hanging files.

Oh, he knew the paperwork was important, but it sometimes took an hour to sort it all into any semblance of useful form. And now reporting requirements on potential health risks posed by donors—from AIDS, hepatitis, malaria, recent surgery, relapsing fevers, polio, over-age, and high blood pressure—had made the system dangerously inadequate. Time for a computer?

Hoping to solve the problem, Paul called Neal Hundt, Valley Blood Center's only programmer/systems analyst, to plead for help. Neal interrupted Paul's harangue by saying, "Write it down, Paul. It isn't a problem until it's on paper!" Well, that response didn't surprise Paul, because of the Blood Center's reputation for bureaucracy.

Paul wrote the required memorandum (Figure 4.10) and sent it along to Neal, with copies to both his and Neal's supervisors.

Shortly after receiving the memorandum, Neal and his boss—the director of Valley Blood Center's MIS department, Kristen Nobles—checked the schedule of other requests and decided that Paul's urgent need justified Neal's performing a quick preliminary analysis.

Reviewing the center's organization chart, Neal decides that he first needs to talk with Paul's boss, Rita Ortiz, the Center's Blood Fulfillment Officer. He telephones Rita, who suggests that she, Paul, and Neal meet in her office the following day.

The next day, Neal and Paul leave Rita's office after she gives her full support to the project. As Neal looks over Paul's filing system, he shakes his head. As a donor himself, and as a father whose son has required several blood transfusions, Neal appreciates the

Figure 4.10 Paul Willey's memorandum to Neal Hundt, asking for help in solving the donor recordkeeping problem.

V A L L E Y
BLOOD CENTER

MEMORANDUM

DATE: April 16, 1993

TO: Neal Hundt, Programmer/Analyst

FROM: Paul Willey, Donor Record Tracker

SUBJECT: Donor donation tracking system

Since the Blood Center opened, we have kept records concerning donors manually in a five-drawer manila folder filing system, but recently, the volume of paperwork has made the manual system impractical and cumbersome. We now track 54,236 donors, as well as their donations. Our files are bursting. Furthermore, we now have to track many potential health threats, such as AIDS, a requirement that could make the present system downright dangerous.

To make matters worse, for 1993 the American Blood Commission (ABC) mandates that every blood center report its tracking of donors, donations, and receiving patients. We cannot easily abide by ABC's guidelines with the present system.

cc: Kristen Nobles
     Rita Ortiz

need for an accurate and timely donor tracking system. How has the center gotten along with this antiquated approach for so many years?

"Can you find my folder and my medical history card in that filing system of yours?" Neil asks Paul.

Opening the G-through-I drawer, Paul starts looking. At the same time, Neal observes the second hand on the clock above the cabinets. Twelve seconds later, Paul fishes Neal's folder from the drawer.

"I didn't know that you kept so much data on donors," Neal says. "What are the other papers in my folder?"

"Forms that we save each time you donate. They tell what happened to the pint of blood you gave. We keep them forever."

"I didn't realize how much paperwork was involved in collecting blood." Neal concludes the fact-finding interview by asking Paul to photocopy Neal's medical history card and send it to him when convenient.

Later the same day, Neal returns to chat with Paul's boss, Rita Ortiz, to find out more about all the paperwork that goes to Paul's office. Rita explains what happens to donated blood. "We send packets of blood products to requesting hospitals every day. We're sort of a warehouse and distribution center, so we have to keep track of a lot of things, such as blood types. At the end of the year, I write up a report for management that summarizes everything we shipped and who received what. That report is my biggest headache; it takes about two weeks to complete."

Neal notes that complaint and asks for sample figures, which Rita briefly sketches. "Valley Blood Center serves a sizable geographic area, about 45 hospitals. We ship maybe 350 units a day, seven days a week."

"When you ship a product, how much paperwork gets generated?"

"We fill out 350 of these a day and send them to Paul Willey to cram into those filing cabinets of his."

Having gleaned all the facts he can from Rita, Neal thanks her for her time and asks if he can drop by again in a week or so to fill in any gaps.

Neal returns to his desk and starts sketching a possible system with a data flow diagram (Figure 4.11). He starts with an overview, then adds more details. Donors complete a medical history form before they make their first donation and receive a donor card afterwards. Hospitals request blood products from Rita, who ships them on a daily basis. The Blood Center's management receives the year-end report. Starting in 1993, the American Blood Commission also receives an annual report.

Figure 4.11 Data flow diagram of the blood bank donation tracking system.

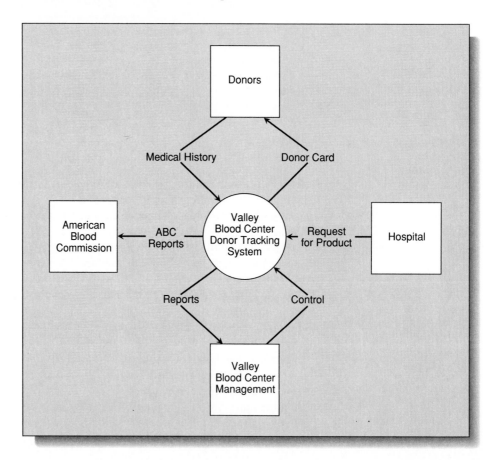

Neal envisions a fairly simple system with which a computer could easily streamline the entire operation, reduce paperwork, eliminate manual files, and make Paul and Rita's jobs more efficient.

The next day, Neal calls up the word processing program on his personal computer and begins composing his preliminary report (Figure 4.12). He agrees that a real problem exists with Paul's filing system and Rita's manual recording of blood product deliveries to hospitals. He also sees a cost-effective potential for computerization.

Figure 4.12 Neal Hundt's preliminary report to Paul Willey and Kristen Nobles.

V A L L E Y
BLOOD CENTER

MEMORANDUM

DATE: April 20, 1993

TO: Paul Willey, Donor Record Tracker
             Kristen Nobles, MIS Director
             Rita Ortiz, Fulfillment Officer

FROM: Neal Hundt, Programmer/Analyst   *NH*

SUBJECT: Preliminary report on donor and product tracking system

This memorandum summarizes my findings with regard to problems in the present donor and product tracking system, and it offers what may be a sensible solution. I hope my recommendations will convince all concerned that we should proceed with a more detailed study. I have estimated the cost and schedule for such an undertaking.

During my interviews, and after examination of existing forms and reports and filing systems, I concluded the following:

1. We are keeping records on over 50,000 donors.

2. An inordinate amount of time is spent filing forms in cabinets.

3. Rita Ortiz spends two weeks preparing the annual management report.

4. Rita Ortiz hand-writes 350 orders per day, all of which Paul Willey files.

5. Paul Willey's job will become much more difficult when the Blood Center adopts the ABC's 1992 guidelines.

6. The possibility for error, misfiling, or loss of paper work is high. In some cases, it takes 3 days to file a form.

7. In a year or so, both Paul and Rita will need additional staff to continue performing their jobs.

8. It takes Paul Willey almost 15 seconds to locate a donor's folder.

9. Paul Willey's three-cabinet system already needs to expand to a fourth or fifth cabinet. The existing cabinets are spilling over.

10. No procedure or plan exists to purge the cabinets on a regular basis.

(continued)

Figure 4.12 *(continued)*

Based on these preliminary findings, I recommend the following course of action.

1. Perform a detailed analysis, with the goal of investigating computerization of the entire donor and product record-keeping system. Computerization would start with the collection of data about donors and continue through delivery of blood products to hospitals.

2. Redesign the donor medical history form to facilitate entry directly into the computer.

I am eager to begin the detailed analysis, which should take about forty hours to complete.

Neal sends his preliminary report to Paul, Rita, and Kristen, then schedules a meeting with all three to discuss it within the week. During the meeting, Neal shows his data flow diagram to Paul, Rita, and Kristen, explaining how the new system would function. He reviews the problems, analyzes possible benefits and costs, and offers his recommendation. Paul and Rita express their enthusiasm for his ideas, and Kristen authorizes Neal to proceed with a detailed analysis (Figure 4.13).

Figure 4.13 Kristen Nobles authorizes the detailed analysis with a memorandum.

V A L L E Y
BLOOD CENTER

MEMORANDUM

DATE: April 21, 1993
TO: Paul Willey, Donor Record Tracker
              Rita Ortiz, Fulfillment Officer
              All department heads
FROM: Kristen Nobles, MIS Director *KN*
SUBJECT: Detailed Study of Donor Tracking System

Neal Hundt has recently researched the feasibility of a computerized blood donor and blood product tracking system. During the next two weeks, Neal will perform an in-depth study, and would greatly appreciate all your help and support. Our first look at this problem has indicated not only that major savings can result from automation, but that with it we might provide better service at lower costs and with fewer mistakes.

For his part, Neal marvels once again about how much he learns as an analy
he's always loved technical details and enjoys working with computers, the reai
job comes from the people he meets and the ways in which he can help them solve pi

SUMMARY

The systems life cycle moves through four phases: analysis, design, implementation, and maintenance. Systems analysis begins when a user or manager requests a study of a problem in either an existing system or a projected one.

Analysis itself breaks down into two stages: preliminary and detailed. During preliminary analysis, the analyst evaluates and researches the user's request, estimates costs to perform a detailed analysis, and lists objectives of the system. These findings come together in a preliminary report.

To arrive at a preliminary report, the analyst interviews key personnel in the organization. Preceding the actual interview, the interviewer develops a list of questions. Effective interviews contain five components: preparation, establishing rapport, questioning, summarizing, and follow-up.

After completing research, the analyst schedules a meeting with users and management. During the meeting, the analyst presents the preliminary report orally—allowing ample time for discussion—and strives toward a decision, which may range from halting the project to proceeding with detailed analysis.

If management approves the preliminary report, the systems analysis phase advances to the second stage, detailed analysis, which is covered in the next chapter.

KEY TERMS

Preliminary analysis	Interview
Detailed analysis	Questioning
Feasibility study	Preliminary report

QUESTIONS FOR REVIEW AND DISCUSSION

Review Questions

1. What is preliminary analysis and what is the analyst's main function during this phase of the systems process?
2. Which topics does the preliminary report to management cover?
3. What are the five stages of an interview?
4. What are the four steps for setting up a meeting?
5. Which four factors will management usually take into account before allowing an analyst to proceed with a detailed analysis?

Discussion Questions

1. List the inputs to preliminary analysis.
2. List the outputs from preliminary analysis.

3. Which of the four key people in Fleet Feet's organization should the analyst interview?
4. What should be reported to management in the preliminary report?
5. Who makes the final decision about continuing the analysis or terminating it?
6. When you interview a user or manager, how should you make the contact?

Application Questions

1. List three factors that the analyst should consider in determining the cost of a detailed investigation.
2. Whom else in the blood bank should Neal interview?
3. Should Neal Hundt have collected any other documents during the preliminary analysis?

Research Questions

1. What are the usual goals of a new or improved system?
2. What outside stimuli do users have to initiate the systems cycle?
3. List three reasons to terminate analysis.
4. The blood donor tracking system appears simple. What other issues do you suppose Neal will face in detailed analysis?

While interviewing other videotape rental store owners and managers for background facts concerning the tape check-out and check-in system, Frank Pisciotta discovers that most stores also rent products such as VCRs, game machines (Nintendos), and game cartridges. Further conversations reveal that renters of machines must leave a cash deposit before they can rent the machine.

Frank's analysis senses tell him that there is a potential problem with his context DFD and that it will need some modification (Figure 4.14). He draws a data flow from customers to the center of the diagram and labels it "deposits." Frank reviews his level 1 DFD and makes another modification, renumbering as he makes his change (Figure 4.15). His research shows that stores want an average of a $300 deposit for a machine. The deposit can take the form of cash or a credit card imprint. When the machine is returned, the cash is returned to the customer or the imprinted credit card charge is torn apart and thrown away.

Frank knows he needs to keep his focus on the tape rental problem and not let his attention become diverted to such other issues as payroll or accounts receivable that View Video will have to solve. Staying focused on the correct problem is something Frank learned a long time ago from his first boss, Susie Mathews, vice president of software development at U. S. Computer Systems. Susie had six axioms that she made all of her staff follow:

Figure 4.14 View Video's modified context DFD. The new data flow shows deposits for machine rentals.

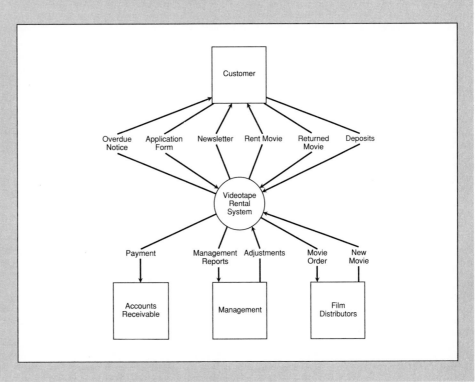

Figure 4.15 The level 1 data flow diagram of View Video's tape rental system now shows the recording of the deposit. Processes are assigned new numbers from the earlier context diagram.

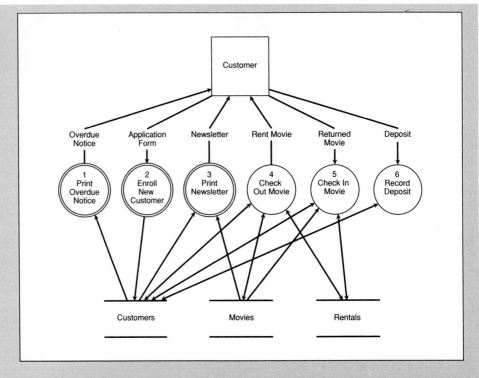

1. Say frequently, "What is the problem that we set out to solve?"
2. Say frequently, "Let me see it for myself."
3. Try to perform the task yourself so that you can experience the problem firsthand.
4. Say occasionally, "That sounds like a solution—tell me the problem."
5. Realize that often the problem that you are addressing is not the same one as described originally.
6. Ask the originator of the problem for more input.

Since leaving U. S. Computer Systems two years ago, Frank sees the wisdom of Susie's six axioms and he follows them with regularity as he works with View Video.

After bumping into the "machine" problem, Frank wonders how many other issues he doesn't fully understand. He decides that he needs to follow the six axioms himself. Does he really understand all the nuances of check-out and check-in? Frank wonders. He realizes that more research is needed. Frank would like to see how a video store really functions, but doesn't know anyone in the business—at least not yet. He picks up the phone.

While at U. S. Computer Systems, Frank also learned the value of keeping in contact with business acquaintances; they often give each other information and advice. After a few phone calls, Frank locates a former co-worker, Richard Donaldson, who owns a video store in a small town about 50 miles away. Because of the distance, his store is not competitive with View Video. Richard is feeling a little isolated, so when Frank asks if he can stop by, Richard readily agrees.

After reminiscing about their former days at U. S. Computer Systems over lunch, Richard proudly shows Frank around the store. Richard invites Frank to rent a videotape so he can see how the system works.

To join up, Frank must complete an application form (Figure 4.16). The questions on the form are easy and Frank answers them without any trouble. Turning in his form, a clerk enters the data into a computer terminal, asking if he wants to add any other authorized borrowers on his account. Frank says that he would and lists the other three members of his family and his

Figure 4.16 An application form for a videotape rental store.

DONALDSON
VIDEO

MEMBERSHIP APPLICATION

1. Name (Last) (First) (Initial)	7. Name of Employer
2. Home Address (Street)	8. Work Phone ( )
3. Apt/Unit	
4. City State 5. Zip	
6. Home Phone ( )	

SECURITY TYPE

Select one of the following:
9. VISA   MASTERCARD   AMERICAN EXPRESS   10. NUMBER _____   11. EXP. DATE _____

Additional family members authorized by applicant to rent videotapes, games, and related equipment.

Name(s) 12. _____   13. _____   14. _____   15. _____

16. Number of additional cards requested _____   (Maximum 3 cards)

Applicant's Signature _____   Date _____

mother-in-law. The clerk tells him that he can check out a tape at any time now; the computer will recognize his account immediately.

Frank browses the aisles, finds a recent release, and takes it to the counter. The clerk picks up his empty box, walks back into the stacks, trades the empty box for the actual tape, and returns to the counter. She asks for Frank's telephone number, keys in the digits of his phone number on the terminal, asks his name, and keys in the videotape's unique tape identification number. That's it!

As he starts to turn away, the clerk tells him that he must return the tape by midnight the following day; otherwise, the tape is considered late and he will receive a late charge. Frank and Richard stand near the clerk, watching the process with other customers. Frank sees that his check-out was similar to others. In many cases, the "return by midnight" warning that he received was not repeated to other customers. Curious, Frank goes back to his clerk and asks why. She tells him that her terminal screen reports the number of tapes that he has checked out and flashes a message to her to remind the customer of the "return by midnight" policy.

In another case, the clerk refuses to check out a tape until the customer pays off back-due amounts. When Frank inquires about this, Richard tells him about their "drop box." Working just like an outdoor mailbox, the "drop box" lets customers return their tapes by simply dropping them in the slot. The next time the customer comes in, they must pay their bill.

Frank takes his tape home and returns it the next morning. He purposefully comes in early to see if any different procedures are used in the morning. He does notice an employee manually checking in a stack of tapes. After returning his tape, he asks Richard what the other employee is doing. Richard reports that the employee is checking in tapes from the drop box.

Frank's check-in is uneventful. The clerk enters the tape identification number, the computer calculates how much is due for the rental, the clerk tells Frank the amount, and Frank pays his bill. The clerk then places Frank's tape on a cart with other tapes that are ready to go back to the rental stacks.

Again, Frank and Richard chat for awhile. Frank stays for a half hour, but does not notice anything new or unusual. He thanks Richard for his assistance and invites him to stop by sometime.

Returning to his home office, Frank knows that he needs to outline his observations. He has ideas for two new DFDs: one for tape check-out and another for returns.

Frank starts to sketch the check-out diagram, identifying three processes (Figure 4.17). Each process is given a number—4.1, 4.2, and 4.3—where the numbers show the hierarchy of ownership, with process 4 as the "parent" of the 3 "child" processes.

Process 4.1 represents selection of a tape from the rental racks. Frank brings his tape to the rental counter. The clerk enters his telephone number, which accesses his data in the CUSTOMERS file, 4.2. The access confirms if this is a valid account and what the account's balance is. If money is owed, the customer must pay in full before proceeding.

With the account balance at zero, the clerk proceeds to actually check out the tape, 4.3. This process must record that a customer has a rental tape and change the "date last rented" field to the current date in CUSTOMERS. It must mark the tape with this unique tape id as "checked out" in MOVIES. Finally, it must create an entry in RENTALS that shows the details of this rental (customer number, tape id, date of rental, and so on).

Frank draws the DFD for checking in a tape in a similar manner (Figure 4.18). He identifies three processes at this level and numbers them 5.1, 5.2, and 5.3.

Figure 4.17 A second level data flow diagram for process 4, check out movie.

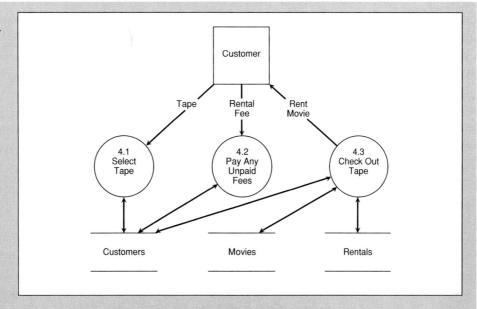

Figure 4.18 A second level DFD for process 5, checking in a movie.

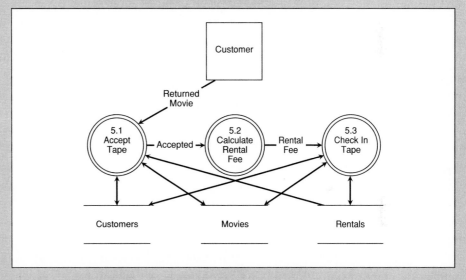

In the first process, Accept Tape, the clerk takes the tape from the customer, enters the tape's identification number, and enters the customer's telephone number on the computer terminal. The process then verifies that this is a legal rental, a valid tape, and a legal customer by searching for the entries in CUSTOMERS, MOVIES, and RENTALS.

With the data verified, the second process calculates the amount owed. This process reads the date that the tape was rented from RENTALS, computes the number of elapsed days, and figures out the amount owed.

The last process, 5.3, completes the check-in. The data about this customer is updated in CUSTOMERS, income earned increased by rental fee, and number of tapes rented incremented by 1. MOVIES is altered to reflect that this tape is back on the shelves and no longer checked out. RENTALS is also updated to reflect that the tape is back.

With his new drawings finished, Frank writes a memorandum to the Stensaases that tells them of his findings (Figure 4.19). The memo outlines what his research has revealed. Frank also estimates the cost for detailed analysis and a projection on the time it will take to finish the last phase of analysis.

After receiving the memo, the Stensaases authorize Frank to continue to detailed analysis (Figure 4.20). This authorization is the green light that Frank needs to proceed with further study.

Working With View Video

As you just saw, preliminary analysis is quite formal. We have diagrams, interviews, observations, collection of sample documents (membership application form), fact finding, and memoranda among all parties. This is the way it should work. We have a business transaction and we want to record the event on paper. Memos and diagrams help improve communication among all parties and let each group know exactly what is expected, time frames, and costs. The formality protects both the user (the Stensaases) and the analyst (Frank Pisciotta). Neither party can later claim "foul" for not knowing what took place.

If preliminary analysis is telephone-based, each party will have selective memory; that is, they will remember what they want to remember, not what actually takes place. Practicing analysts should remember that "if it is not in writing, it does not count!"

Case Study Exercises

To involve yourself in Frank's preliminary analysis, complete the following:

1. Collect an application form from your local video store.
2. Does your store rent machines? If so, what is the rule on deposits? How much is the deposit?
3. Does your local store sell or rent any other services?
4. What is the fee schedule for your store?
5. Level process 1.2 in Figure 4.15, Enroll New Customer. What lower-level process did you identify?

Figure 4.19 Frank
Pisciotta notifies View
Video owners of his
findings.

PISCIOTTA
C O N S U L T I N G

MEMORANDUM

DATE: October 8, 1993

TO: Mark and Cindy Stensaas, Owners, View Video

FROM: Frank Pisciotta, Consultant Analyst   *FP*

SUBJECT: Preliminary Findings

While performing my preliminary analysis on the video cassette check-out and check-in system, I discovered a few missed concepts or ideas that you may want to consider.

Findings:

1. We need to consider a "drop box" where customers can drop off a tape at any time. We can charge a fee for the use of the drop box in addition to the normal rental fee. The drop box is located outside so that customers can drive up to it. We collect the fees at the next check-out. We run a risk of not collecting the fee, but customer satisfaction will offset the risks. Tapes returned via the drop box are entered once per day, as soon as the store opens.

2. Most video stores rent various machines, VCRs, and videogames, such as Nintendo. They require a deposit, averaging $300, which is refundable when the machine is returned. My research shows that most stores charge less than $10 per day for game machine rental and around $5.00 per day for VCR rental. Game cartridge rental is an additional fee.

3. We should consider researching use of bar codes to speed data entry during both check-out and check-in. All the stores I visited require manual entry, which has the associated potential for error of the customer's telephone number and the tape identifier. We can also use bar codes to indicate the type of transaction, rental, return, machine rental/return, deposit, and so on.

4. All the stores I visited use empty tape boxes in the customer area and keep the actual tapes in the stacks. This dual set of boxes is quite inefficient. We need to research how

(continued)

Figure 4.19 *(continued)*

libraries use embedded trigger devices to discourage theft. I think this alone can represent a vast savings in the size of the store that you need to rent and will result in efficiencies in the check-out/ check-in processes.

5. I have not performed any additional research with the interface to BusinessWorks PC, but am convinced that it is still a good way to operate the accounting part of the system.

6. The clerk operating the computer terminal needs a great deal of information at his or her disposal. I saw a case in which the clerk refused to check out a tape because the customer always returned tapes without rewinding them. In another case, the clerk reminded a chronically late customer of the returns policy. In my personal borrowing, the clerk knew I was a first-time borrower and reminded me of the returns rules after I checked out the movie.

7. We should spend time developing the reports that you need from the system. We can produce a variety of information that will help you better manage the business and the customer base. For example, if we keep track of every tape every customer rents, we can establish a viewing history and preference profile on each customer. With the profile, we can selectively advertise to customers when we receive new tapes.

8. Our system needs to minimize clerical check-out/check-in time and allow them to help customers find the tapes they want.

9. I have not researched the tape locator you want. As I understand it, you want customers to have the capability to ask the computer for a shopping list of potential movies. A customer could choose tapes from those offered for rent based on a tape's rating, category, year made, and which actors or actresses appeared in the movie. This part will need to wait until we proceed to detailed analysis.

Recommendations:
On the basis of these preliminary findings, I recommend starting a detailed analysis to determine the feasibility of computerizing videotape check-out and check-in.

Cost and Time Schedule:
I know that both of you are quite busy with setting up View Video and do not have much spare time in which to continue our analysis. I would continue the analysis of both the tape check-out and check-in system and the linking of it to BusinessWorks PC. My time and cost estimates for detailed analysis are

(continued)

Figure 4.19 *(continued)*

1. Two weeks for Frank Pisciotta	$3,000.00
2. Secretarial aid (10 hours)	125.00
3. Travel expenses for 2 days for BusinessWorks PC	
research and discussion	400.00
4. Bar code travel expense	200.00
Total	$3,725.00

If you have any questions or need clarification on any of these points, please feel free to contact me.

Figure 4.20 The Stensaases confirm Frank's findings and authorize him to proceed with a detailed analysis.

|||||||| *View* Video ||||||||

MEMORANDUM

DATE: October 15, 1993
TO: Frank Pisciotta
FROM: Mark and Cindy Stensaas   *MS   CS*
SUBJECT: Preliminary Study

Thank you for your memo of October 8 and for bringing several points to our attention that we had previously overlooked. We very much appreciate your thoroughness on this project.

We have reviewed your preliminary report, data flow diagrams, sample reports, and recommendations, and authorize you to proceed with a detailed analysis. Please continue to keep us informed of your progress.

Detailed Analysis

GOALS

After reading this chapter, you should be able to do the following:

- Describe detailed analysis
- List the analyst's main functions during detailed analysis
- Define fact finding
- Outline methods for gathering data
- State the components of the feasibility study
- Name three different financial analysis techniques
- Explain two reasons for management's investment opportunities
- List the participants in a management review

PREVIEW

Most people in the market for a new car carefully consider whether or not it really suits their needs. Even before they sit behind the wheel to test drive it, they check its specifications. Do they like the way it looks? Does it offer enough room for a family of four? Will the trunk hold luggage for a two-week vacation? Most importantly, can they afford it? If a family feels satisfied after reviewing the specifications for the new car, it will seriously consider making the purchase if it compares favorably to cars offered by other manufacturers.

Preliminary analysis works much the same way. The analyst provides sufficient specifications so that users and management can consider a project's feasibility. If for any reason the proposed system does not seem workable or affordable, the organization halts the investigation; but, if it does seem to offer a viable, practical, and cost-effective alternative, investigation will probably continue to the detailed analysis phase. Preliminary analysis is a quick look at the problem. The intent is on deciding if the problem is really a problem—and if so, how important is it to the overall goals and objectives of the organization.

Detailed analysis, as shown in Figure 5.1, involves thorough consideration of all of the problem's aspects. Its ultimate goal is using a new or revised system to provide a solution. As before, the analyst strives to solve an organization's problems by determining what data an organization collects, how it processes that data into information, what outputs are required, and, specifically, how to improve all three procedures. To do so, the analyst adds detail to the preliminary study by interviewing personnel at all levels within the organization, preparing questionnaires for those not interviewed, studying all relevant forms and procedures, contacting hardware or software vendors for information about their products, calculating costs and benefits of new or improved systems, reporting findings that include alternatives, and recommending a course of action. Those participating in the preliminary analysis—users, management, and the computer services staff—become even more intimately involved in detailed analysis.

Detailed analysis results in a business decision: Should the organization commit funds, resources, and personnel to resolving a specific issue? Because of this, many of the aspects that the analyst must consider are business-related, such as costs, benefits, and potential future revenues. The software and hardware requirements are not ignored, but become an integral part of the overall business decision.

REVIEW AND ASSIGNMENT

The structure chart in Figure 5.1 levels detailed analysis into its components: review and assignment, fact finding, and presentation of the system to management. The two major outputs of detailed analysis are analysis documentation and a feasibility study, which ultimately provide input to systems design.

Figure 5.1 The three major activities of detailed analysis.

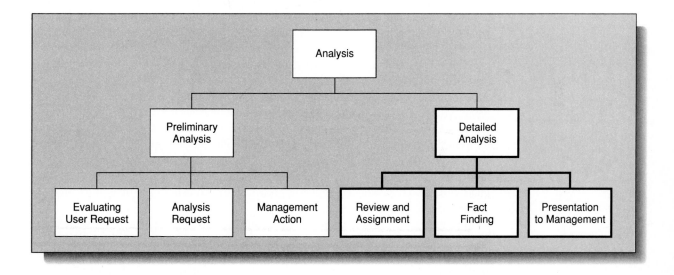

The Preliminary Report

Review and assignment begins with a careful reconsideration of the preliminary report. Within the preliminary report, the analyst reviews the original problem, states his or her findings, makes a recommendation, and estimates costs and a schedule for a detailed study. If someone else takes over at this point or if a lot of time has elapsed since the initial investigation, careful review will acquaint the new analyst with earlier findings or refresh the original analyst's memory.

In large organizations, a complex study may involve a group of analysts that are headed by a team or project leader who divides up the tasks and assigns responsibilities to individuals. In our continuing case involving Fleet Feet, only one analyst, Peggy Adams-Russell, assumes responsibility for detailed analysis.

Scheduling the Detailed Analysis

Aware of the need for planning a project carefully, the analyst or team leader draws up a detailed schedule that outlines anticipated activities. Scheduling may seem like a straightforward task, but given the technical nature of systems analysis and the number of people involved with even a simple system, it can become quite complicated. The analyst should expect both anticipated and unanticipated delays, such as vacations and illnesses. Also, the analyst should understand that an organization's people must continue with their regular meetings and job responsibilities. Taking all of this into account, the analyst must fit users, management, computer services staff, outside vendors or suppliers, and him or herself into a realistic schedule.

Most analysts use a **Gantt chart** for scheduling activities. A Gantt chart (Figure 5.2) lists activities vertically down the left-hand column, while time periods appear horizontally across the top. For each activity, the analyst draws a hollow bar across the chart, indicating when that task should start and stop, then later changes the bar to depict the actual time that was taken. Changing the hollow bar makes it possible to determine visually whether initial time allotments were accurate and whether or not the project is on schedule.

A Gantt chart does not show dependencies among activities—only the time period that each should take and the order of the activities in relation to each other. The bars are drawn to scale so that the reader can quickly spot short- and long-term tasks.

To construct a Gantt chart, start by naming each activity in the project. Next to the activity name, write down your best estimate of the amount of time that it will take to complete the activity. Now, place the activities into chronologic order, deciding which events can overlap each other. Draw the bars horizontally and name the individuals responsible for each activity.

Although Gantt charts show personnel assigned to specific tasks, they do not reveal the degree of a given person's involvement. In the case of Fleet Feet, for example, the Gantt chart will not tell us whether Peggy has devoted 100 percent of her time to the study or whether she has mixed it in with other duties.

Schedules in the form of Gantt charts not only help the analyst plan the work, they also benefit management because they provide such a quick visual check of a project's progress. In this sense, Gantt charts provide managers, even those unfamiliar with computers and the technical aspects of systems, a certain amount of control over the project.

Figure 5.2 A Gantt chart depicts time, measured in weeks, on the horizontal axis; events are measured on the vertical axis. Hollow bars, with black diamonds at both ends, indicate the time frame within which an event is supposed to occur. The gray diamonds show the actual completion dates.

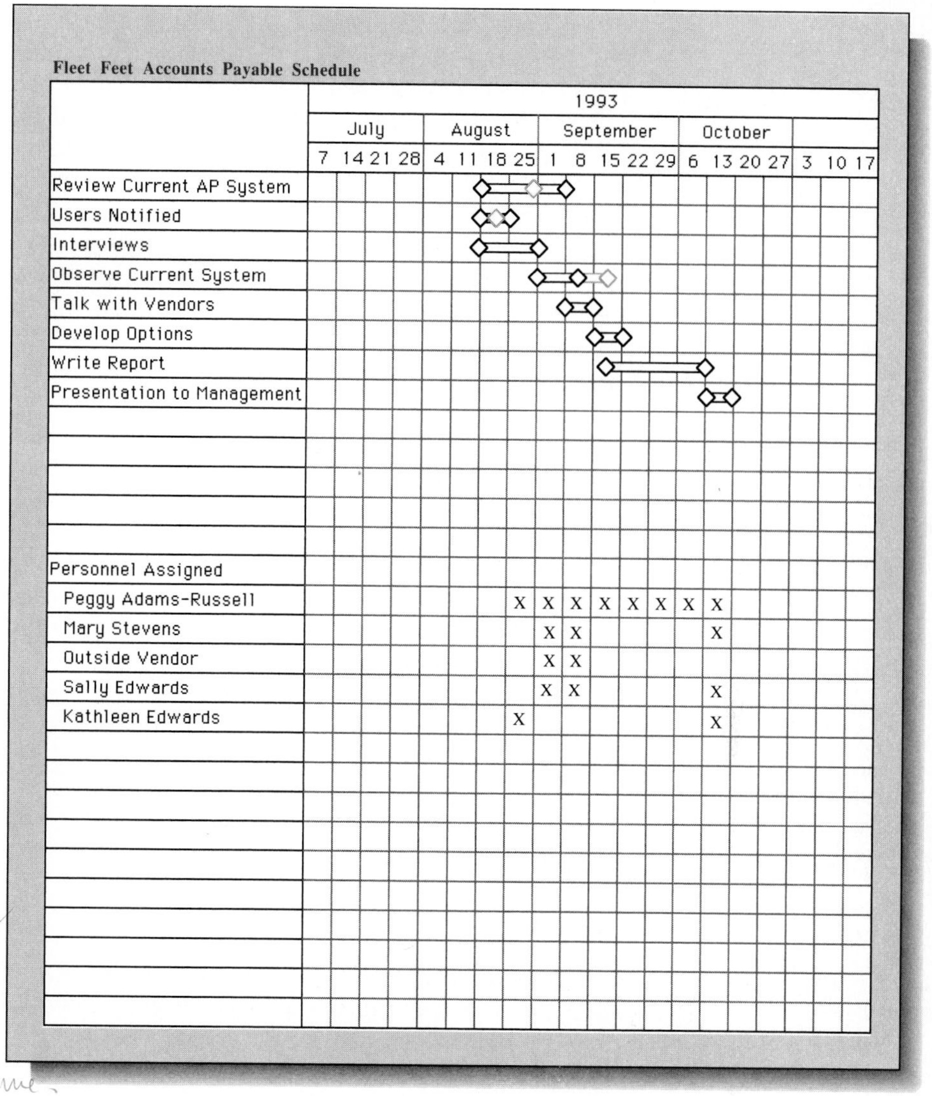

Fleet Feet Accounts Payable Schedule

	1993																		
	July				August				September				October						
	7	14	21	28	4	11	18	25	1	8	15	22	29	6	13	20	27	3	10	17
Review Current AP System							◇	◇	◇										
Users Notified							◇◇												
Interviews							◇	◇											
Observe Current System									◇	◇ ◇									
Talk with Vendors									◇◇										
Develop Options										◇◇									
Write Report										◇			◇						
Presentation to Management													◇◇						
Personnel Assigned																			
Peggy Adams-Russell							X	X	X	X	X	X	X	X					
Mary Stevens							X	X					X						
Outside Vendor							X	X											
Sally Edwards							X	X					X						
Kathleen Edwards						X							X						

work/time

A Gantt chart depicting the major steps of detailed analysis gives a sweeping picture of the whole project to those involved. On a large and complex project involving more than one analyst, each analyst would prepare other specific Gantt charts for specific components of the project. Some versions of the Gantt chart display the percentage of each activity completed, activities completed early, time over-runs, and the cost of each activity. Other versions show resources on the vertical axis and use bars to indicate what each resource will do during a given period.

For instance, we can see from Figure 5.2 that Peggy has scheduled the detailed analysis phase of Fleet Feet's accounts payable system to start the week of August 18. At that time,

Peggy will begin reviewing the current AP system and interviewing users. The chart also reveals that Peggy plans to observe the current system during the weeks of September 1 and 8, followed by discussions with vendors the weeks of the September 8 and 15. Management can expect to see the presentation of her final report the week of October 13.

Personal computer software vendors, as well as mainframe vendors, sell a variety of project scheduling and management software. The Harvard Total Project Management or Microsoft's Project systems will allow you to enter activities, estimated duration of activities, personnel assignments, and priority of activities. Then, the Gantt chart is drawn for you. As the project progresses, you can enter what actually took place or changes in activities, durations, or assignments, and the software will redraw the Gantt chart.

Authorization and Notification

After approving the assignment and scheduling of tasks, management notifies all appropriate personnel via memo (Figure 5.3). The memo outlines decisions made in conjunction with the preliminary report and alerts members of the organization that a detailed system study has commenced.

In our Fleet Feet example, the president authors the memo, which briefly summarizes the results of the preliminary analysis, advises personnel of the impending investigation, and reports an important decision: The controller now has the authority to sign checks.

Figure 5.3 As soon as management approves the assignment of tasks and the schedule, it notifies all involved individuals about the impending detailed analysis.

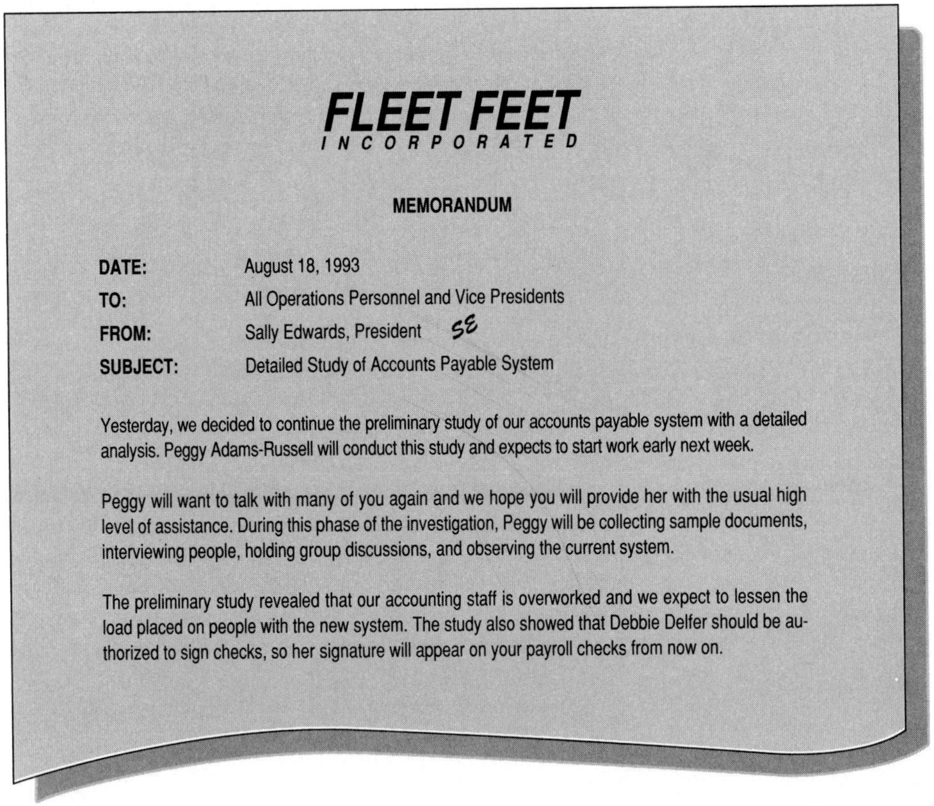

FLEET FEET
I N C O R P O R A T E D

MEMORANDUM

DATE: August 18, 1993

TO: All Operations Personnel and Vice Presidents

FROM: Sally Edwards, President   *SE*

SUBJECT: Detailed Study of Accounts Payable System

Yesterday, we decided to continue the preliminary study of our accounts payable system with a detailed analysis. Peggy Adams-Russell will conduct this study and expects to start work early next week.

Peggy will want to talk with many of you again and we hope you will provide her with the usual high level of assistance. During this phase of the investigation, Peggy will be collecting sample documents, interviewing people, holding group discussions, and observing the current system.

The preliminary study revealed that our accounting staff is overworked and we expect to lessen the load placed on people with the new system. The study also showed that Debbie Delfer should be authorized to sign checks, so her signature will appear on your payroll checks from now on.

WORKING WITH PEOPLE
ACCOUNTING AND ANALYSTS

After graduating magna cum lauda with a degree in computer science, Matthew Reid accepted a job as a beginning systems analyst in the headquarters of Riteway Hardware, a home-building supply chain in the Midwest. Looking forward to his first real assignment, Matthew walked confidently into his meeting with Shelby Ray, Riteway's controller.

As Ms. Ray described her needs for a new accounts receivable system, which should include conversion from balance forward to open item, Matthew found himself sinking deeper into his chair. From his brief exposure to accounting, a one-semester introductory course in his second year in college, Matthew knew that aging was an important concept, but he couldn't recall any of its basic principles. While his mind danced with bits and bytes, Ms. Ray talked casually about debits and credits. Not wanting to appear ignorant, Matthew tried to look attentive, but he left the meeting frustrated and bewildered.

He immediately consulted with Dale Hardt, Riteway's senior analyst. "I spent four years studying COBOL, computer architecture, and database management systems," he mused, "but now I feel like a complete idiot!"

Dale laughed and told Matthew about her own early experience with the company. Like Matthew, she had discovered that her accounting background was embarrassingly weak, but she had done something about it. Approaching an accountant friend, Dale asked him to give her a weekend crash course in accounting in exchange for some free programming on the accountant's new microcomputer system.

"Believe it or not," she told Matthew, "I found the subject fascinating and ended up taking two night courses in it. Not only did I learn to understand people in Shelby Ray's department, but I also gained confidence and expertise by helping to solve their problems." Offering her help, Dale outlined several options for Matthew. He could:

- Quit the accounts receivable project and let another analyst take his place
- Try to bluff his way past Shelby Ray
- Do what Dale had done, perhaps by reading a book or taking a course that she recommended
- Ask that Shelby Ray allow Dale to help Matthew with the project so that he could benefit from her experience
- Find a software firm specializing in accounting to tutor and help him

After debating the pros and cons of each option, Matthew decided to combine two of them. First, he would spend evenings and weekends reading about accounts receivable in the book that Dale suggested. Second, he would try to convince Ms. Ray that together, he and Dale could more quickly convert the accounts receivable system. It worked. Six months later, Riteway had its new system, as well as a much more effective analyst.

With Fleet Feet's staff aware that detailed analysis of their accounts payable system has begun, they can start helping the analyst by gathering and organizing necessary information. The fact that the president authored the memorandum gives it a lot of credibility, letting those involved know that this project has the approval of the highest level of authority.

FACT FINDING AND THE CURRENT SYSTEM

Fact finding means learning as much as possible about the present system. To do this, the analyst interviews personnel, prepares questionnaires, observes the current system, gathers forms and documents currently in use, determines the flow of data through the system, and clearly defines the requirements of the system.

Interviewing

By studying Fleet Feet's organization chart, Peggy can pinpoint key personnel who are involved with the system and schedule interviews with them. Although Peggy conducted interviews during the preliminary analysis stage, now she wants to go into much greater depth. As before, she wants to maintain close rapport with the users because they will not only use the newly developed system, but they may harbor fear or uncertainty about future changes, especially if they think that the computer might replace them.

Like an investigative reporter trying to discover the who, what, where, when, why, and how of a story, the analyst should conduct the interview in a way that stimulates people to provide a full and honest description of their jobs. The following questions can help accomplish this goal:

- What do you do?
- Who else does what you do?
- Where do you do it?
- When do you do it?
- Why do you do it the way that you do?
- How do you do it?
- Do you have any suggestions for change?

Interviews help gather vital facts about existing problems, such as a lack of quality control or insufficient security, and they also allow the analyst to involve people in possible changes. Change is easier when those most affected by it act as change agents themselves. It can cause resentment when changes are inflicted on employees, especially by someone who does not share their daily work.

Questionnaires

Questionnaires can provide an efficient means of gathering facts from both large and small groups of people. Questionnaires are documents that gather data from those involved with the system. Analysts can use questionnaires during both preliminary and detailed analysis, as well as during various tasks of the design phase of the system life cycle.

Properly constructed, questionnaires pose essential questions, require a minimum amount of time to complete, and permit rapid tabulation of statistical results. Developing a questionnaire requires thought and planning, and usually more than one draft.

Analysts should write questions that are short, easy to understand, unbiased, nonthreatening, and specific (Figure 5.4). Suppose, for example, that Question #5 on the Fleet Feet survey asked, "In what ways were Fleet Feet reps discourteous to you?" Worded that way, the question directs the supplier to think only about bad experiences, and it will probably prompt a negative answer that will bias the survey. This question could also be reworded more positively: "Have you received courteous attention from Fleet Feet reps? If not, please explain." Similarly, questions that reflect the analyst's own prejudices or preconceptions will invariably taint the survey. To make sure that the questions will provide the needed information, the analyst can test them with one or two people before widespread distribution. Prepaid return envelopes accompanying questionnaires sent to those outside the company help ensure prompt responses.

When employing the questionnaire approach, the analyst should send forms to everyone involved with the system. A questionnaire works particularly well when gathering

Figure 5.4 A questionnaire used by an analyst to gather data from suppliers.

FLEET FEET *Vendor Survey*

We are in the process of evaluating our accounts payable system in order to speed payments to our suppliers. Please complete the following form within 5 days and return it in the enclosed prepaid envelope.

1. Name of your firm: _____
    _____

2. Name of person completing this form: _____
    _____

3. Title of person completing this form:
    a. Business Manager
    b. Accounts receivable clerk
    c. Controller
    d. Salesperson for our account
    e. Other:_____

4. Have you experienced any late payments from Fleet Feet? If so, when did they occur and how late were they?_____
    _____

5. When you contact a Fleet Feet representative, is he or she courteous? Are you satisfied with the way your problems or questions are handled? (Please circle your response)

    Dissatisfied   Satisfied
          1   2   3   4   5

6. How does Fleet Feet compare to your other customers?
    (a) about the same
    (b) better than most
    (c) worse than most
    (d) no experience

(continued)

Figure 5.4 *(continued)*

7. Would you offer Fleet Feet a better discount rate if payment were made within five days? If so, what might your terms be? _____

_____

_____

8. Who is the person in our organization that you contact most frequently?

_____

9. Comments (please write them here or on the back of this form).

_____

_____

_____

_____

_____

_____

_____

_____

_____

10. Date: _____

facts from a large number of people, when asking everyone the same questions, or when collecting facts from people outside of the organization, such as suppliers.

A questionnaire can include one or more of the following types of questions:

- Multiple choice: Gives the respondent a specific set of potential answers. This format is ideal for computer tabulating (Question #3).
- Open-ended: Respondent must answer the question in his or her own words. Space is provided under each question for the response (Questions #1, 2, 8, 9, and 10).
- Rating: Similar to multiple choice, except that respondents must rate their satisfaction (Questions #5 and 6).
- Rank: Requires respondents to assign priorities, from high to low or on a percentage basis, to their responses.

Peggy Adams-Russell, aware that most people do not enjoy spending a lot of time on questionnaires, decides to mix question formats, including follow-up questions within the original questionnaire to permit elaboration of certain responses (such as Questions #4 and 7). By so organizing the questionnaire, Peggy makes the respondents feel free to express opinions fully, yet quickly.

When the questionnaires come back, Peggy proceeds to tabulate the findings (Figure 5.5). Peggy's survey reveals that the company could save a substantial amount of money by paying bills faster because the discount rate would increase from 0 or 2% to over 4%, outweighing the interest or other income that Fleet Feet might have earned by keeping the money longer.

Figure 5.5 Peggy Adams-Russell's tabulation of Fleet Feet's survey.

Results of *FLEET FEET* Vendor Survey

Survey sent to 127 vendors on September 6. 57 responses received by September 22.

1. Name of your firm: _____
   _____

2. Name of person completing this form: _____
   _____

3. Title of person completing this form:
   a. Business Manager   15
   b. Accounts receivable clerk   12
   c. Controller   9
   d. Salesperson for our account   9
   e. Other   12

4. Have you experienced any late payments from Fleet Feet? If so, when did they occur and how late were they?
   No: 20
   Yes: 37
   Average: 45 days; Longest: 98 days

5. When you contact a Fleet Feet representative, is he or she courteous? Are you satisfied with the way your problems or questions are handled? (Please circle your response)

   Dissatisfied   Satisfied
   1   2   3   4   5
   5   7   8   22   15

6. How does Fleet Feet compare to your other customers?
   (a) about the same   20
   (b) better than most   10
   (c) worse than most   15
   (d) no experience   12

(continued)

Figure 5.5 *(continued)*

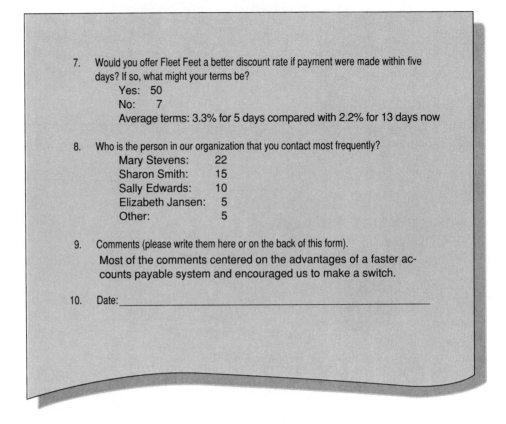

7. Would you offer Fleet Feet a better discount rate if payment were made within five days? If so, what might your terms be?
   Yes: 50
   No: 7
   Average terms: 3.3% for 5 days compared with 2.2% for 13 days now

8. Who is the person in our organization that you contact most frequently?
   Mary Stevens: 22
   Sharon Smith: 15
   Sally Edwards: 10
   Elizabeth Jansen: 5
   Other: 5

9. Comments (please write them here or on the back of this form).
   Most of the comments centered on the advantages of a faster accounts payable system and encouraged us to make a switch.

10. Date: _____

With these findings, Peggy can estimate the possible savings that would result from an automatic accounts payable system. Since purchases amount to well over $1.2 million per year, a 3.3% average savings for payment within five days equals $37,000 per year. Since Peggy's research indicates that discounts for early payment saved $25,000 during the previous year, the company might save an additional $12,000 a year by paying earlier. Fleet Feet can begin paying earlier either by hiring additional staff or by computerizing the payment process.

If the results of a questionnaire survey do not seem conclusive, the analyst may want to contact selected respondents by phone or in person. Of course, this requires tact and an understanding that an analyst's own pressing need may not seem very important to others.

Observing the Current System

The analyst may want to observe the existing physical system by following a transaction, such as an invoice, all the way through it. **Observation** allows the analyst to verify his or her understanding of the system, rather than merely relying on second-hand impressions. Provided that the analyst remains outside the flow as an observer, he or she can experience the actual process without introducing biases or tampering with results. Observing a system requires caution; when people are under observation, they usually behave differently, working more efficiently and at higher speeds to impress the observer a they might otherwise ignore.

In some instances, the analyst may find it useful to visit another o computerized system similar to the one under consideration. Finding a lation may pose a problem, however. Competitive organizations may not want to share their experiences, others may be too large or too small for accurate comparisons, and still others may be unwilling to waste employees' time by demonstrating their system for an outsider. Whenever visiting another organization, an analyst should follow the rules of etiquette: make an appointment, research the organization beforehand, know what you want to see, and write a follow-up thank-you letter.

Hardware and software vendors can also provide an opportunity for observation. Computer sales representatives will gladly demonstrate their products for potential clients, and some of those products may ideally suit the application in question. Though very useful, information gleaned from such sources should be reviewed carefully because vendors will try harder to promote their products than to solve your problems. To minimize this difficulty, try to visit an installation where you can test a salesperson's claim.

THE FEASIBILITY STUDY

At this point in detailed analysis, the analyst has gathered many facts and figures and is now ready to bring everything together. For many analysts, this is a high point at which they use all their skills, moving into a problem-solving mode and out of a problem-understanding mode.

Diagramming the Logical System

have alternatives.

With the interviews completed, questionnaires tabulated, and observations done, the analyst can now describe the current physical system in narrative form, with a data flow diagram, or by using another tool. In the case of Fleet Feet, Peggy chooses to use a data flow diagram to depict the current AP system. She begins by sketching a context data flow diagram (DFD), as in Figure 5.6. A context DFD provides a general view of the accounts payable system under study. In this case, the key ingredients include:

- *Inputs to AP:* packing slips, invoices, canceled checks, purchase orders, and control
- *Outputs from AP:* reports to management and checks to vendors

A context DFD does not include the details of the system, but simply offers a comprehensive overview of it. Given its general nature, it provides an especially useful and accessible picture for management, which cares more about the principles rather than the details of a system. Context DFDs place a boundary around the system under investigation, thus focusing everyone's attention on the problem that the analyst is working to solve.

After completing a context DFD, Peggy turns her attention to the details of accounts payable. In Fleet Feet's case, management reviews inventory reports to help determine

Figure 5.6 A context data flow diagram depicts Fleet Feet's accounts payable system in its broadest perspective, not showing any of the details or internal processes.

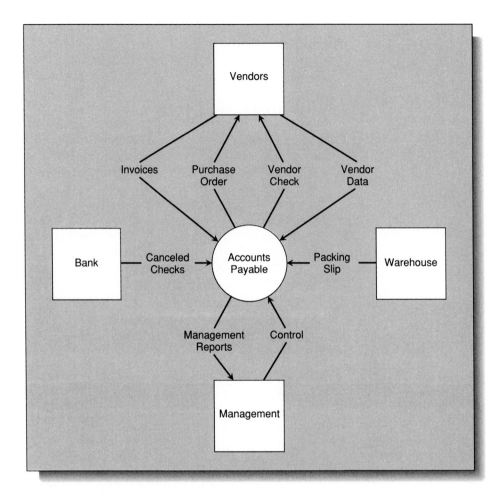

what it should order from suppliers. The accounting department places orders with a purchase order/requisition. Upon delivery, merchandise and packing slips enter the warehouse; the packing slips go to the accounting department, which will receive the invoices themselves directly from suppliers. The merchandise remains in the warehouse until it is shipped to a franchise outlet.

Accounting clerks compare purchase order requisitions with invoices and packing slips to make sure all invoiced items have actually arrived, then they post the purchase to the supplier's ledger card. Around the 15th of each month, the accounting department receives Fleet Feet's canceled checks and checking account balances from the bank. Then, on the 24th of the month, management instructs the accounting department to pay certain suppliers, who will receive a payment check generated by accounting and signed by an authorized individual. At the end of each month, the accounting department prepares a report of balances due suppliers and an inventory report for management evaluation.

These detailed activities by the accounting department, management, warehouse personnel, the bank, and suppliers reveal the following major activities:

- Generating reports
- Ordering stock
- Printing checks
- Posting accounts
- Bank reconciliation
- Payment authorization

Now Peggy can use her research to draw a more detailed DFD, as shown in Figure 5.7. To obtain such a diagram, the analyst follows seven key steps:

Figure 5.7 This data flow diagram depicts Fleet Feet's accounts payable system. Packing slips arrive in the warehouse and then go to the accounting office for filing; the bank sends account balances, and management receives reports as well as controls payments.

1. Look at the system from the inside to the outside
2. Identify important activities
3. Locate the data flows
4. Show the relationships among activities
5. Find the internal inputs or outputs that exist within the system
6. Level complex processes into simpler ones
7. Look for duplication of data flows or files

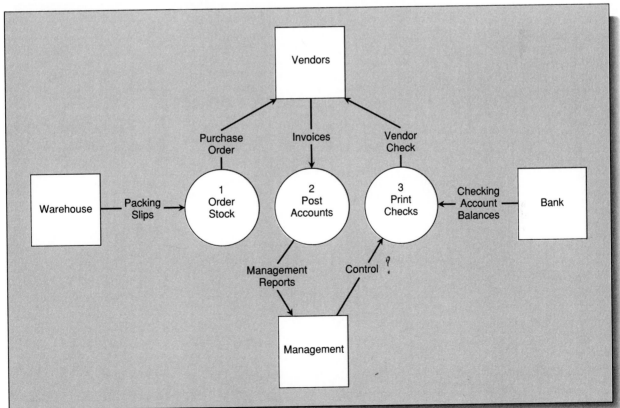

A well-constructed leveled DFD helps reduce the time that it takes to analyze a system. Coupled with context data flow diagrams, they depict the system in a way that users, managers, and technical people can understand. During the design phase of the systems process, Peggy will level the data flow diagram from Figure 5.7 into far more detail, but for now her DFD provides enough information to proceed with her analysis of the system.

Sample Documents and Data Dictionaries

As part of the effort to determine the flow of data through a system, an analyst collects samples of all relevant documents. In our example, Peggy Adams-Russell has collected sample checks, invoices, packing slips, and other accounts payable forms (Figure 5.8). Written in data dictionary format, the list becomes

Sample Documents = Sample Checks +
Invoices +
Packing Slips +
Ledger Cards +
Other Accounts Payable Forms

Figure 5.8 Sample check, invoice, and packing slip that were collected during the fact-finding phase. Checks are used to pay suppliers, invoices show what Fleet Feet owes on a particular purchase, and packing slips detail what actually arrived.

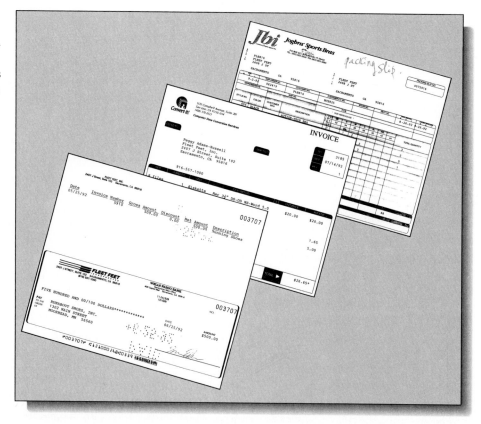

In general, manual systems such as this employ more documents than their au........
counterparts. Fleet Feet's ledger card, for example, could be automatically generated by a computer system.

From the assembled documents, the analyst can better understand what data the new system must collect and process. For example, Peggy can define an invoice as:

Invoice = Supplier Name +
Supplier Address +
Supplier Telephone Number +
Invoice Number +
Invoice Date +
Items Shipped +
Invoice Terms +
Invoice Amount

From the packing slip, she finds:

Packing Slip = Supplier Name +
Shipping Date +
Date Goods Are Received +
Freight Charges +
Invoice Number

Packing slips are carbon copies of invoices that omit certain data not relevant to the warehouse, such as the dollar value of the shipment. The warehouse clerk checks the merchandise received against the packing slip to ensure that the order is accurate and complete, noting any discrepancies. Then the packing slip goes to accounting for comparison with invoices to make sure that the company received what it's paying for.

The ledger card (Figure 5.9) offers two categories of facts: supplier data and purchase/payment history. Peggy again defines the card with a data dictionary:

Figure 5.9 Fleet Feet's ledger card for Nike shoes. A handwritten card is produced for each supplier. When merchandise arrives, the accounting department notes the date and amount, records payment, and calculates a new balance.

NAME FAR WEST NIKE SHOES
STREET 1072 HIGHWAY 72
CITY REDDING, CA 96001   BALANCE FORWARD 1073 28

DATE	VENDOR	DETAIL	DEBIT	DISCOUNT	CREDIT	BALANCE
12/4/92	FWNS		124 22	2 48		1195 02
12/10/92	Payment				973 —	222 02

Ledger Card =   Supplier Name +
                Supplier Address +
                Transaction Date +
                Description of Transaction +
                Amount of Invoice or Payment +
                Discount Amount +
                Balance Due Supplier

Each check sent to a supplier contains the following data:

Check =   Check Number +
          Check Date +
          Vendor Number +
          Check Amount +
          Vendor Name +
          Vendor Address +

Besides these documents, Peggy goes to Fleet Feet's management to secure copies of reports that the accounting department prepares. The monthly supplier balance report lists each of Fleet Feet's suppliers, including the amounts owed them. The weekly inventory report shows the variety of items that Fleet Feet sells and the quantities on hand.

The list of data elements from each document that Peggy builds from the forms, along with the data flow diagrams and other documents of the system, give Peggy the background she needs before starting to plan for the new AP system.

Determining Alternatives, Costs, and Benefits

During fact finding, Peggy researches the problem thoroughly, gathering all pertinent facts, figures, and documents. Peggy must now pinpoint the costs and expected benefits of the new system. She remembers that she is presenting a business plan to a management that is very concerned with the financial impact that the system will have on the organization. She will have to write her report to emphasize the business benefits of her alternatives.

Before settling on her own recommendation, Peggy develops a complete list of **alternatives**, possible solutions to the accounts payable problem. She decides that the viable solutions range from doing nothing to fully computerizing the AP system. In the end, she considers four distinct alternatives:

1. Do nothing; leave the existing system alone
2. Hire more staff, but continue with a manual system
3. Purchase AP software from an outside software supplier
4. Design, program, and install a customized AP system

For each of the alternatives, Peggy considers probable costs, savings, effects, and benefits.

The first alternative, doing nothing, would not capture the $10,000–$12,000 discount savings that Fleet Feet currently loses each year. Thus, it offers no real benefits, except that the company would incur no out-of-pocket expenses by maintaining the status quo. However, inaction would create some long-range liabilities: the AP system will deteriorate as Fleet Feet grows, losing more potential discounts every year; bookkeepers will struggle harder to do their jobs well; and morale of those who come in contact with the system will continue to sag.

The second alternative, hiring more staff, while it would not dramatically change the existing system, would require hiring one new person immediately and another in a year. Peggy estimates costs for this alternative—including salaries, equipment, furniture, and space—at $12,500 for part of the first year and $33,600 for the second. Hiring new people would stabilize the problem with the AP system over the short term and permit the company to take advantage of more discounts. In the long term, however, the problem will recur as Fleet Feet outgrows the manual system, causing the company to again lose most discount opportunities.

Purchasing software, the third alternative, would undoubtedly permit Fleet Feet to capture previously lost discounts. Leafing through two software magazines that she reads regularly, *Interface* and *Interact*, Peggy scans software advertisements for products specifically designed to run on Fleet Feet's Hewlett Packard (HP) computer.

Looking at back issues of the journals she has, Peggy discovers AP systems ranging in price from $500 to $15,000. The lower-priced ones permit a single user at a time (unacceptable in Fleet Feet's situation), while the higher priced ones—although too expensive for Fleet Feet—allow for multiple users, tie AP to the general ledger, and permit multiple corporations. Peggy contacts Micro Computer Business Applications (MCBA), which provided their other accounting systems when Fleet Feet purchased their Hewlett Packard computer. Peggy arranges for a demonstration from an MCBA user, a local educational TV station. Based on the TV station's experience and MCBA price lists, Peggy tallies expected costs:

1. AP software from MCBA — $ 7,500
2. Modifications to software for franchisee — 9,200
3. Equipment—two terminals — 2,000
4. Yearly software update — 2,000/yr.
5. Training of staff — 1,250

In arriving at a total, Peggy separates the one-time costs from the annual costs. The total one-time cost is $19,950, with an annual cost of $2,000. Fleet Feet could probably install the software and bring it on-line within three months, thus allowing the company to begin taking advantage of lost discounts fairly soon. Despite its costs, it would eliminate overtime now paid to bookkeeping employees. Peggy combines the one-time costs with projected savings in overtime and lost discounts:

1. Overtime pay for employees — $ 4,200/yr.
2. Discounts to suppliers — 12,000/yr.

$16,200/yr.

She sees that she can recoup this cost from the savings in a little over a year, a rapid payback.

Peggy's fourth and final alternative, writing customized software, provides a more flexible solution to the problem because it would give Fleet Feet software that was specifically tailored to the needs of a franchise operation. The estimated costs include:

1. System design $ 5,000
2. Programming   24,000
3. Training, installation, conversion   1,200
4. Maintenance of software   900/yr.
5. Equipment, including two terminals   2,000

In estimating the programming costs Peggy predicts that she will need to have at least a dozen new COBOL programs written. Peggy's 12 programs will allow for entering new vendors, changing vendors, printing invoices, printing checks, printing other management reports, and posting invoices. She computes the average program length by looking at Fleet Feet's general ledger programs, finding that the 20 programs in this group average 1,000 lines each. Since both systems will use the Hewlett Packard database manager, are accounting related, and operate interactively, she estimates that the difficulties are somewhat equal.

Multiplying the number of programs (12) times their average length (1,000 lines), she comes up with 12,000 lines. Peggy divides the needed 12,000 lines by 100, an industry average number of lines per day that a typical programmer can write. She comes up with 120 days of programming effort. Since experienced programmers earn about $50,000 per year (including fringe benefits) and work an average of 250 days per year, Peggy computes the daily rate of a programmer at $200 (50,000 divided by 250 equals $200/day). Total programming costs are, therefore, $24,000 (120 days times $200 per day equals $24,000).

Again, Peggy totals the one-time and annual costs: $32,200 in one-time costs and $900 in annual costs. This alternative still gives her $16,200 per year in overtime savings and lost discounts. The payback period is quite different from the third alternative, around two years.

This fourth alternative will allow Fleet Feet to pay suppliers within the five to ten day discount period. Some staffing changes would have to occur in the warehouse and in the bookkeeping area, but Peggy estimates that she can have the new system in place within six months.

With all the alternatives' costs, savings, and benefits figured out, Peggy can now compare the advantages and disadvantages of each. Some costs and benefits would occur one time only, while others would recur annually. Of course, some other costs would crop up, such as supplies and terminal maintenance, but those would not dramatically affect the comparisons.

Some costs, benefits, and savings are not quantifiable, but Peggy knows they exist. For instance, she cannot place a dollar value on better service to vendors, staying ahead of Fleet Feet's competition, more fully utilizing the HP computer, increased morale/job satisfaction, or better management information. Peggy will have to ignore these intangible costs in her calculations, but will document them in the Feasibility Study. It is these intangible costs that make a custom solution better than a prepackaged one.

From her past experiences, Peggy knows that most systems attain a four- to six-year life-span. Because the first alternative does not solve Fleet Feet's discount problem, can never capture the $12,000 per year discount potential, and cannot be maintained for 4–6 years, Peggy rejects this alternative. Using a five-year planning horizon as an average, Peggy then compares the costs and benefits of the remaining three alternatives:

Category	Alternative 2 (Hire staff)	Alternative 3 (Buy software)	Alternative 4 (Write custom software)
Total one-time costs	$ 0	$ 19,950	$ 32,200
Total yearly costs	-46,200	-2,000	-900
Total yearly benefits	0	16,200	16,200
Annual yearly savings	$ -46,200	$ 14,200	$ 15,300

Studying this comparison chart, Peggy rejects alternative 2 (hire staff) because of its high yearly losses, $46,200. However, alternatives 3 (buy software) and 4 (write custom software) both achieve annual cost savings and therefore could make sense for Fleet Feet.

Can Peggy simply multiply the annual cost savings for each of the two remaining alternatives by five to estimate total cost savings for the five-year life of the new AP system? Recalling her college business mathematics course, Peggy knows she can't, because the benefits will occur in the future, while costs will occur now. Future benefits require conversion into present-day dollar amounts with a type of financial manipulation called **present value analysis**.

Peggy looks up the present value formula for converting or discounting in her old college business math textbook:

$$P = \frac{F}{(1 + I)^N}$$

Two things to concern, 1) eat when we break even.

P represents the present value of the benefits, F the future value of the benefits, I the interest rate, and N the number of years. Using 10% as the interest rate (Fleet Feet earns that rate of profit on their least profitable product, socks), Peggy builds another table for the two remaining alternatives under consideration:

Year (N)	(1 + .10)^N	Alternative 3 (Buy software) Future value of benefits (F)	Present value of benefits (P)	Alternative 4 (Write custom software) Future value of benefits (F)	Present value of benefits (P)
1	1.10	$ 14,200	$ 12,909	$ 15,300	$ 13,909
2	1.21	14,200	11,736	15,300	12,645
3	1.33	14,200	10,699	15,300	11,495
4	1.46	14,200	9,699	15,300	10,450
5	1.61	14,200	8,817	15,300	9,500
Total			$ 53,829		$ 57,999

With the present values of the future benefits computed, Peggy subtracts the one-time costs to obtain the net present value of each alternative:

Net present value for alternative 3: $53,829 - 19,950 = $33,879
Net present value for alternative 4: $57,999 - 32,200 = $25,799

Thus, alternative 4 costs about $8,000 more net present value obtainable profits than alternative 3 and represents the "worst" financial solution to the AP problem. It would cost Fleet Feet appreciably more to write the accounts payable software than to purchase it from MCBA. However, both alternatives solve the problem of lost discounts. Also, both alternatives permit Fleet Feet to recover its costs in a reasonable amount of time.

Tax considerations would also affect the ultimate decision. For private profit-oriented businesses, the tax laws allow companies to depreciate hardware and software, which are legitimate business expenses.

In addition to a present value analysis, Peggy considers two other factors: the break-even point and the return on investment. Break-even point computes how long it takes a project to repay costs. In the case of alternative 4, Peggy can compute the payback period from the following table:

Year	One-time costs	Annual costs	Yearly benefits	Total costs	Total benefits
0	$ 32,200	$ 0	$ 0	$ 32,200	$ 0
1		900	16,200	33,100	16,200
2		900	16,200	34,000	32,400
3		900	16,200	34,900	48,600
4		900	16,200	35,800	64,800
5		900	16,200	36,700	81,000

This table, easily built with a spreadsheet program, shows the total costs at the end of year 1 as $33,100 with total benefits of $16,200; obviously, a negative return. However, by year 2, total costs amount to $34,000, with benefits of $32,400. The new system will break even early in the third year.

The **return-on-investment (ROI)** projects the profitability of various alternatives. ROI is a percentage rate measuring the amount that the enterprise *receives* from an investment, versus the amount it *pays* for the investment. Businesses calculate ROI with the formula:

$$ROI = \frac{\text{Total benefits minus total costs}}{\text{Total costs}}$$

The ROI for alternative 4 is:

$$ROI = \frac{\$81,000 - 36,700}{\$36,700} = \frac{\$44,300}{\$36,700} = 120\%$$

Thus, alternative 4 returns 1.2 times more benefits than costs. If the calculation had taken into account the time value of money (present value analysis), the ROI would drop somewhat.

Looking further into alternative 3, Peggy finds a severe potential drawback. While the purchased AP software interacts with Fleet Feet's general ledger, the cost of converting the software for franchise operation may, according to a colleague Peggy met at a trade meeting of franchises, be uncomfortably high.

After careful consideration, Peggy recommends alternative 4, writing customized software. While many computer professionals today advocate buying accounting software, Peggy views Fleet Feet's franchise operation and its Washington, D.C. to California base of operation sufficiently unique to merit a customized approach.

Reporting Findings

Once the analyst has gathered all necessary facts from interviews and questionnaires, collected relevant documents, constructed data flow diagrams, written data dictionaries, observed the system and performed cost-related calculations, the analyst brings it all together in the feasibility study (also known as the requirements document or final report). The format of the feasibility study parallels that of the preliminary report prepared earlier. As you can see in Figure 5.10, it starts with a restatement of the problem and its importance, continues with a list of the study's objectives, reviews the analyst's findings, tallies expected costs and savings, and concludes with the analyst's recommendations.

In most situations, management will want to see a formal proposal before deciding to invest in a new system. Even in small organizations, a verbal discussion alone can lead to eventual misunderstandings. The proposal must take the form of a written document.

When management considers any one proposal, it will compare it to proposals for any other decisions and expenditures. Departments within a company usually vie for the same monetary resources, and most companies cannot support every new idea. Other proposals might include purchasing sophisticated word processing software, renovating manufacturing facilities, or developing bar code inventory systems. Management must decide which is the best investment to make at this particular time.

Even the most experienced analyst will have proposals rejected by management. However, rejection does not necessarily mean that the analyst has failed to do the job well or that management has lost confidence in the analyst. Regardless of the reasons for a rejection, the analyst must treat it as a learning experience: an opportunity to learn more about the business and the way management make decisions.

In our Fleet Feet example, Peggy suggests that the company design, program, and install its own AP system because, after weighing all the options, she determines that doing so will cost less in the long run, increase efficiency, and provide more flexibility.

In a large organization, the analyst may use a standardized form for the final report; in smaller organizations, the analyst may have the freedom to select the most logical format. In any case, the analyst distributes copies of the printed or typed report to the managers who will decide to adopt, reject, or modify the recommended solution.

Figure 5.10 Final report and main product of detailed analysis: the feasibility study.

MEMORANDUM

DATE:	October 17, 1993
TO:	Sally Edwards, President
FROM:	Peggy Adams-Russell, Systems Analyst *PAR*
SUBJECT:	Final Report of the Detailed Analysis of the Accounts Payable System

Executive Summary:

We have completed the detailed analysis of our accounts payable system. This report outlines our objectives, findings, and recommendations. In brief, we recommend the design and development of our own accounts payable system, which would ultimately result in a cost savings of over $16,000 a year.

System Objectives:

How can Fleet Feet speed payments to vendors? Can we accomplish this goal and still lessen the heavy workload in the bookkeeping department?

Findings:

1. Processing time of invoices has dropped by two days since Debbie Delfer was authorized to sign vendor and payroll checks. Unfortunately, this does not save enough time to permit us to take advantage of discounts.

2. The monetary loss of not taking discounts is significant. For the calendar year 1992, Fleet Feet could have saved 8,234 -- or around $700 per month.

3. The usual discount offered by vendors is 2 percent for payments made within ten days. More than half our vendors reported in a telephone survey that they would give a 3.3 percent, five-day discount. Our calculations show that this could boost our savings to over $12,000 for last year.

4. The present manual system includes excessive paperwork. Invoices are hand-posted to both the ledger card and the stub portion of the check to the vendors.

(continued)

Figure 5.10 *(continued)*

5. We currently generate no reports that could automatically trigger payments. Debbie Delfer depends on her memory to know which to pay when.

6. We face rising costs. Any new franchises will require us to hire additional personnel. Our current pay rate for a clerical position in the accounting department exceeds $1,200 per month. We'll need one new person within the next three months, a second person within a year.

7. We do not adequately identify our vendors. We presently list them by name only. Frequently, we post invoices to the wrong ledger card and suffer a finance charge for late payment.

Alternatives:
Four possible solutions could eliminate the problems in our accounts payable system and expedite payments. Each offers differing costs, savings, and effects on our operation. None of the alternatives takes into account effects of investment tax credits for equipment or depreciation as a savings. All costs are pre-tax costs.

Option I.   *Do nothing.* Leave the system alone.

    Costs: Loss of potential discounts ($10-12,000 per year)

    Savings: None

    Effects: The manual AP system can only get worse. Payments to vendors will get slower as our business volume grows. Bookkeeping will struggle to keep up and morale will suffer.

Option II.   *Hire staff.* One additional person to start now, another in a year. Both people would work in bookkeeping to maintain the workload at its current level. Note second year cost is high due to first person working remainder of this year.

    Costs: Loss of potential discounts: $10-12,000/year
                  Two employees: First year $12,600
                             Second year $33,600

    Savings: None

    Effects: The AP system would stabilize, but not improve. In four years, we would be back where we are now.

(continued)

Figure 5.10 *(continued)*

Option III.   *Purchase AP software.* An accounts payable software system written by MCBA (Micro Computer Business Applications) is available and would fulfill most of our needs. This system does not have the ability to deal with franchises, but is compatible with our HP computer. This software is installed in 40 businesses in the United States and Canada.

Costs: MCBA accounts payable software $7,500
          Modifications to MCBA system for franchises   9,200
          Two terminals (in warehouse and bookkeeping)   2,000
          Update of MCBA on a yearly basis   2,000/yr.
          Training of staff   1,250

Savings: Overtime pay for bookkeeping employees $4,200/yr.
          Discounts to vendors   12,000/yr.

Effects: Vendors could be paid within the five- to ten-day discount period. Some changes would be necessary in the warehouse and office areas. The system can be installed and operating in three months. Good training manuals are part of the system.

Option IV.   *Design, program, and install our own AP system.* This alternative gives us the most flexibility: We can tailor our reports to our needs and accommodate franchise operations.

Costs: System design $5,000
          Programming   24,000
          Training and installation   1,200
          Maintenance of system   900/yr.
          Two terminals (in warehouse and bookkeeping)   2,000

Savings: Overtime pay for bookkeeping employees $ 4,200/yr.
          Discounts to vendors   12,000/yr.

(continued)

Figure 5.10 *(continued)*

Effects: Vendors could be paid within the five- to ten-day discount period. Some changes in the warehouse and office would have to be made. The system can be installed and operating in six months. The problems associated with rising costs, excessive paperwork, workload, and vendor identification can be alleviated.

Options three and four do not show the cost of the computer, since we own our computer. Neither of these two choices requires us to modify or upgrade the computer, nor will they damage the system's performance on existing applications. We have sufficient processor time and disk storage space to add the accounts payable system to the computer.

Recommendations and Rationale:

I recommend management approve option four for the following reasons:

1. The cost of a customized system is not appreciably greater than the MCBA software.
2. It will solve the problems in the current manual AP system.
3. MCBA's estimate may be low because they want our business. Other users have found actual costs to run higher than MCBA's prediction.
4. The new system would interact with our current general ledger system.
5. The payback for the new system is a little more than two years.

Intangible Benefits:

Option four also provides Fleet Feet with the following nonmonetary benefits:

1. Better service to our vendors.
2. Information not now available to management will arrive in a monthly report, which will help us make better decisions.
3. Our firm will continue to stay ahead of the competition.
4. The bookkeeping department will feel higher job satisfaction.
5. We can more fully use our current HP computer.

WORKING WITH TECHNOLOGY
WRITING SUCCESSFULLY FOR MANAGEMENT

Writing a document for management review is quite different than many other types of reports. Managers want a slimmed-down "just the facts" report, not one filled with rhetoric. Here are some guidelines for writing to this special audience:

1. Start the report with an executive summary. Write a brief description of the project, along with a financial summary.

2. Next, describe the project in detail, including objectives, scope, level of effort required, and anticipated time frames.

3. Define the problem and the solution. Make each definition short and specific. Avoid long-winded historical reviews of how the organization got into the situation. A short definition gives managers confidence that you have done your research.

4. Give the financial impact of the project. A detailed financial report (typically created on a spreadsheet)— showing cash flow requirements, costs, benefits, and savings—is easily read and understood by managers.

5. List all possible risks and potential pitfalls.

6. Show opportunity costs. Since few businesses enjoy sufficient financial resources to seize all opportunities, management sets priorities. Projects that achieve little return on investment will fall low on the list.

7. Make a quality presentation. The depth and quality of the proposal and its presentation strongly influence management, which proceeds most confidently when management believes that all factors have been considered.

8. Include your track record. Management places great emphasis on historical performance. Past projects that have met their goals and were delivered on time and within budget are good indicators of a future project's success.

9. Don't worry about early project termination. Approval comes more easily for projects that the enterprise can halt before incurring unacceptable unrecoverable losses. Unanticipated changes in the business climate or ownership sometimes requires drastic action. In large projects, it may help to list possible termination points.

10. Give details about vendors or suppliers. Many proposals utilize outside vendors or suppliers. Tell managers about the vendor, including financial strength and references. An appendix is a great place to put corporate literature.

11. Make a recommendation. After all, you are the expert on the project and you must have formed an opinion. Tell people about it. Make your recommendation short, concise, and justify it.

PRESENTATION OF FINDINGS TO MANAGEMENT

After management has thoroughly considered the feasibility study, Peggy calls a meeting to discuss the study and, if all goes well, to choose a course of action. Such a decision-making meeting usually occurs a few days after the study's distribution. Responsibility for leading it typically falls on the manager of the computing services department or the person who requested the analysis.

The analyst plays a major role in this meeting and should prepare thoroughly in order to answer questions and supply needed information. In fact, the analyst should rehearse the presentation a few times in an effort to identify and improve upon weak points in the proposal.

If the analyst leads the meeting, he or she must exercise control. The following rules can help ensure a successful meeting:

- Never read the feasibility study aloud. Instead, summarize it, guiding the audience toward the study's recommendations.
- Use visual aids, such as chalkboards, flipcharts, slides, photographs, videotapes, and overhead transparencies.
- If appropriate, demonstrate equipment or software to show how it will work.

Often, one key individual requires convincing, and this person should receive respectful attention. If all runs smoothly, the meeting may end with a decision to implement the analyst's recommendations.

After the meeting, management notifies all appropriate staff members of its decision. If management has decided to proceed to the design stage, the notification memo explains the plan briefly and establishes an overall schedule, as in Figure 5.11. Even if management decides to maintain or modify the current system, it should still issue a memo, or else people will wonder why the company wasted time with a study that produced no results.

After receiving a decision to proceed to the design phase, the analyst organizes all the memoranda, questionnaires, interview documents, forms, data flow diagrams, data dictionaries, entity relationship diagrams, financial spreadsheets, and reports from both the preliminary and detailed analysis into one file, which becomes the **analysis documentation**. Now the design phase can start.

Figure 5.11 Management communicates its decision to the staff through the use of a formal memorandum. This notice should go to all those affected by the action. In our example, Sally Edwards sends a memo to everyone involved with the accounts payable system.

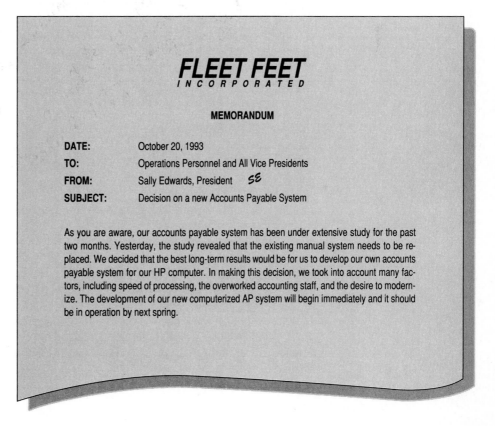

FLEET FEET
INCORPORATED

MEMORANDUM

DATE: October 20, 1993
TO: Operations Personnel and All Vice Presidents
FROM: Sally Edwards, President   *SE*
SUBJECT: Decision on a new Accounts Payable System

As you are aware, our accounts payable system has been under extensive study for the past two months. Yesterday, the study revealed that the existing manual system needs to be replaced. We decided that the best long-term results would be for us to develop our own accounts payable system for our HP computer. In making this decision, we took into account many factors, including speed of processing, the overworked accounting staff, and the desire to modernize. The development of our new computerized AP system will begin immediately and it should be in operation by next spring.

CHOOSING CASE TOOLS

In the early days of the personal computer market, customer options were limited to only a few choices, such as an Apple II or a Radio Shack TRS-I. Today, it is quite different. Apple, IBM, NeXT, Sun, and a wide variety of other suppliers offer an overwhelming number of products. The same situation exists in the CASE marketplace. In the early days, few products existed; now, there are many CASE suppliers. So how do you pick the right CASE tool?

Some CASE tools are designed for creating new applications and not for maintaining or enhancing existing ones. If an organization is in a maintenance mode, it must have a CASE tool that supports the maintenance aspect of software development and may limit its reviews to products strong in systems maintenance. Conversely, if an organization is heavily involved in design, it needs to make sure that the CASE tool supports many design activities—including report prototyping, database definition, screen generators, and process outliners in which the logic of the system can be described. If an organization is heavy in analysis, the CASE product must support diagramming tools, data dictionaries, and financial modeling.

All CASE tools support one or many philosophies or methodologies for software development. Some examples are Gane/Sarson, Ward/Mellor, Shlaer/Mellor, or Yourdon/DeMarco. If the CASE tool does not support the methodology an organization currently uses, it is doomed to failure.

Many large projects are so involved that no single person can do it all. Instead, the CASE tool must allow partitioning of the project so that each person on the team can tackle his or her assignment. With a team environment, a good CASE tool keeps track of the status of each part and provides detailed reporting, such as percentage completed on assignments. It also promotes team communication.

Good CASE tools collect a wide range of facts, diagrams, rules, report layouts, and screen designs. The CASE tool must format the data collected into a meaningful document that managers and users can understand. This is one of the "claims to fame" of CASE tools. Make sure that whatever tool you pick generates good documentation that people can read and use, not just lists of facts and figures.

Most CASE software suppliers are proud of their product and the organizations that have bought them. To get a well-rounded description of them, you should ask for references of customers who use the product and talk with the customers about their successes and failures. A CASE supplier that will not give you references is one to avoid.

Many CASE tools generate COBOL, C, C++, or SQL code. The tool should translate logical process models into physical realities. In the database area, a CASE product should produce SQL statements to create the tables that you have defined in entity relationship diagrams, as well as views and indexes.

What about support after the sale? Besides a cartridge tape, reel of tape, or a set of floppy disks and a few manuals, the CASE supplier must offer training on the tool and techniques. The supplier should also have a consulting arm to help the user through his or her first experience, and a user's group where people meet to discuss the reality of the product versus the hype from the vendor.

Regardless of which tool the user selects, management involvement and support at the start are essential. All too often, CASE is seen as the panacea for all the ills of a management information system. Not true. Gains in productivity and quality will come over time, but a steep, expensive learning curve and the costly purchase of CASE software occur first. Managers must allow for adequate training and experimentation to take place. Bottom-line results won't seem significant during the first year, but the long-term benefits could be substantial.

SUMMARY

If management approves the results of a preliminary analysis, the analyst conducts a detailed analysis, gathering facts about the old system, outlining objectives for a new one, estimating costs, listing possible alternatives, and making a recommendation.

Before finalizing a recommendation, the analyst may interview people in the organization, talk with hardware or software vendors, collect typical forms currently in use, personally observe the existing system or similar systems outside of the organization, and perhaps send questionnaires to other people—even outsiders—involved with the system. The Gantt chart depicts scheduled events in a way that even nontechnical people can grasp.

Since the organization is making a business decision, the analyst evaluates costs and benefits. Costs and benefits are divided into one-time, periodic, and intangible categories.

Management evaluates the analyst's final report or feasibility study and decides whether to halt the project or proceed with it. In the latter case, the analyst ends the analysis phase by assembling all documentation and initiating the design phase of the systems life cycle.

KEY TERMS

Gantt chart
Fact finding
Questionnaire
Observation

Alternatives
Present value analysis
Return-on-investment (ROI)
Analysis documentation

QUESTIONS FOR REVIEW AND DISCUSSION

Review Questions

1. What is detailed analysis?
2. What are the analyst's main functions during detailed analysis?
3. Which methods does an analyst use to gather data?
4. What is fact finding?
5. What are the components of the feasibility study?
6. Who participates in a management review?
7. What are three different financial analysis techniques?
8. What are two reasons for management's investment opportunities?

Discussion Questions

1. List the inputs to the detailed analysis phase.

2. List the outputs from the detailed analysis phase.

3. Make a list of three dos and three don'ts that you should follow when interviewing a user.

4. What questions should an analyst ask the person during the fact-finding phase of the detailed analysis?

5. List three types of questionnaire formats.

6. What problems does an analyst face when observing an existing system?

Application Questions

1. List four vendors that an analyst might to contact to gather information about terminals.

2. Why is a vendor number better to use than a vendor name when identifying invoices?

3. What is the purpose of an invoice number on the packing slip?

4. Why would Fleet Feet keep the name of the contact for each vendor?

5. What will happen if Fleet Feet's management decides on alternative 3?

6. Match the following techniques:

   a. Present value analysis   1. A percentage

   b. Return-on-investment   2. A year

   c. Payback analysis   3. Time value of money

Research Questions

1. Make a list of audio or visual aids that an analyst can use when presenting the feasibility report to management.

2. Using spreadsheet software, build a payback analysis table for alternative 3.

3. Using spreadsheet software, modify the payback analysis for alternative 4 to take into account present value analysis. Note: Most spreadsheet systems have the net present value formula built in.

The memo from the Stensaases to Frank Pisciotta was the authorization that he needed in order to proceed to detailed design. Frank started by reviewing tasks that he felt he needed to perform. At the top of the list was further research into the existence of a software package for videotape check-out/check-in. Second was the interface to BusinessWorks PC, the integrated accounting software the Stensaases bought. Next came a more detailed definition of the files that the system would require and an expansion of the data flow diagrams and data dictionaries. Finally, Frank would have to write up his findings in a report to Mark and Cindy.

Looking for an existing package was straightforward. The Stensaases subscribed to an industry journal, *Video Store Owner*, and every issue was packed with advertisements from software developers. Frank also called his friend, Richard Donaldson, and asked for the name of the package his video store used. Frank then contacted seven vendors and asked for literature on each of their products. They sent him materials within a few days and he started his analysis.

Packages were priced between $8,000 and $10,500 for a four-station network version, were used on IBM PCs, and were in operation in many stores across the country. None of them had an interface to an outside accounting software system or supported bar codes on boxes. Also, none of them could report facts about individual videos, such as rating, year made, names of stars, category, and so on. Each package required an average of $1,000 per year for a maintenance agreement, and on-site training was priced at $750 per day plus expenses, with a minimum of 3 days recommended.

In spite of all the apparent deficiencies, Frank looked over each package's literature, called for references of local installations, and went to a computer store to view his top two contenders. The first package, Video In/Out, looked great. It had color windows, mouse support, and a series of very nice reports that the system produced. The second package was a flop. It didn't offer color and looked like a system designed for the casual user, not one for high-volume transactions. Frank ruled out this package and called the software supplier of Video In/Out, the first package. Answering his inquiry on software modifications, the firm responded that they were willing to make them, but charged a $2,000 fee—plus the cost of the actual modification. What Frank deduced was that they were not really interested in making changes, but would make them for a large fee.

Frank was discouraged. He had hopes of finding a package that would meet the Stensaases' needs "off the shelf." Unfortunately, that was not the case. At this point, it appeared to Frank that he would have to design and write a system just for Mark and Cindy Stensaas.

While he was researching existing software packages, Frank re-established contact with Mark Havener at Manzanita Software Systems. His original contact had been a few months ago and now his interest was even greater. He wanted to know the exact specifications for a bridge or link between BusinessWorks PC and a video check-out/check-in system. Mark's first question centered on the type of bridge or link. Did Frank want a real-time link? Or would a batch update once a day suffice? Mark explained that BusinessWorks PC is capable of updating a customer record immediately after the data are entered or could wait and take the data periodically, perhaps at the end of every business day. If Frank's customer could wait for daily batch updates, no modifications to BusinessWorks PC were necessary, since that capability was present now with the software. All Frank's new video software system needed to build was a file fitting a pre-agreed format and BusinessWorks PC would do the rest.

A real-time link was not much harder, but did require additional software. Two years ago, Manzanita gave BusinessWorks PC the ability to link to a cash register and Mark did not see any reason that the interface would not work with their videotape system. Mark envisioned that the video system would issue a request for customer data; the system would send the data to the terminal for display purposes; the clerk would enter charges or payments amounts; and the system would then transmit the data to BusinessWorks PC, which would update the customer's account balance. BusinessWorks PC would not maintain data on tape check-out or check-in, but would keep accounts in balance.

Frank agreed that this plan looked great and asked for prices and timeframes. Mark responded a few days later that Manzanita would write all the needed software for $1,200 and could have it ready 2 weeks after signing a contract. Manzanita would retain ownership of the software, paying the Stensaases a royalty of 10% if they ever resold the software, and wanted Frank's customers to serve as reference accounts for future sales. When Frank inquired about resales, Mark responded that Manzanita wanted to sell their product and that this looked like an ideal opportunity for them to enter the video store market. Mark and Frank agreed that while they still needed to resolve many issues, they could delay these discussions until the design stage. The link was possible, not too costly, and within a timeframe that each organization thought acceptable.

Frank had two issues resolved and was ready to work on the files needed for the video half of the system. The BusinessWorks PC half would maintain all of the financial aspects of the system. He felt that his earlier design of a file of customer and rental data was still logically sound, but needed expansion based on the facts that he had uncovered during the preliminary design.

The Stensaases' requirement that the system maintain data on the tapes and the major actors and actresses appearing in each videotape mandated a modification to his entity relationship diagram (Figure 5.12). Frank saw the need for five entities or files:

- CUSTOMERS: customer number, name, address, date joined
- MOVIES: movie number, title, rating, year made, category
- STARS: star number, name
- RENTALS: customer number, movie number, date rented
- PERFORMANCES: movie number, star number

Each customer would have a unique customer number, his or her telephone number. Every movie would receive its own movie identification number. Each movie star would receive an arbitrarily assigned unique star number.

The rentals file would allow the system to cross over between which customers rented which movies and which movies were rented by each customer. In a similar fashion, PERFORMANCES would track which stars appeared in which movies and in which movies each star appeared.

Frank remembered that a customer could have multiple borrowers on an account and that this would require that CUSTOMERS have multiple name fields. BusinessWorks PC had a problem here since it kept a single customer name and this single requirement would mean

Figure 5.12 The modified ER diagram for the video system.

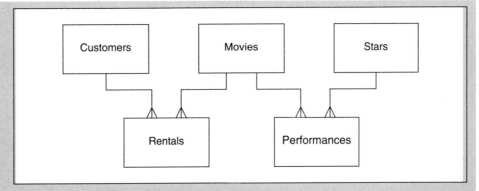

keeping names in both halves: BusinessWorks PC and the video system. How he would keep multiple names was another issue that Frank would delay until the database design phase of the systems process.

Each file would need a complete data dictionary definition, including entity names, list of data elements, data types, lengths, and all the other specifics about entities and data elements. Frank decided to put these off until database design.

Frank put together his findings and wrote his feasibility study or requirements document (Figure 5.13). Purchasing software is an alternative that Frank must consider in the report. He states all the facts about his observations, interviews, and questions. However, this alternative does not solve Mark and Cindy's problem of wanting an integrated accounting system with tape check-in/check-out, including data about the stars appearing in the videos.

The final alternative is easy to estimate because of his prior experience. He knows he'll need some design time, probably 20 programs, equipment, documentation, and some user training.

Frank writes his feasibility study in the form of a short memorandum and sends a copy to Mark and Cindy. Both agree with his rationale and authorize him to proceed to designing a customized videotape check-out/check-in system.

Working With View Video

Frank's role in detailed analysis mimics that of a detective trying to solve a mystery. He interviews a variety of people, gathers his thoughts, writes reports, and finally makes a recommendation. The primary goal is to make sure that Frank really understands the problem that he needs to solve. Notice that very little of Frank's efforts are focused on the solution to the problem, but on understanding it. Finally, near the end, he puts together his findings—but he does not design the screens, database, or reports. This will come later during design, if the decision leads to this phase of the systems process.

Many analysts want to jump into the solution before getting a firm grip on all the issues. A programming axiom states that "the person who starts coding first finishes last." Frank's need to understand dramatically affects his solution. For example, suppose that during design, he discovers that multiple people can borrow with the same account number. This little tidbit dramatically alters his database design, data collection screens, and reports. Now suppose that each borrower can have a limit placed on the type of video he or she rents; for example, parents can rent any type of video, but they may want to restrict their children to "G" rated videos and prohibit them from music videos. All of these facts impact the solution and must be known *before* starting design.

Figure 5.13 The requirements document for View Video.

MEMORANDUM

DATE: October 22, 1993

TO: Mark and Cindy Stensaas, Owners

FROM: Frank Pisciotta, Consultant Analyst   *FP*

SUBJECT: Detailed Study of View Video's Tape Check-out and Check-in System

I have completed my detailed study of your videotape check-out/check-in, movie stars, and link to an accounting system in less time than anticipated. As you know, my prior employer assigned me to analyze and program their accounts receivable system, so I had that experiance to draw upon. My greatest fear was that View Video's accounting needs differed from my prior employer's, but after spending time discussing needs with Mark Havener at Manzanita Software, we discovered that both systems were very much alike.

My study revolved around three alternatives. I feel that one is superior to the others, but would be happy to discuss all of the alternatives with you.

Alternative #1: Use a manual paper-based system.

    This alternative does nothing for us.

    Errors will occur in tape check-in/out.

    We will still not know who owes us what and how old the debt is.

    I did not pursue this option further because it does not solve the tape check-in/out problem.

Alternative #2: Buy Video In/Out software.

    Video In/Out software is an off-the-shelf package.

    It is expensive: $10,500.

    Does not interface to BusinessWorks PC.

    Is available within 2 weeks after receipt of order.

    References are positive.

(continued)

Figure 5.13 *(continued)*

Management reports are plentiful, as well as useful.

Is PC-based and we will need to buy equipment. My estimate is $5,900 for equipment.

Uses color, windows, and mouse.

Does not have bar scanner capability.

Does not have movie star cross-references.

Training is $750 per day plus expenses, 3 days recommended.

Software modifications require a $2,000 fee plus the cost of the modification, estimate of $2,000.

Annual maintenance charge of $1,000 per year.

Alternative #3: Write our own software.

Writing our own software is the most expensive and time-consuming of the alternatives. My cost estimates:

System Design:	$3,500
Programming:	8,500
BusinessWorks PC Link:	1,200
Training, Documentation, Installation:	900
System Maintenance:	800/year
Equipment:	5,900
Total One-Time Cost:	$20,000

Will have an interface to BusinessWorks PC.

Will have bar code readers.

Can have movie star cross-references.

We will own the software.

Should take 4 months to have in operation.

Can have customized reports, meeting your needs.

Possible royalties from Manzanita if they market vertically.

Recommendation:

Alternative #3, Write our own software. Picking this alternative gives you the system you want, not somebody else's system. You can tailor it to the method of operation you want and are not held to a fixed method of doing business. It will provide better service to our customers, which will assist them in finding the tape they want to rent. We can meet and stay ahead of the competition with a custom solution.

Case Study Exercises

To help you with detailed analysis and understanding the video problem, complete the following:

1. Visit three video stores and see if they can tell you who stars in which videos.
2. Some video stores have customers pay for their videos at the time they are borrowed; others, when the video is returned (called pre-pay and post-pay). What new issues does this bring out?
3. How do pre-pay stores bill their customers if they paid too little during check-out?
4. Do you see any advantages to either pre- or post-pay?
5. Do your stores have a different fee structure for mid-week rentals than weekends or holidays?
6. Do the stores have a different fee structure for recent releases rather than older videos?
7. Will different fee structures have any effect on systems design?

Prototyping and Fourth-Generation Languages

GOALS

After reading this chapter, you should be able to do the following:

- Differentiate between a third-generation and a fourth-generation language
- Define a productivity tool
- Explain why prototyping is becoming the preferred way to build systems
- Explain the uses of prototyping tools
- Cite three advantages of 4GLs
- Name the four parts of most 4GLs
- Name three disadvantages of 4GLs
- Cite the features found in most CASE products
- Cite the two primary differences between 4GLs and CASE
- Name three disadvantages of CASE

PREVIEW

Finally, it's time to start building a system. Assuming that the analyst correctly identifies the problem that prompted the user's original request, lists alternative solutions plus attendant costs, and receives the green light from management, work can now proceed on actually designing the new or enhanced system.

Before going ahead, however, the analyst must decide whether to follow the standard system development life cycle (SDLC) and use conventional approaches to building the system, or to employ rapid system development techniques. The pressures to get systems done quickly are increasingly pushing analysts toward choosing rapid system development.

THIRD-GENERATION LANGUAGES (3GLS)

Most departments in an organization have ongoing demands for new computer applications. The personnel department needs a new system in order to monitor the company's efforts to meet federal equal opportunity employment laws. The marketing division desperately wants a new report that shows sales by geographical area. And the manufacturing group requests a better method for tracking machine performance. All of these requests may make sense, but how can a manager of information services fulfill each one while still maintaining all current applications? In most cases, something has to give, especially if an organization relies on **third-generation languages (3GLs)** such as COBOL, Pascal, C, C++, Ada, FORTRAN, or BASIC (see Figure 6.1).

In the past, it usually took many systems revisions before users felt happy with an application. Users often did not really know what they wanted until seeing what they *didn't* want. In the end, users either accepted an imperfect system or rejected it outright.

Traditional third-generation methods of system development are anything but rapid. Most users and analysts agree that the old methods are agonizingly slow. This slow, careful approach was considered necessary to ensuring that the detailed logical specifications for a system were absolutely correct before writing the programs. Systems designs were frozen at the time programming was started because changes were difficult and expensive after that point.

The problem with this 3GL approach is that long periods of time pass between finalization of the logical system specifications and the completion of programming. For larger systems, it was not unusual for a year or more to pass between the time a user agreed to a screen's final layout and the time he or she actually saw that screen on a monitor.

Figure 6.1 Programming languages have evolved in four generations.

First generation:
Machine language, hardwired instructions, numeric instructions and addresses, machine-dependent programming.

Second generation:
Symbolic instructions and addresses, translation of program with an assembler, machine-dependent programming. Typical languages include IBM's BAL and Autocoder.

Third generation:
Problem-oriented languages, translation with compilers or interpreters, structured programming, database management systems, on-line program development. Typical languages include COBOL, FORTRAN, Ada, Pascal, C, BASIC, PL/I, and C++.

Fourth generation:
Nonprocedural languages, integrated data dictionaries, dynamic relational databases. Typical languages include Oracle, FOCUS, NOMAD, Natural, POWERHOUSE, RAMIS, and INQUIRE.

PROTOTYPING: RAPID SYSTEM DEVELOPMENT

To resolve this predicament, today's systems developers turn to **rapid system development** and productivity tools. A productivity tool is any technique, usually software-based, that enables analysts to produce systems faster. The general idea is to help the analyst to work more productively.

The most important productivity tools are **fourth-generation languages (4GLs)** and CASE. Compared to traditional 3GL languages, 4GLs and CASE dramatically reduce programming efforts. Functionally as powerful as their 3GL counterparts, 4GLs and CASE can get a system up and running ten times faster, and since they involve fewer instructions, the system is easier to maintain after it becomes operational.

Both of these productivity tools classes support prototyping. The word suggests an approach that an analyst uses to render a quick, though unpolished, handmade version of a desired end-product destined for subsequent volume production. Prototyping software lets the user quickly see a system, and it allows anyone to suggest or make immediate revisions (Figure 6.2). Because prototyping recognizes that numerous changes will occur between the first and final versions of any application, it permits such changes with unprecedented ease.

CASE and 4GLs are tools to help the analyst, programmer, and user develop the system right the first time. Analysts still do the design work, programmers still program, and users still must describe their wants and needs. These new tools help automate the systems process; they do not replace them.

With the advent of newer CASE tools coupled with 4GLs, the traditional systems analyst may find his or her job evolving. Instead of solving only computer-related problems, the new analyst will need more skills in solving complex business problems and procedures. The analyst of the future will require as much in business expertise as they do in analysis acumen.

Figure 6.2 CASE and prototyping create a small working model from user specifications. The model changes as users work with it.

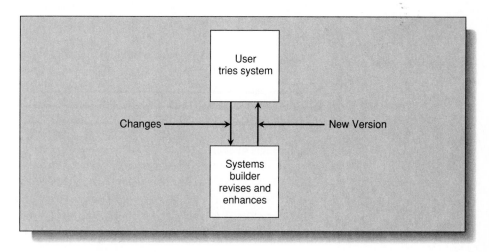

In the interim, a lot of things could go wrong. Aspects of the business could change, making parts of the screen design or many other features of the system obsolete or incorrect. Promotions and other forms of user turnover mean that new people might end up using the system, not just those who originally helped design it. These new users might have different perceptions about what the system should do. Budget priorities might change so that a system project that was "hot" and well-funded at its initiation might become underfunded and less important by the time it is completed. The "window of opportunity" for improving business competitiveness or introducing a new product or service might close by the time the system finally comes out. Laws could change. And worst of all, users might decide, after seeing the actual system in operation, that they don't like the design after all, and want to change it.

Naturally, analysts are very nervous when users sit down to see a new system for the first time because they usually request changes. But these user-requested changes are normal. It is not unlike buying a new sofa, setting it in your living room, and deciding that you don't really like it after all. It looked great in the store, but your mental visualization of what it would look like at home wasn't very accurate.

Analysts ask users to do impossible amounts of mental visualization. In the traditional system development life cycle, we ask them to make hundreds of detailed decisions about screen designs, report layouts, screen dialogs and sequencing, business rules, and many other abstract concepts. That is, we give them system blueprints—paper specifications—and ask them to pretend, or visualize, what the system would look like and how it would function if it were real. Then we ask them to commit to these detailed specifications, to agree not to request any changes, and even to sign specification acceptance documents. Then, after completing the programming, we call them in to see the system work.

It is natural that they ask for changes. It is much easier to notice ways to improve an information system once you can actually see and experiment with it. Working only from the blueprints is hard. You can't accurately visualize the sofa in your living room. Likewise, users can't perform the much more difficult task of remembering what a system looks like that they haven't seen for a long time—and might not have fully understood even then.

You can think of the life cycle that accompanied this traditional approach to system development as a long, straight line. Projects start with a preliminary scope and feasibility study; move to logical design; then to physical design, programming, testing, and installation. The steps come one after another, in sequence, and there is no doubling back to "tune up" the specifications once programming is underway. The traditional approach to system-building is a long, linear life cycle.

Rather than this long, linear life cycle, both analysts and users would prefer a life cycle that consists of a series of repetitive, tight circles. In this life cycle, analysts and users would sit down and rough out a preliminary design for, perhaps, a screen. They would then have the opportunity to actually see and manipulate that screen in a couple of days, rather than a couple of years. After experimenting with the screen, they'd next develop ideas for changes and improvements, make the modifications, and quickly see and use that new screen. This circular process would continue for screens, reports, and all the other system features until a complete system emerges, fully consistent with the users' needs. This prototyping approach would result in the delivery of systems much more rapidly, and represent a welcome change from the traditional linear system life cycle.

Tools now exist that make system prototyping a reality. Gradually, the traditional approach to system development is losing out to new techniques for rapid system development. The first prototyping tools developed were fourth-generation languages. Both 4GLs and CASE complement the newly shaped life cycle of tight circles, and both support rapid system development.

[handwritten margin note: & there's a text function — good for application.]

FOURTH-GENERATION LANGUAGES (4GLS)

Fourth-generation languages emerged in the late 1970s, and most came from third-party software companies, rather than from mainstream computer manufacturers such as IBM. Campaigns and products have come and gone, and today there are only a handful of 4GLs that survived the product shakeout process to dominate the market.

Regardless of the particular language or vendor, 4GLs share several common characteristics. Most 4GLs are easy to learn. The software companies clearly targeted 4GLs at the computer user market, rather than at general business professionals. Their syntaxes and commands are so simple that software companies claim that non-programmers can learn to use most of the key features of a 4GL in just a few hours. In reality, this is mostly hype, but they do have fewer rules to memorize.

Most 4GLs (as does COBOL, a 3GL) have built-in "report writers" that make it easy to extract data from files or databases, summarize them, and produce reports in a wide range of formats. Most 4GLs have built-in statistical analysis functions (regression, correlations, and so on) and financial modeling routines (such as present value and annuities). Some 4GLs can convert data to graphics formats (pie, bar, line, and scatter charts).

To understand the advantages of fourth-generation languages, you must first understand the chief drawback of third-generation languages (Figure 6.3). 3GLs such as COBOL, C, and Ada are procedural. Programs in these languages must instruct the machine not only in what to do, but in exactly how to do it. Writing even a file update program in a language like COBOL becomes quite a job because you must keep detailed track of many nitty-gritty details.

Fourth-generation languages are non-procedural languages. The programmer need only specify what to do and the language figures out the procedural details of how to do it. This greatly simplifies the programmer's tasks, making it end-users' most popular feature. A lot of work becomes easy to do with only a few lines of 4GL code. For example, the following three lines of SQL code will produce a sales analysis report.

[handwritten margin note: In 4GLs, Design so nontechnical people can understand.]

```
SELECT PROD_NAME, UNITS_SOLD, SALES_REVENUE,
    FROM SALES_SUMMARY
        ORDER BY REGION, MONTH
```

[handwritten annotations: example of SQL; expecting table name.]

The SQL software automatically decides how to arrange the titles, column headings, and detail lines on the printout, and will monitor for an end-of-page, page numbering, and other details of the job. This same task would require at least 200 lines of COBOL code, and the logic in the COBOL program is also more intricate. The 4GL represents a dramatic productivity improvement over the 3GL.

[handwritten note at bottom: SQL = Structure query LANG]

Figure 6.3 Operating environment of 3GL and 4GL systems.

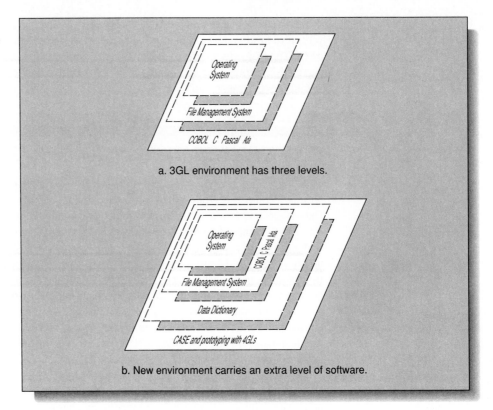

a. 3GL environment has three levels.

b. New environment carries an extra level of software.

Some drawbacks, however, are common to most 4GLs. Many products do not generate 3GL output, but instead create machine language. A few 4GLs don't execute efficiently. Finally, 4GLs have the drawback that each is unique; no standardization exists. Users familiar with one 4GL cannot easily converse with users of another 4GL, nor can the systems they develop move easily between 4GLs.

Most industry observers do not believe that 4GLs will ever gain wide use for developing large transaction systems. Since their focus is on direct implementation, taking users from a perceived need to functioning code quickly and directly, they bypass the modeling and logical design activities circles that are so essential to the development of conventional systems.

Weighing these pros and cons, it is apparent that 4GLs are best suited for systems that must be developed as quickly as possible. Generally, they are not used for development of large, high-transaction volume systems.

In contrast to our evolutionary prototyping, which focuses on rapid application development, we have throwaway prototyping. This version more closely follows the traditional system life cycle with the prototype discarded but used as a model for 3GL implementation.

CASE AND 4GLS

Amazingly, CASE technology is a classic example of the "cobbler's children have no shoes" syndrome. The analyst's job is one of the last ones automated in most organizations. We have historically automated everybody else in our organizations but not ourselves; for example, most secretaries got their word processors before most analysts received CASE.

Some CASE products support only small parts of the system life cycle, while others cover the entire life cycle (Figure 6.4). Code generators that can automatically write COBOL and other types of source code are examples of tools that target only the development phase of the life cycle. CASE tools fall into four main categories: Analysis/Design Tools, Implementation Tools, Maintenance Tools, and Project Management Tools.

Similarly, designs for reports (printouts) and screens are quickly laid out with CASE and refined successively until users are satisfied. These designs then support prototyping of the system, in which users can "test drive" the system, moving from screen to screen just as they would in the real system. Other CASE tools can take these screen layouts and use the specifications to automatically write code that will generate these screens in the environment of the real system.

Figure 6.4 CASE tools support single or multiple activities within the systems process.

Implementation tools can generate program and database code, automatically test systems, and support requirements traceability. Code generators use the diagrams and logical

VENDOR	PRODUCT	ANALYSIS	DESIGN	IMPLEMENTATION	MAINTENANCE	DOCUMENTATION
Intersolv	Excelerator	X	X			X
Ken Orr	Design Machine	X	X	X	X	
Chen & Assoc.	E-R-Modeler	X				
Arthur Anderson	Design/1	X				
KnowledgeWare	IEW/WS	X	X		X	
Texas Inst.	Info. Eng. Fac	X	X	X	X	X
Yourdon/DeVry	Analyst/Designer Workbench	X				
Cadre Tech.	Teamwork	X	X			X
IDE	Software Through Pictures	X	X			X
Oracle	CASE * Designer and CASE * Dictionary	X	X	X	X	X
SoftStruct	TurboCASE	X	X			

specifications prepared by an analyst to automatically create the program code (usually COBOL or C) to implement those logical specifications.

CASE can help in system testing through tools such as test data generators. These generators can create thousands of records for input to a new system in order to test its accuracy and speed. Analysts specify the range of values allowed in each field of the test records, the data types, and the dispersion of values desired. The CASE test code generator does the rest automatically. This is a tremendous improvement over the old method, in which analysts and programmers made up test data manually, often requiring months of effort for a large system.

CASE maintenance tools are used primarily for revitalizing old programs that are slow, difficult to understand, or poorly documented. For example, Peat Marwick offers a CASE tool called Pathvu that accepts an old, disorganized program as input, and produces a description of that program's logical structure and an analysis of its quality as output.

Retrofit, a companion product, will then restructure the program, making it "logically" cleaner, with better structure, and producing documentation in the process. The new program will preserve the features and functions of the old one, but is more easily modified. CASE tools such as these are often referred to as **reverse engineering** products because they accept program code as input and produce specifications and structure charts as output. This is the reverse of the usual direction—hence the name.

Project management CASE tools assist analysts by providing automation of scheduling, task completion and milestone analysis, and estimating. These tools can compare preliminary estimates with actual expenditures, report on system status, and trace quality control. Most CASE tools start with a repository (Figure 6.5), which becomes the heart of the CASE software. Similar to a data dictionary, the **repository** collects, stores, and maintains a complete set of facts about the system, including data elements, data structures, processing rules, screen and report layouts, and documentation. This computerized repository allows the analyst to quickly review and update the dictionary, list dictionary entries (on the screen or printer), or update any item stored in the dictionary.

In some CASE tools, the database manager surrounds the repository. Analysts can create user views of the database, which may differ markedly from the physical reality of the database. In fact, different users may access different versions of the database, tailored to their own individual needs. The database manager provides for security, recovery in the case of a system failure, multi-user access, and integrity controls to ensure consistency.

Figure 6.5 The focal point of all CASE systems is the repository.

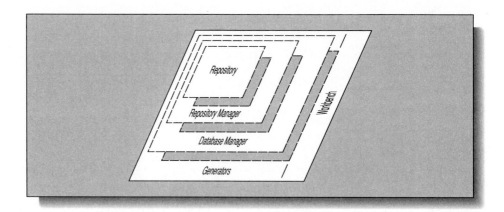

Handwritten margin notes:
Take the code & develop / To build new system.
(Can Recode designed system / and more to structure system)
Bad → Current system is / not a good way to / start.

The **workbench** provides necessary tools for manipulating the specifications dictionary. With the workbench, the analyst can modify, add, or delete applications in the repository.

The **generator** is used by the analyst to develop the system prototype. With the generator, analysts can design screens and reports, develop programs, and write documentation. In fact, the generator allows the analyst to set field and label attributes—such as reverse or blinking video—and define help messages.

The prototyping function allows the developer and user to build the system, review it when actually running, change it, review it again, and approve it when completed. Since the prototype simulates the running system, the analyst or user can preview the results of design changes at an early stage. By repeating this prototyping process as often as necessary, those involved can perfect the application even while it continues to evolve through the systems process.

Prototyping also permits such attractive features as function keys, color screens, windows, and mice so that the user can gain some familiarity with the system before it is finalized. Once again, to ensure consistency, the specifications dictionary captures the final version of the prototype, guaranteeing that the last version of the system will match the prototype.

The screen painter and report writer portion of the generator eases the construction of data entry and reporting screens, as well as any printed reports. Again, the user and analyst work hand-in-hand, building and refining the required screens. Once completed and approved, the screens and report definitions also go into the specifications dictionary.

After the analyst describes the processing rules, files, database manipulations, error messages, and so on, the program generator automatically creates the necessary programs. In certain situations, special routines written in a 3GL language can interact with the CASE software. If so, the processing logic of those routines will also go into the specifications dictionary.

The final part of the generator, the **documenter**, provides end-user, program, operations, test, and system documentation. Documentation typically includes narrative descriptions, data flows, screen formats, data entry editing rules, and lists of allowable values for inputs. Since each documentation item comes from a single source, the specifications dictionary, all items are by definition consistent in content.

CASE TRENDS

Even though CASE is a new technology, some aspects of the future of the systems industry in the CASE era are clear. Industry experts make three predictions:

1. CASE will impact software maintenance much more heavily than new system development
2. CASE tools are nothing more than methodology companions
3. CASE will fundamentally change the system life cycle and the job roles of the programmer/analyst

System maintenance is presently difficult and expensive because few old systems were well-documented, and no standard methodologies were used when these systems were built. As a result, it often took an analyst hours or even days to find the section of program code needing modification. Once located, the modification might only take a few minutes or it could take hours. CASE-developed systems feature full documentation and use standardized techniques, so modifications such as this are faster and easier. Some observers think that CASE will eventually drop the proportion of dollars spent for maintenance from its present rate of about 80% to approximately 20%.

Some experts predict that organizations trying to adopt CASE without first adopting a structured development methodology will fail with CASE. Most CASE software products conform very tightly to a particular methodology, and successful CASE implementation depends heavily on consistency between the CASE tools and the underlying methodology.

Finally, forecasters say that CASE will lead to a fundamental, breakthrough change in the way programmers and analysts work. The system life cycle, through prototyping, will change dramatically because users can see their systems evolve piece by piece. This means that users will have a much closer involvement with the design details than in the past. Analysts will take on a consultative role, moving them closer to the business aspects of their systems projects.

CASE METHODOLOGIES

The CASE approach parallels the system life cycle methodology. During CASE analysis, the repository collects rules and procedures, data elements, screen layouts, data flows, and report formats. At the end of analysis, the CASE software documenter combines the information collected in the repository and produces a functional specification. Similar to the feasibility study, the **functional specification** provides a clear and complete summary that those involved with the system can consider for approval. Often, the functional specification includes data flow diagrams (context and leveled), structure charts, entity relationships, state transition diagrams, presentation graphs, and data model diagrams.

During CASE design, the CASE software refines the repository built during analysis by building data and program structures and organizing data into logical records, files, and databases. In addition, the documenter creates complete program specifications, including processing instructions, data validation rules, updating requirements, and structure charts. Given ongoing user interaction with the developing system, the CASE design should deliver what users want, rather than what they thought was needed when their initial request was written.

During CASE development, the software builds the complete system from the repository. The documenter produces program, user, and operations documentation. Some CASE systems produce programs in 3GLs, while others produce programs with their own non-procedural language.

The greatest benefit of CASE software arises during system maintenance. During this important phase of a computer application's lifespan, CASE software can continually manipulate the repository with the workbench, thus making quick and timely modifications that are responsive to users' needs. After making changes, the CASE software recreates the entire system automatically, regenerating affected programs, screens, repoɪ bases, and documentation with computerized accuracy.

Most CASE software operates on personal computers, but can easily link to mainframes. When employing such CASE systems, analysts can conduct all of their analysis, design, and development on a PC and then upload their work to the host mainframe.

Contrasting 4GLs and CASE

Significant differences exist between 4GLs and CASE tools. A clear contrast between 4GLs and CASE is that 4GLs are relatively user-friendly and as a result, a training class of a few hours (or at most, a few days) can enable a user to work productively with most 4GLs.

While users can readily understand the modeling diagrams, screen prototypes, and other outputs of a CASE tool, few can actually use the tool themselves. Full life cycle CASE tools such as IEW from Knowledgeware and IEF from Texas Instruments are complex, and extensive training is required in their use. Since an understanding of the technical aspects of programming, database, human interface, security, control, and project management connected with large systems projects is necessary to use CASE tools effectively, usually just practicing systems analysts are CASE users.

Currently, larger systems development is a full-time occupation, not a part-time one. Even if users have the needed technical experience, their other duties would prevent them from developing large systems on their own.

Another difference is that 4GLs seldom are rooted in a structured system development methodology. 4GLs normally support only the system implementation part of the life cycle, while some CASE products can support all phases.

Analysts refer to both CASE and 4GLs as productivity tools because they speed up the process of building systems. Both are needed and both are useful to have in an analyst's toolbox.

OBJECT-ORIENTED ANALYSIS

Like ice cream, CASE software systems come in many flavors, but some consistencies exist among all of them. Every CASE tool allows you to draw diagrams that match those we saw for the pictorial view of the View Video Cassette Tape Rental System, Fleet Feet's accounts payable system, and the Valley Blood Bank donor system.

A new methodology called **object-oriented analysis (OOA)** and design is in its infancy. This methodology asks the analyst to determine what the objects of the system are, how they behave over time or in response to events, and what responsibilities and relationships an object has to other objects.

Before progressing any further, let's look at two simple example objects: a box and a cylinder. We are all familiar with boxes and see them every day as packing cartons, holding ice cream, or containing presents. What is a box and what are its elements or attributes?

USING 4GLS TO CUT SOFTWARE MAINTENANCE COSTS

In many cases, corporate users of computer systems spend too much time on software maintenance and not enough time on devising new ways to decrease the need for maintenance. Some estimates suggest that half of the total cost of computing, including hardware, software, and personnel, goes directly or indirectly into maintenance. On the average, maintenance consumes up to three-quarters of the total cost of software over its lifetime, and thus claims considerable resources that organizations could better invest in application software development.

Maintenance, which modifies existing operational software while leaving its primary functions intact, can take many forms. Corrective maintenance remedies failures in the implementation of the system design; preventative maintenance enhances the software by making it more efficient and maintainable; and adaptive maintenance enables the software to accommodate environmental changes, such as new hardware or new data types. Finally, functional maintenance alters the function of the software by adding or deleting features.

All types of software maintenance involve three stages. First, the programmer assigned maintenance responsibility must understand the software itself and the nature and scope of the requested change. Second, the programmer must make the necessary changes. Finally, the programmer should revalidate the software to ascertain that it works correctly.

That may sound simple enough, but so much older software, written in unstructured code and lacking consistent standards, resists easy modification. In such cases, maintenance becomes bothersome and time-consuming.

That's where fourth-generation languages and application generators may come to the rescue. Their built-in consistency can make all three stages of maintenance much swifter, smoother, and more accurate. Most importantly, 4GLs may eliminate corrective maintenance if their program generators do, in fact, produce the error-free code that they promise. Enhancement modifications will also become easier because enhancement only involves changing the system specifications, rather than the source code. Adaptive maintenance, too, will benefit because a 4GL can contain so many environmentally dependent routines and permit migration of system modifications to other environments using an unmodified 4GL.

4GLs pose a number of other interesting maintenance ramifications. By using 4GLs, designers and programmers must conform to a standard methodology, thus ensuring easy verification of specifications. Automatically produced documentation provides further assistance because it helps those involved in maintenance to more easily grasp and master the system.

On the human relations side, 4GLs can help developers improve rapport with users because they afford the developer the opportunity to provide a quick prototype, which can elicit more exact specifications from users. Early and constant user involvement in the process helps overcome much of the fear and anxiety that often accompany computerization of a function. Of course, management must exert close control over 4GL application development or run the risk of too much unsupervised "seat of the pants" application development.

If companies are to stay profitable and competitive, they must manage software in a cost-effective manner. 4GLs introduce structure and reliability into the design process and elevate maintenance to its role of software asset protection.

Most boxes have sizes that we measure with length, width, and height. A box is colored and is made from some material, usually cardboard. Boxes usually have the same shape, rectangular, but can also have a different orientation. A box may sit on its bottom, side, or end. Further, we can view the box from an angle: above, from the left, from the right, or from the bottom (when placed on a glass table, for example). We must know all these factors about the object called *box* in order to properly describe it.

Now consider a cylinder. Cylinders have diameters, lengths, and colors, plus material, orientation, and a view. How different are the two objects? Not very much at all! Colors, material, orientation and view are common; the physical descriptors (cubical versus cylindrical) differ.

Object-oriented analysis has the analyst look at all the objects in a system, their commonalities, differences, and how the system needs to manipulate the objects. Determining a system's objects seems difficult at first, but CASE software simplifies it with different types of drawings: traditional data flow, entity relationship or object life cycle diagrams. During object-oriented analysis, the analyst determines the objects. For each object, the diagram provides you with a tool to analyze how an object is created, how it changes states in response to external factors, and how it can cease to exist.

An object becomes the building block on which systems are designed. Analysts think of the systems they want to design as a collection of objects that have data and functions or actions that need to take place on the data. From a mathematical perspective, we could state:

$$\text{Object} = \text{Data} + \text{Functions}$$

The combining of data and functions into an object is called **encapsulation**.

As an example of an object, let's go back to our accounts receivable system. Our data flow diagram revealed a data flow object of the system: the payment. Some of the data for a payment include the customer's account number, the amount of money the customer sent, the date the payment was received, and the date the payment was due. Some functions that we need to perform on a payment are

- Determining if the payment is late
- Assessing any finance charges
- Associating this payment with the correct customer
- Posting the payment to the customer's account
- Reducing the customer's account balance

When we combine all these functions together with their associated data, we have a payment object. We really don't care what the specific data values are; we just need to have our object manipulated as the functions dictate. The concept of divorcing the data and functions from the user of the object is part of encapsulation and is often called **information hiding**.

The second object-oriented viewpoint is the concept of **inheritance**. Inheritance allows software developers to order objects so that an object at a higher level can use some or all of the objects at a lower level. An analogy is a family tree in which people inherit the characteristics of their prior family members. Inheritance allows software developers to add new objects or modify existing ones to suit their needs.

Polymorphism is the third form of abstraction in this philosophy of software development. Polymorphism is the ability of two or more classes of objects to respond to the same stimulus, each with their own set of operations. Polymorphism comes from a Greek word meaning "many shapes."

As an example of encapsulation, inheritance, and polymorphism, consider a collection of graphic objects, lines, circles, squares, arcs, or points. What is different among the objects is their physical shape—the way each one looks. A point is a circle with a very small radius and an X and Y coordinate location. A circle is a form of an arc that comes back on itself with an X and Y coordinate. A line is an arc that does not bend. Here, we see the inheritance of the objects. What we also need is the ability to "show" each graphic object on the screen. Polymorphism in this example is the "show" concept. We want one word, "Show," but when it is invoked along with an object, it will display many shapes.

Many programmers are now favoring object-oriented programming (OOP) languages such as C++, Object-C, and Smalltalk because they allow programmers to write software as definitions of and operations on objects. All languages, even COBOL, are jumping into the OOP world by adding features to allow encapsulation, inheritance, and polymorphism. Further, OOP tends to make maintenance easier, systems that more easily mirror actual components, and software that is more portable across a wide range of applications, as well as hardware platforms.

Once an object is defined, it becomes easy to prototype. When in prototype mode, the designer can study it along with the user to make sure the object works as it is supposed too.

Will the object-oriented philosophy dominate software development in the remainder of this decade, or is this merely a passing fancy? Time will tell, but early results are very promising as object-oriented software helps projects get delivered on time and within budget—meeting user and customer needs.

SYSTEMS DEVELOPMENT USING CASE AND 4GLS: AN EXAMPLE

To illustrate the power of these new software tools, consider how the Valley Blood Bank system might use them. We must assume, of course, that the CASE and 4GL approach would incur equal or lower costs and match or exceed the capability of a system developed with traditional life cycle methodology and the current manual system.

Analyst Neal Hundt and primary user Paul Willey meet in Neal's office with Neal's personal computer, which ties into the Blood Bank's mainframe. While Paul describes his desired application, Neal enters the description into the data specification dictionary, and after a few minutes, Neal calls on the CASE software to produce a picture of the system. The data flow diagram for the system shows data flowing into it in the form of medical history forms, donation receipts, product requests and control (Figure 6.6). Outputs take the form of donor identification cards, the American Blood Commission's annual report, and periodic management reports.

Paul and Neal continue to refine the picture until they think that it's right (Figure 6.7). They add two new data flows: the receipt the hospital sends the Blood Bank when a blood shipment arrives, and the notice the Blood Bank sends to the hospital with every unit of blood it ships.

Neal knows that the Blood Bank's CASE software will automatically:

- Construct files
- Produce printed reports
- Collect and verify data
- Sort data into order
- Process data appropriately
- Print documentation

Figure 6.6 Data flow diagram of the Blood Bank donation tracking system.

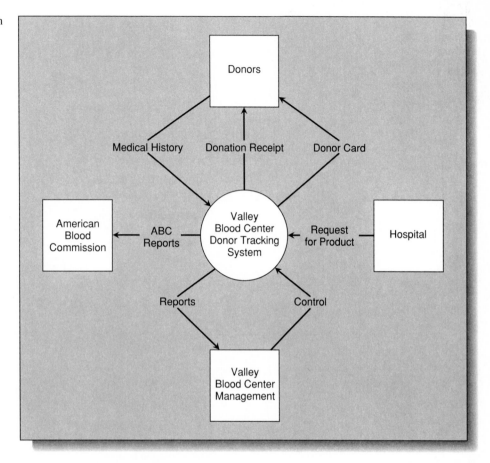

Despite the seemingly "automatic" nature of such a system, however, Neal must still provide it with a certain amount of direction. Specifically, Neal must describe each

- Data element
- File (data store)
- Report (printed and screen)
- Transaction processing activity
- Data collection screen

However, Neal does not have to rely on the complex procedures of traditional 3GLs. He can write quite simple and straightforward descriptions that the software will convert into the desired results.

The computerized repository lies at the heart of Neal's new system (Figure 6.8). The report generator, the data collection screen generator, the file builder, and the transaction processor all draw from the data dictionary. From initial development through actual operation and maintenance of the final system, the data dictionary provides the links among these other components.

Figure 6.7 Modified data flow diagram of the Blood Bank donation tracking system, showing two additional data flows.

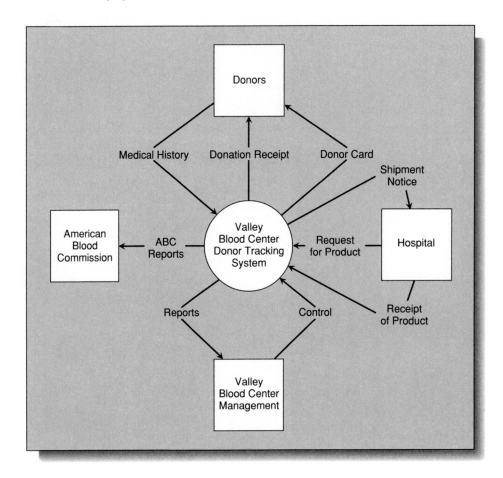

Entering Data Elements

Neal starts by entering the data elements and all relevant facts about them into the system. The software includes default values; on-line help if the user gets stuck; the ability to add, change, or delete definitions; error messages when users make mistakes during data entry; and the ability to print lists of the data dictionary. Of course, every commercial CASE and 4GL system, with its own unique language structure, syntax, and other idiosyncrasies, works differently, so we'll follow a *hypothetical* one for the Blood Bank.

Because "DONOR-ID" is the most important data element in the blood donor tracking system, Neal Hundt defines it first (Figure 6.9). The data dictionary will consistently employ this definition during data entry (validating maximum and minimum values), report generation (edit requirements and headings), and data storage (length and type). To complete the definitions, Neal must enter all the descriptions of all the various data elements. Fortunately, if he overlooks one, the CASE/4GL software will let him add it easily at a later date.

Figure 6.8 The specifications data dictionary forms the foundation of the prototype-4GL software.

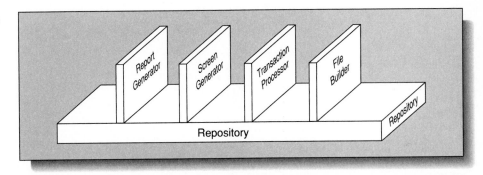

Figure 6.9 The screen for entering a data element allows places for a complete description of the element.

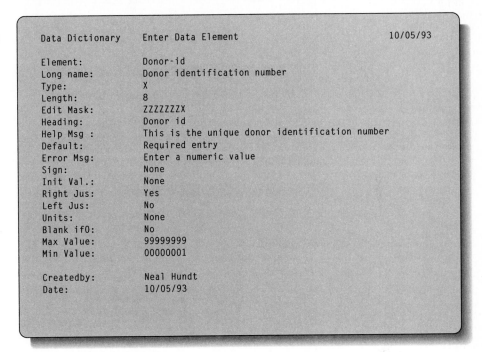

```
Data Dictionary      Enter Data Element                        10/05/93

   Element:           Donor-id
   Long name:         Donor identification number
   Type:              X
   Length:            8
   Edit Mask:         ZZZZZZZX
   Heading:           Donor id
   Help Msg :         This is the unique donor identification number
   Default:           Required entry
   Error Msg:         Enter a numeric value
   Sign:              None
   Init Val.:         None
   Right Jus:         Yes
   Left Jus:          No
   Units:             None
   Blank if0:         No
   Max Value:         99999999
   Min Value:         00000001

   Createdby:         Neal Hundt
   Date:              10/05/93
```

Creating Files

After defining every data element with the data dictionary software, Neal organizes the data elements into records and files with an interactive, CRT-based file builder. The donor tracking system will need two files: one for each donor and one for each donation made by a donor.

Since each file needs a unique name, Neal might select DONORS and DONATIONS (Figure 6.10). The interactive screen builder requires Neal to enter his name as the file creator, the file creation date, file name (found in the OUTPUT statement), and, in some cases, the key (located in the KEY statement) for the file.

Figure 6.10 The screen for defining files requires the user to enter the data element names.

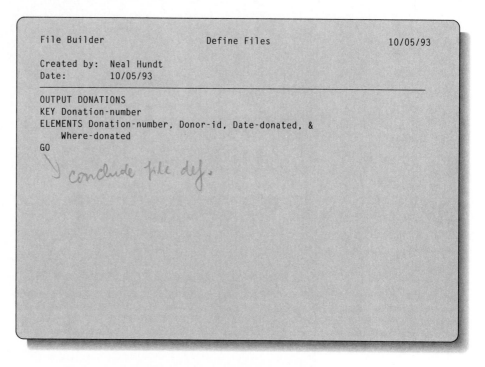

```
File Builder                    Define Files                    10/05/93

Created by:  Neal Hundt
Date:        10/05/93
_____
OUTPUT DONATIONS
KEY Donation-number
ELEMENTS Donation-number, Donor-id, Date-donated, &
    Where-donated
GO
```

conclude file def.

Having made these entries, Neal uses ITEM statements to identify the data elements that make up a record in the file. An ampersand (&) indicates that this line will continue on the next line. To create the actual record structure, the file builder stacks the element descriptions from the data dictionary like building blocks.

Some unique elements, such as Donation-number, disallow duplicates. Two or more records may reside in DONATIONS with the same Donor-id, but no two records could possess the same Donation-number. To distinguish unique elements from those that can appear more than once, the system relies on the unique key field's element name.

The GO statement concludes file definition, telling the software that an operator has finished the task. EXIT gets the user out of the software and returns control to the operating system. After entering the EXIT command, the user should see the operating system prompt.

Figure 6.11 illustrates a file with four data elements. To create the DONORS file, Neal would repeat the procedure. With the data dictionary files in place, he can then begin using the other CASE components to enter, manipulate, and retrieve information.

Figure 6.11 Record structure for DONATIONS.

Constructing Data-Entry Screens

All applications begin with collecting, verifying, and storing data. Any subsequent changes in data, processing of data, and reporting of information derived from data all depend on proper data collection.

The Blood Bank needs two primary data collection screens. One screen will collect data about each donor, while the other will collect data about each donation.

The interactive, CRT-based **screen generator** component of the software also is tied to the data dictionary (Figure 6.12). It constructs data collection screens complete with data error checking and error messages. Anyone wishing to design a screen enters his or her name, the date, the name of the new screen, and what file the screen should reference with an ACCESS statement. AT 3 CENTERED specifies that the title should appear on the third line and centered on the screen.

On the bottom portion of the screen, the designer lists each data element. SKIP 1 specifies a blank line. To obtain data elements, the software consults the data dictionary for length, type, format, and validation criteria. For example, Donation-date must always appear in the form MM/DD/YYYY with MM between 01 and 12, DD between 01 and 31, and YYYY this year or prior ones, and with slashes between the month number, day number, and the four-digit year number.

When entering actual data into the system, data entry operators can never skip REQUIRED data elements. The LOOKUP phrase tells the system to refer to the DONORS file to make sure this donor actually exists, and GO tells the screen generation software that the task has concluded. GO gets the user out of the software and back to the operating system.

Figure 6.12 Building a data collection screen follows the pattern of constructing the data dictionary and the files.

```
Screen Builder                                          10/05/93
Created by   : Neal Hundt
Date         : 10/05/93
Screen Title : Collect and Post Donations AT 3 CENTERED

OUTPUT DONATIONS
SKIP 1
ITEM Donation-number REQUIRED
ITEM Donor-id          REQUIRED LOOKUP DONORS
ITEM Date-donated      REQUIRED
ITEM Where-donated
GO
```

To identify the data elements that make-up a record in the file.

With screen descriptions completed, the software fabricates a real screen (Figure 6.13). The Xs show the size of each field and the bottom line of the screen remains reserved for data entry error messages.

The MODE command, which tells how the screen will function, appears on the top line and reveals the great power of this type of software. E = Enter a new record, F = Find a specific record, U = Update a record after altering it, and D = Delete a record. With these rapid development languages, one screen permits all these functions, while with a 3GL, a programmer would be forced to write different screens (and complex programs in 3GL) for each function.

Screen generators are extremely flexible and allow users to do the following:

- Perform and display calculations
- Execute summing and balancing
- Specify security for screens and fields
- Highlight important fields
- Skip data entry fields if a predetermined condition exists
- Build screens for entry of multiple file transactions
- Create screens that are automatically called up when the entry sequence reaches a certain field on the active screen
- Design menu screens to connect other screens together
- Draw lines or boxes to separate data
- Set certain fields to automatic default values
- Prevent users from changing data in important fields
- Prohibit users from deleting records

Available on most screen generators, these features speed screen construction, thus increasing productivity, decreasing costs, and reducing errors. But their real advantage is that they are interactive. This allows the analyst and user to work together, perfecting the application, while seeing the results of their work immediately.

Producing Printed Reports

To management and users, nothing about a system matters more than the reports it generates because the information contained in those reports supports crucial decision-making. Just as a system relying on 3GLs requires different programs for every report it generates, so do CASE and 4GL systems—but with nowhere near the number of instructions.

Before we watch Neal use the report generator component of his CASE/4GL software, review the "12-Month Donor List" report in Figure 6.14. The **report generator's** interactive CRT-based screen generator permits Neal to write a description list, Figure 6.15, beginning with his name, the date, the title of the report, and the file from which the report should extract data (with an ACCESS statement). Unless instructed otherwise, the software assumes that the report will end up on a user's terminal or personal computer screen.

The SORT DONOR-ID statement specifies the order in which the data should appear; in this case, in ascending order by DONOR-ID. REPORT identifies what each line in the

Figure 6.13 The actual screen fabricated by the 4GL software looks like a normal data collection screen.

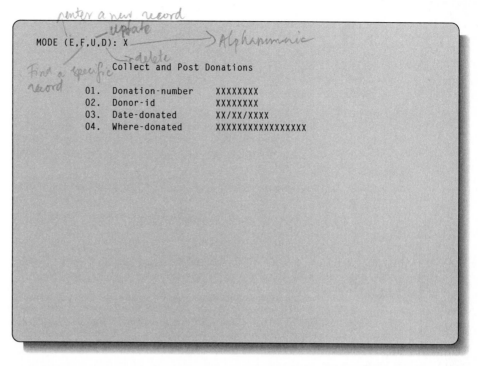

```
                              enter a new record
                                  update            → Alphanumeric
       MODE (E,F,U,D): X
                                  delete
       Find a specific Collect and Post Donations
       record
              01.  Donation-number    XXXXXXXX
              02.  Donor-id            XXXXXXXX
              03.  Date-donated        XX/XX/XXXX
              04.  Where-donated       XXXXXXXXXXXXXXXX
```

Date: 10/05/93		12 Month-Donor List						Page 1
Donor Id	Donor Name	Telephone Number	Date Last Gave1	Date Last Gave2	Date Last Gave3	Date Last Gave4	Date Last Gave5	Total Given
00001234	Joye, Sol	(408)823-9144	5/93	1/93	6/92	12/91		4
00002354	Leeper, Matt	(415)823-8124	2/93	7/92	6/91	1/90	3/89	8
00007823	Lee, Sarah	(408)782-6789	8/92	5/91	1/91	9/90	2/90	52
00009812	Lundberg, Andrew	(408)823-3210	8/93	2/93	4/91	2/90	4/89	18
00009912	Haverberg, Mary	(408)823-0098	5/93					1
00009923	MacFarlane, Josh	(916)823-1819	4/93	1/93	10/92	7/91	1/91	16
00010875	Belajic, Daniel	(408)823-7473	12/92	6/92	1/92			3
*	*	*	*	*	*	*	*	*
*	*	*	*	*	*	*	*	*
*	*	*	*	*	*	*	*	*
Number of Donors on File								5,234
Total Units Given by All Donors								45,342

Figure 6.14 The list of donors shows all those people who donated blood during the prior 12 months.

body of the report should contain. Data elements appear in left to right order. Since column headings are a part of the data dictionary definition of each data element, the software will "know" to place them over the top of each column. The software also "knows" how much space each data element will consume, and how to format each element. TAB 65 overrides normal horizontal spacing and begins TOTAL-GIVEN in position 65 on that line.

Figure 6.15 This description list produces the "12-Month Donor List."

```
Report Generator                                        10/05/93

Created by:   Neal Hundt
Date:         10/05/93
Report Title: 12-Month Donor List
_____
ACCESS DONATIONS
SORT DONOR-ID
REPORT DONOR-ID, DONOR-NAME, TELEPHONE-NUMBER, &
     DATE-LAST-GAVE1, DATE-LAST-GAVE2, DATE-LAST-GAVE3, &
     DATE-LAST-GAVE4, DATE-LAST-GAVE5, TAB 65, &
     TOTAL-GIVEN
FINAL FOOTING SKIP 2, COUNT, "Number of Donors on File",
     SKIP 1, & "Total Units Given by All Donors", TAB 65,
     TOTAL TOTAL-GIVEN
GO
```

FINAL FOOTING identifies what should appear at the end of the report, SKIP 2 advances the paper two lines, and SKIP 1 prints one blank line before the footing line. COUNT tallies records automatically; then, after printing the record count, it displays the caption "Donors on file." The word "Totals" starts this line and the word TOTAL causes the software to sum the data element "TOTAL-GIVEN" and begin printing it in position 65.

Built-in page headings do not require any programming. A well-designed page heading includes a date, title, and running page number. The software centers the title of the report across the top of each page.

While some reports need only appear on a CRT screen, others should go to the printer. Most report writers include statements such as:

```
SET REPORT DEVICE PRINTER
SET REPORT DEVICE TERMINAL
```

These commands allow the user to change the destination of the report without any fancy operating system commands.

The Blood Bank's simple report takes only six commands. A comparable COBOL program would take at least 200. Report writers can produce extremely complex reports: those with control breaks; those that must access data from multiple files; and those that require subtotals, averages, maximums, minimums, and so on. Most report writers allow the designer to override the default values, reset page numbers, change the column headings, alter the format of the data from those specified in the data dictionary, avoid orphaned and widowed headings and footings (those at the top or bottom of a page that contains no data), and print an initial heading different from other headings.

Computers store data. Report writers turn that data into meaningful information and save months of programming time in the process.

Transaction Processing

The **transaction processor** component of the system allows the designer to change the data in individual records, in selected groups of records, or in the entire file in one sweep. Like all the other features, the transaction processor works closely with the data dictionary. Both powerful and easy to use, the transaction processor deserves respectful treatment because a user can wipe out entire files full of data in a few seconds.

As we saw, the Blood Bank system employs two files, DONORS and DONATIONS. DONORS keeps track of personal data on individual donors, such as name, address, age, weight, and the dates that the donors made their last donations (Figure 6.16). DONATIONS tracks specific blood donations contributed by a single donor. In the DONATIONS file, we would find the donation number, donor's identification number, and the date and location of the donation. When a donor gives blood, the system must update his or her DONOR record, adding 1 to the total number of donations. Similarly, the computer must "age" the donations:

Figure 6.16 The data dictionary definition of DONORS. This file chronicles all the data about a donor.

```
Name of file: DONORS                                      Date:10/05/93
Analyst: Neal Hundt

        DONORS = Donor-id +              {    8    Alphanumeric   }
                 Donor-name +            {   32    Alphanumeric   }
                 Donor-address +         {   60    Alphanumeric   }
                 Donor-phone +           {   10    Numeric        }
                 Blood-type +            {    4    Alphanumeric   }
                 Height +                {    4    Numeric        }
                 Weight +                {    4    Numeric        }
                 Sex +                   {    2    Alphanumeric   }
                 Race +                  {    2    Alphanumeric   }
                 Date-of-birth +         {    8    Numeric        }
                 Date-of-application +   {    8    Numeric        }
                 Date-last-gave-1 +      {    8    Numeric        }
                 Date-last-gave-2 +      {    8    Numeric        }
                 Date-last-gave-3 +      {    8    Numeric        }
                 Date-last-gave-4 +      {    8    Numeric        }
                 Date-last-gave-5 +      {    8    Numeric        }
                 Total-given    +        {    5    Numeric        }
                 Medical-history-code    {    4    Numeric        }

    Key field:              Donor-id.

    Order of file:          Indexed by Donor-id.

    Length:                 Approximately 40,000 records.

    Media:                  Disk.

    Security:               Internal use only. Telephone numbers,
                            race, date of birth present in file.
```

Date-last-gave5 becomes date-last-gave4.
Date-last-gave4 becomes date-last-gave3.
Date-last-gave3 becomes date-last-gave2.
Date-last-gave2 becomes date-last-gave1.
Date-last-gave1 becomes today's date.
Add 1 to Total-donated.

With the transaction processor, the system can create, delete, and change records quite easily. These types of definitions form the rules that the software records in the repository.

As with the other components of the software, Neal need only describe events that should occur (Figure 6.17). As usual, the software prompts for the name of the designer and the date. Once the "File Name" line identifies the Blood Bank's donation record program as BBUPDONAT, any user can execute it by running BBUPDONAT.

An ACCESS statement names the input file that holds the transactions, and an OUTPUT statement identifies a file that the user wishes to update. With the KEY statement, the software cross-links records in each file that contain the field "Donor-id."

The UPDATE command, followed by a description of desired changes, prompts modification of an existing record in the output file. For example, the statement:

Date-last-gave4 = Date-last-gave3, &

Figure 6.17 The transaction processor adds new data to files, updates existing records, and deletes records.

```
Transaction Processor                                    10/05/93

Created by:   Neal Hundt
Date:         10/05/93
File Name:    BBUPDONAT
_____

ACCESS DONATIONS
OUTPUT DONORS
KEY Donor-id
UPDATE Date-last-gave5 = Date-last-gave4, &
       Date-last-gave4 = Date-last-gave3, &
       Date-last-gave3 = Date-last-gave2, &
       Date-last-gave2 = Date-last-gave1, &
       Date-last-gave1 = Todays-date, &
       Total-given = Total-given + 1
GO
```

indicates that the value of Date-last-gave3 should replace the current value of Date-last-gave4. Similarly, the statement:

Total-given = Total-given + 1

adds 1 to the current value of Total-given. Finally, GO concludes the transaction processor description of tasks and returns control to the operating system.

Transaction processors can perform many additional operations. Tied to data collection screens or printed reports, they can process data immediately or in a batch mode overnight. In either case, the transaction processor keeps files up-to-date.

In complex applications, analysts and users can easily and quickly develop a working prototype or even a complete system. CASE/4GLs especially benefit users who do not know exactly what they want from the system. They also work nicely for one-time applications in which users don't have to worry about specific report or screen formatting.

CONTRASTING 4GLS AND CASE

The primary focus of 4GLs and CASE software is toward development of new applications, starting from analysis. What about existing systems that came from other—perhaps older—traditional software development methodologies?

Reverse Engineering

Attacking the problem from the other end is the goal of reverse engineering. Reverse engineering software takes an operational system and deduces the system's database design and data elements. It also examines which programs use specific data elements. Once the system is analyzed, the software builds a new description as if it had come originally from a traditional CASE software tool.

The advantages of reverse engineering on existing software are found in maintenance. With the system now under the control of a CASE tool, all changes to the system are recorded automatically, documentation materials are a by-product, system data flow diagrams are easily drawn, and the CASE tool can check the system for flaws.

4GLS AND CASE: ADVANTAGES AND DISADVANTAGES

Productivity tools such as 4GLs and CASE offer analyst, user, programmer, and management an attractive alternative to the traditional life cycle approach to systems analysis, design, and development. Among the significant advantages brought about by CASE and 4GLs are the following:

WHEN IT ABSOLUTELY, POSITIVELY, HAS TO BE DONE

Eldon Rowe, sales manager for Tekchem Pharmaceuticals, couldn't believe his ears. The president of Tekchem, Jack Yokote, had just ordered Eldon to compile a new report breaking down sales of new versus old products geographically, and he wanted it first thing tomorrow morning. "Can't be done," Eldon protested. "We've never had that report before. It'll take a week to get the computer people to write the program and test it."

Jack just frowned. "I don't care how you do it. I've got to have that report in my hands when I meet with the Syntonics people at 9 tomorrow morning. This whole contract depends on it."

Since it was already 2:00 in the afternoon, Eldon considered his options: assemble the data manually (he'd have to find 20 helpers), forget the assignment (and possibly lose his job), or beg the Management Information Systems people to perform a miracle (they'd bailed Eldon out of a few tight spots before). Seeing no other viable choice, Eldon trudged to the office of Beverly Berg, the analyst who had designed the new sales forecasting system. As Beverly listened to Eldon's problem, a mischievous grin crept across her face.

"What's so funny?" Eldon asked. "If you laugh me out of your office, I'm a dead duck."

Beverly's eyes twinkled. "I was thinking more about rabbits. Like the kind you pull out of hats. I've been tinkering with some new software on our VAX. I won't bore you with the details, but it's almost like magic. It uses computer-aided software engineering and a fourth-generation language to speed up application development. The R & D Department has used it for a few months and now they can do in one day what normally would take 10."

Eldon's hopes rose. "But can you really design and program this new report overnight?"

"Let's look it over." Beverly scanned Jack Yokote's memo. "Not too bad. Yeah, we can do it. We have all the necessary data; it's just spread out over three files. All we have to do is design a new database, copy the data into it, and write a program to print the report."

Eldon's heart sank again. "That's what I figured. I remember when we designed the monthly field sales report. It took about five weeks."

"That was with COBOL programming," replied Beverly. "This new software skips all that. It's all built in. Come on, let me show you."

At 7:00 that evening, Eldon called out for pizza to be delivered to the MIS department. At 9:00, Beverly announced that the system was ready to run, and at 10:00 the report began streaming off the line printer in exactly the format Jack had requested. By midnight, Eldon was home in bed, with the new geographic sales report on Jack's desk.

The next morning, Eldon dropped by Beverly's office to thank her again for her help. "You were right about those rabbits," he said. "That new CASE and 4GL stuff you used last night *is* magic. Why haven't we used it before now?"

"It's funny," she replied. "But if you hadn't had an emergency problem come up, you probably would have resisted a new approach. I've already seen it with R & D. They know about our plans to go to CASE and 4GL for systems development; with their technical training, they should appreciate this advancement. But it took a crisis with that new cat-worming product to get them to put away their C and COBOL. I wish I could convince people that these new tools do not take the place of the analyst; they merely supplement the person and help automate the task."

1. Increased productivity of analysts and programmers.

2. Improved communication between user and analyst.

3. Higher quality software with fewer errors.

4. More satisfied users.

5. User involvement throughout the entire process.

6. Reduced maintenance. (One financial institution reported a maintenance staff reduction from 12 to 2.)

7. Reduced time spans for development and maintenance.

8. Complete and consistent user and technical documentation.

9. Enforced standards for design, program structures, and documentation for all applications.

10. Ability to evaluate application changes before implemention.

11. Decreased testing time.

12. Increased flexibility and quicker response to rapidly changing business needs.

13. Improved organizational effectiveness.

14. Common interface and documentation for all systems.

One large CASE user actually measured the many differences between the CASE software and 3GL approaches and found that for a personnel system, CASE took 4 months compared to the 3GLs' 25 months; an inventory system took 5 months rather than 28; a quality control system 16 months rather than 56; and a work order management system 4 months instead of 18. In another example, CMS (Club Management System, the world's largest telemarketers) reports that maintenance times have shrunk from 12 days for COBOL programs to 4 days for CASE/4GL systems.

Nevertheless, CASE/4GL software does have disadvantages:

1. Additional expense of CASE/4GL software.

2. No standardization among different CASE/4GL software as products (3GL languages do enjoy a higher degree of standardization).

3. Retraining of staff and associated expense, plus lost time.

4. Difficulty of matching new CASE/4GL software development to existing software.

5. Less efficient use of hardware with 4GLs versus 3GLs.

With so many advantages and so few disadvantages, why have organizations resisted them? In two words, *change* and *trust*. Even MIS managers and their staffs resist change, especially one as bold as this. And the CASE and 4GL products are so new. MIS managers remain unconvinced that CASE and 4GL software will really deliver on its bold promise.

As with most aspects of computing, advances in CASE or 4GL technology should increase its benefits and decrease its liabilities over time. Still, even today, CASE/4GLs provide tangible benefits for many organizations that are developing new computerized information systems. They work especially well in organizations:

- With large backlogs of new application or system enhancement needs

- Whose current set of tools do not add to programmer productivity

- That need stronger control over their data

- Where reduced delivery time is more important than hardware utilization efficiency

SUMMARY

Most of the applications developed since the beginning of the computer age involved third-generation languages (3GLs), such as COBOL, FORTRAN, and Pascal. However, productivity tools such as fourth-generation languages (4GLs) and computer-aided software engineering (CASE) reduce programming efforts, increase communication between users and analysts, decrease systems maintenance, and allow working systems to become operational quickly. Additionally, these tools make rapid system development practical, changing the shape of the system life cycle from a long straight line to a series of tight, iterative circles. Prototyping, the ability to see parts of a system in operation very early in the life cycle, is one of the key features of rapid system development.

Fourth-generation languages and CASE contain certain similar components: data dictionaries and repositories, file builders, screen generators, report writers, and transaction processors. At the core of the 4GL/CASE tool lies the repository, or specification dictionary, which defines the system. These definitions specify the name of the data items in files, their lengths, data types, formats, reports, and data collection screens—plus all the rules for processing.

Drawing from the data dictionary, the file builder constructs physical disk and tape files. The user or analyst merely needs to specify which data elements belong to which files, name the key fields, and determine the order of the records (ascending, descending, or random).

Screen generators ease the creation of the data collection process. This component consults the data dictionary for field definitions, validation requirements, and error messages. It also stores collected data in the files.

Report writers organize the data to place information at the users' fingertips. The report writer accesses the data collected by the screen generator, then formats either printed or CRT reports. Reports allow users to analyze, organize, and synthesize data.

The transaction processor permits changes in the system's files. Converting data into information implies some logical manipulation of the data, and the transaction processor does this with impressive ease.

In addition to the similarities, significant differences exist between 4GLs and CASE. 4GLs are user-friendly and noncomputer professionals can quickly learn to use them for developing systems. CASE is not user-friendly and is intended for use by professionals. 4GLs are useful for small- to mid-sized systems that will not have heavy processing volume. CASE is appropriate for all systems. There are other differences in terms of life cycle stages, methodology support, and human interfaces. Both CASE and 4GLs are highly useful to analysts when applied to the proper type of project.

CASE and 4GLs allow organizations to develop and test prototypes and to get workable systems up and running quickly. They especially benefit users who do not know precisely what they want in the early stages of the systems process.

KEY TERMS

Third-generation language (3GL)
Rapid system development
Fourth-generation language (4GL)
Reverse engineering
Repository
Workbench
Generator
Documenter
Functional specification

Object-oriented analysis (OOA)
Encapsulation
Information hiding
Inheritance
Polymorphism
Screen generator
Report generator
Transaction processor

QUESTIONS FOR REVIEW AND DISCUSSION

Review Questions

1. What is a productivity tool?
2. Why is prototyping becoming the preferred way to build systems?
3. What are two uses of prototyping tools?
4. What are the four parts of most 4GLs?
5. What features are found in most CASE products?
6. What are two primary differences between 4GLs and CASE?
7. What are three advantages of 4GLs?
8. What are three disadvantages of CASE?
9. What are three disadvantages of 4GLs?
10. What is the difference between a third-generation and a fourth-generation language?

Discussion Questions

1. Name four commercial CASE software packages.
2. Make a list of five essential words that the 4GL uses.
3. What are the major functions of any application?
4. List four 3GLs.
5. How many generations of computer languages have we experienced?
6. Why would an organization buy a 4GL?
7. What is the central component of a 4GL?

Application Questions

1. Who writes the 4GL programs?
2. Name three methodologies CASE software follows.
3. What types of applications are best suited for 4GLs?
4. List three disadvantages of 4GLs.

Research Questions

1. Make a list of two commercial 4GL software systems.

   a. How much does each cost?

   b. What types of computers are required?

   c. Does each use the data dictionary as the central component?

2. Make a list of two commercial CASE software systems.

   a. How much does each cost?

   b. What types of computers are required?

3. Write a 4GL report program to print a list of all donations in ascending order by "Donor-id."

Prototyping View Video's System with CASE and a 4GL

In later chapters, we will watch analyst Frank Pisciotta follow traditional system design and implementation procedures, but let's suppose here that he can design the videotape check-out/check-in system with a prototyping-4GL software system called Powerhouse. Installed in over 15,000 locations around the world and operating on IBM, Digital Equipment Corporation, Data General, and Hewlett-Packard platforms, Powerhouse employs the repository approach.

Frank Pisciotta starts from the familiar context data flow diagram (Figure 6.18). He reviews the system inputs (new movies, returned movies, application forms, adjustments) as well as outputs (movie orders, reports, AR payment details, overdue notices, newsletter, and movie rental). The highest volume events are movie returns and rentals, while the lowest is probably the monthly newsletter.

Figure 6.18 The context data flow diagram of View Video's tape check-out/check-in system.

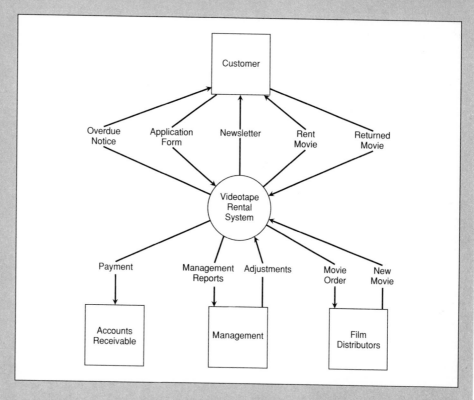

Before he starts Powerhouse, Frank decides to use the computer's word processor to list entities (Figure 6.19). Frank's earlier analysis revealed the need for six entities: CUSTOMER, RENTAL, TAPE, PAYMENT, STORE, and CLERK. For each entity, Frank lists the attributes (fields) in the logical order he wants them. Frank's STORE entity will hold

Figure 6.19 Frank Pisciotta's data dictionaries list the entities and attributes.

Entity Name:	CUSTOMER
Attributes:	Telephone number
	Last name
	First name
	Middle initial
	Street address
	City
	State
	ZIP
	Income generated
	Date joined
	Date last rented tape
	Number of tapes rented
	Credit card number
	Credit card type
	Expiration date

a. The CUSTOMER entity.

Entity Name:	RENTAL
Attributes:	Telephone number
	Tape identification number
	Date rented
	Date returned
	Rental-fee

b. The RENTAL entity.

Entity Name:	TAPE
Attributes:	Tape identification number
	Title
	Year made
	Rating
	Category
	Date purchased
	Date last rented
	Times rented
	Cost
	Income generated

c. The TAPE entity.

Entity Name:	PAYMENT
Attributes:	Telephone number
	Date returned
	Rental-fee

d. The PAYMENT entity.

(continued)

Figure 6.19 *(continued)*

Entity Name: STORE
Attributes: Store name
                   Street address
                   City
                   State
                   ZIP
                   Rental-fee

e. The STORE entity.

Entity Name: CLERK
Attributes: Clerk initials
                   Clerk last name
                   Clerk first name
                   Clerk middle initial

f. The CLERK entity.

global data that all screens and reports may want, such as the store's name, address, and telephone number. The CLERK entity keeps track of each clerk who works for the store, their initials, and their name.

Ready to start his Powerhouse prototype, Frank logs onto his HP 9000 HP/UX computer system. He sees the operating system prompt, a dollar sign ($) that is equivalent to MS-DOS's A> or C> prompt. Next, he activates Powerhouse's data dictionary maintenance program, called QDD (Figure 6.20). QDD requires him to start with a TITLE statement. This statement names the dictionary and will automatically appear on every report that the system produces.

FILE statements name the files and describe their organization. Indexed files allow rapid and random retrieval of data, something his system must have so that customers will not stand in line to check out or check in a tape. Sequential files store data in the order in which they are entered and for PAYMENT, this is the ideal method.

Each attribute is defined with an ELEMENT statement. Now, Frank must decide on the data type for the element: alphanumeric (X), numeric (9), or date (DATE). 9(05)V9(2) means that the element is numeric with 5 digits before the decimal point (implied with the V) and 2 digits after the decimal point. Likewise, X(12) means the element is alphanumeric with 12 characters. PICTURE statements specify special editing that Frank wants performed on the data whenever it is printed or shown on a CRT screen. The caret (^) in the picture will be replaced by a numeral, starting from the right. DATE attributes are entered in MMDDYY format and are displayed according to our cultural convention, as MM/DD/YY. RECORD statements group elements into logical record structures.

The BUILD statement completes the data dictionary statements. SAVE stores them as a file named VTDD (Video Tape Data Dictionary) in case Frank wants to make later additions, changes, or deletions.

EXIT leaves QDD and returns control to the HP/UX operating system. QUTIL activates the Powerhouse file creation utility, and the CREATE ALL statement constructs the physical files on the disk drive. EXIT ends the file creation utility and takes the user back to the operating system.

Figure 6.20 Using the Powerhouse 4GL software, Frank builds the data dictionary .

```
$ QDD
> TITLE "Video Tape Check In/Out"
>
> FILE CUSTOMER      ORGANIZATION INDEXED
> FILE RENTAL        ORGANIZATION INDEXED
> FILE TAPE          ORGANIZATION INDEXED
> FILE PAYMENT       ORGANIZATION SEQUENTIAL
> FILE STORE         ORGANIZATION SEQUENTIAL
> FILE CLERK         ORGANIZATION INDEXED
>
> ELEMENT AMOUNT-PAID 9(05)V99       PICTURE "^^,^^^.^^"
> ELEMENT APPROX-AGE 9(02)
> ELEMENT CUSTOMER-BALANCE           9(05)V99    PICTURE "^^,^^^.^^"
> ELEMENT CATEGORY                   X(12)
> ELEMENT CHECKIN-INITIALS           X(03)
> ELEMENT CHECKOUT-INITIALS          X(03)
> ELEMENT CLERK-INITIALS             X(03)
> ELEMENT CITY                       X(14)
> ELEMENT COST                       9(05)V99    PICTURE "^^,^^^.^^"
> ELEMENT CREDIT-CARD 9(16)
> ELEMENT CREDIT-CARD-TYPE           X(06)
> ELEMENT DAILY-RATE 9(05)V99    PICTURE "^^,^^^.^^"
> ELEMENT DATE-JOINED DATE
> ELEMENT DATE-LAST-RENTED           DATE
> ELEMENT DATE-PURCHASED             DATE
> ELEMENT DATE-RENTED DATE
> ELEMENT DATE-RETURNED              DATE
> ELEMENT EXPIRATION-DATE            DATE
> ELEMENT FIRST-NAME X(20)
> ELEMENT INCOME-GENERATED           9(05)V99    PICTURE "^^,^^^.^^"
> ELEMENT LAST-NAME                  X(20)
> ELEMENT MIDDLE-INITIAL             X(01)
> ELEMENT PHONE-NUMBER               9(10)       PICTURE "(^^^) ^^^-^^^^"
> ELEMENT RATING                     X(04)
> ELEMENT RENTAL-FEE 9(05)V99        PICTURE "^^,^^^.^^"
> ELEMENT STATE-ABB                  X(02)
> ELEMENT STORE-NAME X(40)
> ELEMENT STREET-ADDRESS             X(30)
> ELEMENT TAPE-ID                    9(05)
> ELEMENT TAPES-RENTED               9(05)
> ELEMENT TIMES-RENTED               9(05)
> ELEMENT TAPE-STATUS X(01)
> ELEMENT TITLE                      X(30)
> ELEMENT YEAR-MADE                  9(02)
> ELEMENT ZIP-CODE                   9(05)
```

(continued)

With his files created, Frank can now build the system's data collection screens with Powerhouse's screen builder, QDESIGN (Figures 6.21 and 6.22). Before designing the screen, Frank invites Mark and Cindy Stensaas to join him and provide user input while he creates the rental and return data collection screen.

Figure 6.20 *(continued)*

```
>
> RECORD CUSTOMER
>     ITEM PHONE-NUMBER UNIQUE KEY
>     ITEM LAST-NAME
>     ITEM FIRST-NAME
>     ITEM MIDDLE-INITIAL
>     ITEM STREET-ADDRESS
>     ITEM CITY
>     ITEM STATE-ABB
>     ITEM ZIP-CODE
>     ITEM DATE-JOINED
>     ITEM INCOME-GENERATED
>     ITEM DATE-LAST-RENTED
>     ITEM TAPES-RENTED
>     ITEM CREDIT-CARD
>     ITEM CREDIT-CARD-TYPE
>     ITEM SEX-CODE
>     ITEM CUSTOMER-BALANCE
> RECORD RENTAL
>     ITEM PHONE-NUMBER UNIQUE KEY
>     ITEM TAPE-ID
>     ITEM DATE-RENTED
>     ITEM DATE-RETURNED
>     ITEM RENTAL-FEE
>     ITEM AMOUNT-PAID
> RECORD TAPE
>     ITEM TAPE-ID UNIQUE KEY
>     ITEM TITLE
>     ITEM YEAR-MADE
>     ITEM RATING
>     ITEM CATEGORY

>     ITEM DATE-PURCHASED
>     ITEM DATE-LAST-RENTED
>     ITEM TIMES-RENTED
>     ITEM COST
>     ITEM INCOME-GENERATED
>     ITEM TAPE-STATUS
> RECORD PAYMENT
>     ITEM PHONE-NUMBER
>     ITEM DATE-RETURNED
>     ITEM RENTAL-FEE
> RECORD STORE
>     ITEM STORE-NAME
>     ITEM STREET-ADDRESS
>     ITEM CITY
>     ITEM STATE-ABB
>     ITEM ZIP-CODE
>     ITEM PHONE-NUMBER
>     ITEM DAILY-RATE
> RECORD CLERK
>     ITEM CLERK-INITIALS UNIQUE KEY
>     ITEM LAST-NAME
>     ITEM FIRST-NAME
>     ITEM MIDDLE-INITIAL
>
> BUILD
> SAVE VTDD
> EXIT
$ QUTIL
> CREATE ALL
> EXIT
$
```

They talk about the check-out/check-in process to make sure they all know what data they need to have collected. They agree that rentals and returns are two different events. During a rental, the clerk will need to enter the customer's telephone number and tape number; the computer will use today's date as the date rented. When the tape is returned, the clerk will locate the rental record, the computer will use today's date as the return date and calculate the rental fee, and the clerk will collect what is owed.

Logging onto the HP 9000 again, Frank activates the screen builder by typing QDESIGN at the operating system dollar sign prompt. QDESIGN responds with a "greater than" prompt (>) and Frank types the word "Rent," thus naming the data collection screen.

FILE statements tell QDESIGN which files the screen will need to access. The word REFERENCE defines CUSTOMER and TAPE as "reference only" files, meaning they will not store data; RENTAL will store data.

TITLE places a caption across the top of the screen and centers it. GENERATE tells QDESIGN that the entries that follow will name the fields that should appear on the screen.

FIELD statements cite the specific fields for data collection during data entry. REQUIRED fields identify data items that the user cannot skip when entering data. The LOOKUP clause

Figure 6.21 The terminal session for QDESIGN builds a data collection screen for a videotape rental by a customer.

```
$QDESIGN                        → store data
>SCREEN Rent                    → (Tells which file the screen will need to access)
>FILE RENTAL
>REFERENCE TAPE                 → Defines elements, Tape, Customer are reference only
>REFERENCE CUSTOMER               meaning they will not store data
>TITLE "Video Tape Rental" CENTERED
>                               → Place caption at Center
>GENERATE                       → (The entries follow will name the fields that should appear on screen.
>FIELD PHONE-NUMBER   OF RENTAL REQUIRED LOOKUP ON CUSTOMER     Check to see if no exist
>FIELD TAPE-ID        OF RENTAL REQUIRED LOOKUP ON TAPE
>FIELD DATE-RENTED    OF RENTAL REQUIRED USE TODAYS-DATE        user can not skip when entering data
>
>UPDATE TAPE WHERE TAPE-ID IN TAPE = TAPE-ID IN RENTAL         &
>      SET TAPE-STATUS = "R"    → rented or check-out        cont on
>                                find tape w/this tape ID. in the Tape File    next line
>BUILD
>SAVE VTRENTAL                   store a list of QDesign statements as a
>GO                              file name "VTRENTAL"
>EXIT
$                               → Tell powerhouse to construct the screen named "Rent"
                                → Return to operating system.
```

Generate the actual screen

Figure 6.22 The data entry screen Frank built to collect data about a tape rental.

```
MODE: X  ACTION: XXXXXXXXXX


                  Video Tape Rental

    01 PHONE-NUMBER              XXXXXXXXXX
    02 TAPE-ID                   XXXXX
    03 DATE-RENTED               MM/DD/YY
```

tells the software that it should check each phone number entered on the screen to see if that number exists in the master list of customers. If it doesn't, an error message at the bottom of the screen will automatically alert the user. The USE TODAYS-DATE tells the software to assign the current date as the default value for this field and that the operator may override this entry if he or she chooses.

The UPDATE statement finds the tape with this tape identification number in the TAPE file and sets the status field to "R" to indicate the tape is rented or checked out to a customer.

The BUILD statement instructs QDESIGN to generate the actual screen. Screen generation itself takes place after Frank EXITs QDESIGN. A SAVE command stores the list of QDESIGN statements as a file named VTRENTAL (Video Tape Rental). GO tells Powerhouse to construct the screen named Rent. After typing EXIT, control returns to the operating system and the prompt appears.

When he wants to see the screen that he has designed, Frank types "Rent" to call it up. Mark and Cindy take turns trying it out, while Frank explains how to tell the software what they want it to do:

E : Enter data

F : Find a record previously entered

U : Change or update a previously entered record

D : Delete a record

^ : Terminate record addition

^^: Terminate the screen

As Mark and Cindy test the screen, they decide to make a change. When a tape is rented or returned, they want to collect the initials of the clerk that checked out the tape.

To remedy that problem, Frank simply accesses the VTRENTAL file and enters the new fields:

```
>  RECORD RENTAL
>      ITEM PHONE-NUMBER
>      ITEM TAPE-ID
>      ITEM DATE-RENTED
>      ITEM DATE-RETURNED
>      ITEM RENTAL-FEE
>      ITEM AMOUNT-PAID
>      ITEM CHECKOUT-INITIALS
>      ITEM CHECKIN-INITIALS
```

Frank saves his changes to the QDD dictionary, brings up the definition of the rental data collection screen, and repairs it:

```
>SCREEN Rent
>FILE RENTAL
>REFERENCE TAPE
>REFERENCE CUSTOMER
>REFERENCE CLERK
>TITLE "Video Tape Rental" CENTERED
>GENERATE
>FIELD PHONE-NUMBER OF RENTAL REQUIRED LOOKUP ON CUSTOMER
>FIELD TAPE-ID OF RENTAL REQUIRED LOOKUP ON TAPE
>FIELD DATE-RENTED OF RENTAL REQUIRED USE TODAYS-DATE
>FIELD CHECKOUT-INITIALS OF RENTAL REQUIRED LOOKUP ON CLERK
>BUILD
>SAVE VTRENTAL
>GO
>EXIT
$
```

Within minutes the system is up and running again, which is an advantage of using a 4GL and prototyping.

Next, Frank turns his attention to the return data collection screen and uses QDESIGN to construct it, as shown in Figures 6.23 and 6.24. Again, TODAYS-DATE specifies the default value for the return date and the operator can enter a different value if he or she chooses to do so. Since the TAPE file has no unique key, the LOCATE statement tells the software that it should use the entries for the phone number and the tape identification to find the record in TAPE.

RENTAL-FEE is quite different from the other entries; it is a calculated field. Fields of this type require Powerhouse to calculate the value and then display it on the screen. The user can override the value if he or she so chooses, but the fundamental value is determined by the software. The CALCULATE statement gives the rule on how RENTAL-FEE is found. A straightforward operation, the amount of money owed is the difference between the date rented and the date returned multiplied by the daily fee that View Video charges for a tape rental. For example, if a tape is rented on January 13th, returned on January 15th, and the daily rental fee is $4.00, the amount owed is $8.00 ($4.00 times 2 days equals $8.00). Powerhouse can subtract two DATE fields, find the number of elapsed days between them, and use this fact—an ideal operation for this application. Powerhouse takes into account leap years, months with different numbers of days, and rental days that cross over different years.

The UPDATE statements allow Powerhouse to alter data values in other files. The WHERE portion of UPDATE selects the specific record and the SET statements specify how the data fields in this record need altering. In the example on TAPE, the SET statements manipulate DATE-LAST-RENTED, TIMES-RENTED, INCOME-GENERATED, and TAPE-STATUS using familiar arithmetic operators.

To complete his data collection screens, Frank would need one to collect data about a customer and about new movies purchased from the film distributors. He'd also need the capability to enter the initials for each clerk and build a single record for STORE. As you can see, the definitions are not difficult and Frank can do them in a few hours, reviewing each with Mark and Cindy Stensaas.

Figure 6.23 Frank's videotape check-in definition.

```
$QDESIGN
>SCREEN RETURN
>FILE RENTAL
>REFERENCE TAPE
>REFERENCE CUSTOMER
>REFERENCE STORE
>REFERENCE CLERK
>TITLE "Video Tape Return" CENTERED
>GENERATE
>FIELD PHONE-NUMBER      OF RENTAL REQUIRED LOOKUP ON CUSTOMER
>FIELD TAPE-ID           OF RENTAL REQUIRED LOOKUP ON TAPE
>FIELD DATE-RETURNED     OF RENTAL USE TODAYS-DATE
>FIELD CHECKIN-INITIALS  OF RENTAL REQUIRED LOOKUP ON CLERK
>FIELD RENTAL-FEE        CALCULATED
>FIELD AMOUNT-PAID       OF RENTAL REQUIRED
>
>LOCATE USING TAPE-ID IN TAPE LINK RENTAL
>CALCULATE RENTAL-FEE = DAILY-RATE LOOKUP ON DIRECTOR         &
>                           *(                               &
>                           DATE-RETURNED  OF RENTAL          &
>                           - DATE-RENTED  OF RENTAL          &
>                           )
>UPDATE CUSTOMER WHERE PHONE-NUMBER OF CUSTOMER =             &
>                           PHONE-NUMBER OF RENTAL            &
>         SET BALANCE = BALANCE - AMOUNT-PAID                 &
>         SET INCOME-GENERATED = INCOME-GENERATED + RENTAL-FEE &
>         SET DATE-LAST-RENTED = DATE-RETURNED                &
>         SET TAPES-RENTED = TAPES-RENTED + 1
>UPDATE TAPE WHERE TAPE-ID OF TAPE = TAPE-ID OF RENTAL        &
>         SET TIMES-RENTED = TIMES-RENTED + 1                 &
>         SET DATE-LAST-RENTED = DATE-RETURNED                &
>         SET INCOME-GENERATED = INCOME-GENERATED + RENTAL-FEE &
>         SET TAPE-STATUS = "I"
>
>BUILD
>SAVE VTRETURN
>GO
>EXIT
$
```

[handwritten annotation: Calculate value and then display on screen]

To produce management reports, Frank calls on Powerhouse's report writer, QUIZ. This software can build complex reports and can generate this report with few statements. Frank decides to tackle the customer list report first (Figure 6.25). This report exhibits many elements of most common reports—a title, date, page number, column headings, detail lines, and totals—all within QUIZ's capabilities.

Frank starts creating the report by typing QUIZ at the HP/UX prompt. At the operating system prompt (>), he identifies the files from which his report must use with an ACCESS

Figure 6.24 The data entry screen that Frank built to collect data about a return.

```
MODE: X  ACTION: XXXXXXXXXX


                    Video Tape Return

        01 TAPE-ID              XXXXX
        02 DATE-RETURNED        MMDDYY
        03 CHECKIN-INITIALS     XXX
        04 RENTAL-FEE           99,999.99
        05 AMOUNT-PAID          99,999.99
```

Figure 6.25 Management report lists View Video customers in ascending order by customer number. The last line in the report gives the grand totals of the columns.

```
Date:  12/20/92              View Video, The Movie People                    Page: 1
                               C U S T O M E R    L I S T


                                                                Date
Phone            Customer        Customer   Tapes    Date       Last       Income
Number           Name            Balance    Rented   Joined     Rental     Generated

(415)122-5463    Macauley, Tammi     5.00      142   01/23/91   11/29/92       388.00
(415)122-6782    Nancebo, David                 56   01/12/92   12/12/92       150.00
(415)333-4182    Laki, Michelle     25.00       14   11/21/92   12/19/92        28.00
(415)333-8111    Hakimoto, George   50.00       35   03/05/91   12/23/91        80.00
(415)652-1140    Lanning, Floyd                  2   06/19/92   06/23/92         5.00
      *               *              *        *        *          *             *
      *               *              *        *        *          *             *
      *               *              *        *        *          *             *

Grand Totals                     1,230.00    9,267                          15,789.00
```

statement (Figure 6.26). This version of ACCESS picks data from a single file, CUSTOMER. Other versions of the ACCESS allow two or more files to link together. For example, if we want a report showing all the data about a specific customer and all the tapes that customer has ever rented, the ACCESS would read:

```
> ACCESS CUSTOMER LINK TO RENTAL LINK TO TAPE
```

Figure 6.26 Powerhouse report writer statements to produce the list of purchase orders needing payment.

```
$QUIZ
>ACCESS CUSTOMERS
>SELECT IF INCOME-EARNED NOT EQUAL 0
>SORT ON PHONE-NUMBER
>REPORT PHONE-NUMBER, concat (LAST-NAME, ', ', FIRST-NAME),      &
>        BALANCE, TAPES-RENTED, DATE-JOINED, DATE-LAST-RENTED,    &
>        INCOME-GENERATED
>FOOTING SKIP 2, TAB(27) "======= =======", TAB(20),             &
>        "=========="
>TOTAL "Grand Totals", CUSTOMER-BALANCE, TAPES-RENTED,           &
>        INCOME-GENERATED
>
>SAVE CUSTLIST
>GO
>EXIT
$
```

The ACCESS statement allows the report to cross over among the three files, extracting data from each. Powerhouse checks to make sure the linkage is possible. It determines the linkage by means of the PHONE-NUMBER in CUSTOMER to the PHONE-NUMBER in RENTAL then via the TAPE-ID in RENTAL to the TAPE-ID in TAPE.

The SELECT statement tells QUIZ to retrieve only customers that have rented tapes. SORT tells QUIZ to arrange the data selected in ascending order by phone number. Without a SORT statement, the report would merely list the records in the order in which they exist in the file.

REPORT instructs QUIZ to display the necessary data elements in left to right order on the report. All of the data come from the CUSTOMER file. Column headings are the data names for each field stacked on top of each other. The CONCAT calls for Powerhouse to join together (concatenate) the customer's last name, a comma, a space, and the customer's first name into a single field that minimizes blank spaces.

FOOTING produces the underscores of equal symbols (=========) after skipping two lines at the end of the report. TAB causes correct horizontal spacing of the underscores. TOTAL prints the words "Grand Totals" and then accumulates the sum of each field listed (CUSTOMER-BALANCE, TAPES-RENTED, INCOME-GENERATED) automatically.

SAVE stores the QUIZ statements as a file named CUSTLIST, and GO initiates the execution of the QUIZ report. EXIT terminates QUIZ and returns control to the operating system, where we again see the HP/UX prompt.

Mark and Cindy express amazement over the speed with which Frank has designed the system, constructed files, built the screens, and produced the printed reports. Both agree that with a little practice, they could even write "programs" themselves, producing a wide variety of reports, such as lists of tapes in order by tape identification number and alphabetic by tape title.

While not a complete videotape check-out/check-in system, the prototype does give the two primary users of the system an early look and feel of the system. More importantly, they do not feel threatened by the proposed system and find it friendly and easy to use. Wisely, Frank got his users involved at an early stage and exploited their expertise to spot errors before he went too far with the new system.

Working With View Video

Our look at a theoretical version of the actual Powerhouse software was quite short and the actual version has many more capabilities than you saw. We wanted to show you the speed with which a skilled user can place a system into operation, as well as how it allows users to try out a system before the design is finalized.

Case Study Exercises

Complete the following:

1. Write the QDESIGN statements to collect data about clerks. Sketch the screen you think your statements would produce.
2. Write the QDESIGN statements to enter data for the STORE file.
3. Write the QUIZ statements to list all the videotapes in the TAPE file sorted by tape title.
4. Write the QUIZ statements to print a list of customers that have never rented a tape.
5. Write the QUIZ statements to list all the clerks.
6. Write the RECORD statements in QDD to define another file to store data about each movie star.

Systems Design

Output Design

GOALS

After reading this chapter, you should be able to do the following:

- Establish the criteria for specifying a printer
- Summarize the characteristics of at least three kinds of printers
- Describe the basic report formats
- List the five types of reports
- Explain three design criteria for CRT reports
- Cite two reasons to prototype a system's reports and screen designs
- List three control techniques for monitoring system outputs
- Write a data dictionary definition for a report

PREVIEW

System design transforms a logical representation of what a given system is required to do into the physical specifications. The specifications are converted into a physical reality during development. The design forms a blueprint of the system and how the components relate to each other.

The design phase proceeds according to an orderly sequence of steps, beginning with review and assignment of tasks and ending with package design. During the first stage, output design, an analyst determines what data the application produces and how to organize and present that data (Figure 7.1). Paradoxical as it may seem on the surface, output design precedes input design. Just as a model plane builder wouldn't purchase materials and start assembling them until after designing the plane, a systems analyst wouldn't collect and try to process any data before designing the output of the system.

When designing a system's output, the analyst must make several interrelated and interdependent decisions. For example, the analyst must select the best output medium and has a wide range of choices, including printers, cathode-ray tubes (CRTs), tapes or disks, audio devices, and microfilm.

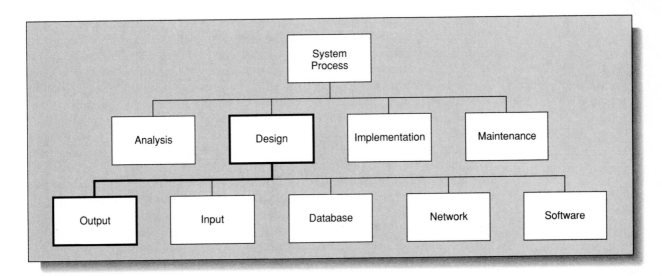

Figure 7.1 Output design is the first phase of the systems design phase of the systems process.

This chapter will explore the major output media, their uses, and how you format reports for each. It concludes with an explanation of how to prepare a report data dictionary, which is a major product of output design.

GUIDELINES FOR OUTPUT DESIGN

You encounter computer output all the time: the morning newspaper, utility and other bills, application forms, and course grades, to name a few. The usefulness and clarity of such commonplace outputs depend a great deal on the care with which an analyst designed them, keeping their major purposes in mind.

Every kind of business and government computer system produces some kind of report, and many systems produce a lot of them. No matter what the content of a report, though, its design should always take into account the audience and purpose to which users will put it.

The following guidelines apply to any report:

- The information should be clear and accurate, yet concise, and restricted to relevant data.
- Reports should have titles, the date, descriptive headings for columns of data, numbered pages, and so on. If printed, they should also appear on a standard size of paper.
- The report's contents should be in a logical arrangement so that users can easily locate what they need.
- The report should come on an output medium that best suits the user's needs (a stockbroker who needs instant information may require a CRT, while a sales manager who consults monthly figures for reference may require a printout).

Business reports fall into two broad categories: internal and external. Managers and other decision-makers within an organization use **internal reports** to track performance and to make a variety of business judgments. Some internal reports are very simple (such as alphabetic lists of customers and vendors), while others are extremely detailed (such as analyses of items ordered from vendors, the amounts owed, and their due dates). Still others contain confidential and sensitive information such as outstanding bank loans/investments or employee pay rates. See Figure 7.2 for an example of a well-designed internal report.

Because internal reports remain within an organization, accuracy and timeliness are as critical as their appearance and style. People who receive them must glean information from them quickly and easily. For example, although an accounts payable report should clearly indicate due dates so a person responsible for decisions in this area can take advantage of purchase discounts, the report does not need to be printed on expensive bond paper.

External reports circulate among customers, clients, stockholders, vendors, government agencies, and others outside of an organization. Here, appearance and style matter as much as clarity and accuracy. Since such reports present the company's image to the outside world, they must make the company look well-organized, stable, and professional. Figure 7.3 shows one type of external report.

Figure 7.2 Internal reports remain within the organization. Usually produced on plain paper, they provide the information needed to perform a particular job. This report contains all the characteristics of good report design: headings, date, time, page number, detail, and totals.

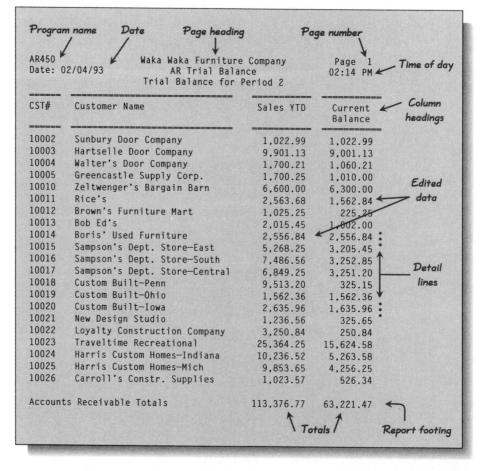

Some reports that begin as computer output eventually become input for a subsequent computer operation. Such **turnaround documents** provide information for outsiders (perhaps items ordered and the amount due sent to a retail store's customers), but they also contain information for use by the organization (payment received). As you can see in Figure 7.4, a designer can accomplish this dual use by creating a report with two easily separated sections: one for the customer's files and one for return to the organization. The organization usually preprints the customer's account number on the turnaround portion

Figure 7.3 External reports circulate outside of the organization. They are normally printed on standard or preprinted forms with column or category headings that apply to all situations. Spaces under the headings will be filled with appropriate data when the form is printed. The familiar W-2 federal income tax form comes to most of us in January of each year.

1 Control number **22222**	For Official Use Only ▶ OMB No. 1545-0008						
201							

W-2 Wage and Tax Statement form:
- 1 Control number: 201 — **22222** — For Official Use Only ▶ OMB No. 1545-0008
- 6 Statutory employee / Deceased / Pension plan / Legal rep. / 942 emp. / Subtotal / Deferred compensation / Void
- 2 Employer's name, address, and ZIP code: WABASH FURNITURE COMPANY INDIANA DIVISION NAPPANEE, IN 46500
- 7 Allocated tips
- 8 Advance EIC payment
- 9 Federal income tax withheld: 2,976.10
- 10 Wages, tips, other compensation: 33,170.45
- 3 Employer's identification number: 36-3367380
- 4 Employer's state I.D. number: 36-336780
- 11 Social security tax withheld
- 12 Social security wages: 33,170.45
- 5 Employee's social security number: 305-77-6579
- 13 Social security tips: 1,999.15
- 14 Medicare wages and tips
- 19a Employee's name (first, middle, last): Christinia Suzanne Munroe, R R 3 Box 42, Nappanee, IN 46501
- 15 Medicare tax withheld
- 16 Nonqualified plans
- 17 See Instrs. for Form W-2
- 18 Other
- 19b Employee's address and ZIP code
- 20
- 21
- 22 Dependent care benefits
- 23 Benefits included in Box 10
- 24 State income tax: 562.80
- 25 State wages, tips, etc.: 33,170.45
- 26 Name of state: IN
- 27 Local income tax: 281.85
- 28 Local wages, tips, etc.: 33,170.45
- 29 Name of locality: C-50

Copy A For Social Security Administration   Dept. of the Treasury—Internal Revenue Service

form **W-2 Wage and Tax Statement 1991**

For Paperwork Reduction Act Notice, see separate instructions.

16-0331690   Do NOT Cut or Separate Forms on This Page

Figure 7.4 Turnaround documents create both a record for the customer and a means for the customer to communicate with the organization.

YES! Send my first issue **NOW!**   ☐ Check ☐ Bill me   ☐ VISA ☐ MasterCard

☐ 1 year (12 issues) $29.95   Card # _____ Exp. _____
☐ 2 years (24 issues) $56   Signature _____

Foreign: 1 year – $54 Canada/Mexico; $65 overseas
2 years – $89 Canada/Mexico; $123 overseas
Canadian orders must add 7% goods and services tax.

The **C** Users Journal™
1601 W. 23rd Street, Ste. 200
P.O. Box 3127
Lawrence, KS 66046-0127 USA

The **C** Users Journal™

- Practical, sophisticated C development information
- more features, columns, and code
- Only $29.95 for 12 issues
- Money back Guarantee

PERRY EDWARDS
711 W. 40TH ST.
SACRAMENTO, CA 95816

Our Guarantee

If you're not completely satisfied with your first issue of The C Users Journal let us know. We'll refund your entire subscription price. No questions asked.

of the document to facilitate cost-effective data collection. A good example of this type of report is the form used by most monthly book clubs, part of which a customer must return each month to indicate whether he or she wishes to buy an advertised book.

SELECTING THE BEST MEDIA

Early in the design phase, the analyst must select the best medium for presenting information to users. In recent years, a wide range of media has become available, including many different kinds of printers and terminals, personal computers linked to central data banks, CD-ROMs, microfilm, tape, disks, audio or sound generators, data communications links, optically scannable forms, and plotters. Each offers special capabilities, speeds, and limitations, and the analyst should carefully weigh these against the needs of the organization.

For a payroll application in one organization, an analyst might select a floppy disk to transport data from the employer's computer to the bank that credits net pay to an employee's account. For another organization, a telephone data communication link between the computers may make more sense. Whereas a telephone data communication link would provide for instant crediting of employees' accounts, it would cost more than sending a floppy disk through the mail. However, mailing would delay crediting. As usual, the successful analyst must juggle a number of trade-offs: cost, speed, expandability, machine readability, and compatibility with existing equipment.

Two of the most popular output media today are printers and screens. Most users now access their reports in the form of either hard-copy printout or a screen display. Some applications mix the media: a screen for rapid inquiry and a printed report with historical account activity for periodic reference.

Printers

Before selecting a printer, the analyst must consider all of the system's output and input requirements, as well as the company's budget. Some printers may satisfy most requirements, but fall short with respect to some critical functions. Again, trade-offs come into play, and the ultimate choice depends on the system's requirements and the company's capabilities, such as:

- *Amount of output per week or month*, which affects the decision in terms of the device's speed. Here, you would take future growth into account.
- *Quality of output*, which affects the decision in terms of cost, especially when a large number of copies is needed. Quality laser printers and bond paper cost much more than dot-matrix printers and stock paper.
- *Existing printers available*, which may make newer acquisitions unnecessary. Still, you must make sure that the old equipment will attain the desired quality and quantity of output.
- *Need for specific function* of multiple copy printing or bar coding.

Consider the trade-offs encountered by Fleet Feet analyst Peggy Adams-Russell. Peggy needed to select the printer for her firm's new computerized office system, which would send out bills and process payroll. Before she even thought about the ideal printer, Peggy first listed the requirements of the system:

■ Except for the bills sent to customers, the reports need not be elegant. Most would circulate internally.

■ The system would print 12,000 pages, or 480,000 lines (12,000 pages × an average of 40 lines per page) a month during a typical month.

■ The organization did not need a printer that could draw graphs or other visual aids.

■ Payroll forms could be oversized (132 characters wide).

■ Billing forms had to permit 6 lines per vertical inch on pre-formatted paper.

After Peggy began researching printers, she found, to her delight, that a number of good ones could nicely meet Fleet Feet's needs. Using this information, Peggy outlined the major differences and trade-offs among the candidates:

■ *Character formation:* fully formed (similar to a typewriter) or dot matrix (composed of a series of dots instead of connecting lines)

■ *Speed:* one character at a time, one line at a time, or one page at a time

■ *Print method:* impact (uses ribbons to imprint letters, but can print carbon copies) or nonimpact (very fast and quiet)

■ *Characters per inch:* most common paper widths are 80 or 132 characters per line, placing 10 or 12 characters per inch

■ *Lines per inch:* 6 or 8 lines per vertical inch of paper

■ *Cost:* ranges from $250 for an impact, dot-matrix printer that produces one character at a time to $150,000 for a nonimpact, letter-quality printer that can produce 100 or more pages at a time

Like most organizations, Fleet Feet followed a fairly strict budget. Though letter-quality or laser printing would enhance the appearance of reports, a draft mode or near letter-quality mode dot-matrix printer could produce acceptable output at a much lower cost. Internal reports did not need great beauty, nor did external reports to customers and state or federal agencies.

Given the need to print approximately 480,000 lines per month, a device printing a maximum of 200 characters per second would take too long to get the job done (426 hours to print 480,000 lines). So Peggy settled on a line-at-a-time printer. But how many lines per minute did the application demand? Using 200 lines per minute as a starting point, she found that it would take 40 hours to print 480,000 lines. That would handle this year's volume, but since Fleet Feet expected strong future growth, Peggy doubled the number of lines printed per minute, thus cutting production in half:

480,000 lines ÷ 400 lines per minute = 1,200 minutes or 20 hours per month

That allowed plenty of margin for growth or unanticipated extra printing.

Fleet Feet did not need a printer that could create graphs or pie charts, but it did sometimes use oversized forms (132 columns wide). Since the billing forms combined six lines per vertical inch, Fleet Feet would not need an eight line per inch printer. Considering the clinic's overall needs, Peggy recommended a 400 line per minute, dot matrix, impact, 132 column, 6 line per inch printer.

Paper

Having selected a printer, an analyst next chooses the type and quality of paper. Important considerations include cost, number of copies needed, and the purpose or audience for the reports.

Based on these factors, the analyst selects:

- Weight and bond of the paper
- Plain paper or preprinted forms
- If preprinted forms, the number to order at a time
- A duplication method

Regardless of its quality or use, some printer paper is continuous fanfolded, with sprocket holes on the left and right edges that enable the printer to pull the paper through the printing mechanism. The pages are perforated at the top and bottom so an operator can easily separate or burst them and strip off the sprocket hole edges. If it takes too much time to burst pages manually, a company can buy a machine that automatically performs this task. Other machines can automatically decollate forms and remove carbon paper.

Paper quality varies dramatically by both weight and bond. Paper weight is the number of pounds per ream (500 sheets) of 17" × 22" paper. Onionskin paper weighs 9 pounds per ream, standard paper about 40–50 pounds, and heavy card stock weighs 120 pounds. **Bond** refers to the percentage of fiber in the paper. The higher the percentage bond, the higher the paper's quality: a 20 pound, 40% bond paper is a medium weight, fairly high-quality paper; 10 pound, 10% bond is a light weight, lower quality paper. If an organization will use a report internally, the latter may suffice; if appearance matters, the former is a better choice.

Plain paper costs less than preprinted forms of the same weight and bond. If an organization requires certain preprinted forms, vendors will gladly provide prices for set-up and production. Set-up involves making the plate required for printing and it usually represents a one-time cost, unless the design changes. Production costs include typesetting, printing, handling, and delivery. The per-page price for production depends on quantity. A thousand forms cost more per page than fifty thousand, so it makes sense to gauge with some accuracy the quantities required over a reasonable period of time, perhaps six months to a year. Allowing for longer time makes sense only if the form will not eventually change. Laser printers can provide savings in this area since the printer can produce the form at the same time that it prints the report as long as color or very small print are not needed.

Since most applications demand the duplication of reports and other printer outputs, the analyst must determine the costs of various duplicating methods by considering the number, frequency, and quality needed. Although impact printers can print carbon copies, carbon paper is expensive and requires decollating the carbon sets and removing the carbon paper. From a legal perspective, 3 copies made from a laser printer are not considered true copies. A true legal copy is a photocopy of an original and not a reprint of the original since it is theoretically possible to make them different.

CRTs

Cathode ray tubes (CRTs) have gained popularity in recent years. They come in many shapes and sizes and offer many different features. The most important options currently available include:

- *Appearance of material on the screen:* monochrome or color
- *Styles and optional equipment:* users who spend many hours in front of the CRT usually prefer those with tilting and non-reflective screens plus detachable keyboards
- *Personal computer and terminal to computer hook-up capabilities:* direct connection between the terminal and the computer costs the least
- *Speeds:* the productivity of expensive high-speed terminals may offset the lower cost of less productive low-speed ones
- *Screen sizes and resolution:* diagonal measure in inches (for example, 14") and pixels (e.g., VGA or SVGA).

Like printers, personal computers and terminals share several characteristics. The screens on most devices manufactured since the mid-1970s can display 1,920 characters arranged in 24 rows of 80 characters each. When designing reports for terminals, analysts can plan for this 24 by 80 window, no matter what type of device is selected for output.

Some terminals display a twenty-fifth or **status line** at the bottom of the screen, while others offer a wider screen or more rows. A status line tells users what the device is doing at any given moment: ready for entry, sending data to the computer, awaiting a response from the computer, or displaying an error message. Wider screens, up to 132 characters per line, permit the terminal to display wider forms. Additional rows, useful for word processing, allow the user to see more data at a time.

An organization can use personal computers and terminals to collect, display, and disseminate information. However, the analyst must still consider format, the possibilities for which include:

Today's analyst faces a wide variety of font options: Times, Helvetica, Courier, Chicago, and hundreds more. Therefore, the analyst needs to know some of the typesetter's vocabulary. Figure 7.5 shows 15 key concepts to type.

1. Point size: A measurement of height, from the top of the ascender to a fixed depth below the descender. There are 72 points in an inch, so 36-point type occupies about 1.2" including shoulder.

2. Cap height: A measure of height, from the baseline to the top of a capital character. The highest swell in curved characters can reach past this height.

3. Shoulder: The fixed space below the descender. This ensures that the descenders from a line of type don't contact ascenders and capitals on the line below when no leading, or space between the lines, has been added.

4. Baseline: An imaginary horizontal line on which the base of every letter aligns. Descenders fall below it, and the deepest swell in curved letters overhangs it slightly.

5. Serif: A small finishing stroke, roughly perpendicular to the main stroke of a character. The presence or absence of serifs is a major type classification.

6. Counter: The white area enclosed by the curves or lines of a character, like the holes in a D, e, or B.

7. x-height: The height of lowercase characters without ascenders or descenders, such as the x or e. Every letter occupies this minimum depth, making the x-height the most visually significant dimension in a type design.

8. Set width: The total horizontal white space that a character occupies. Additional space is added before or after the character and determines the default letter spacing.

9. Ascender: The portion of a lowercase character that extends above the median line. It may be quite tall relative to x-height, or very shallow.

10. Descender: The portion of a lowercase character that falls below the baseline. Usually much shallower than the height of an ascender.

11. Mean line: An imaginary horizontal line that falls along the top of lowercase letters without ascenders to define the top of x-height.

12. Roman: Type that is not italic or bold; often called regular. The root design in a type family is roman, and

- *Blinking video:* Flashing data entry points. Blinking helps the user to rapidly locate the cursor on the screen.

- *Inverse video:* White letters on a black background instead of black letters on a light background. Useful for showing data-entry errors to the terminal user.

- *Secured video:* Data input does not appear on the screen, thus allowing confidential entries. Most often used for entry of a user's password.

- *Scrolling:* Information rolls off the top of the screen and a single new line is inserted at the bottom.

- *Paging:* Information is brought to the screen in blocks of lines (usually 24) at a time. This especially helps programmers who must list long programs and want to see them as a complete unit.

- *Forms or block mode:* Data are sent to the computer in groups of characters at a time. Users enter all the data about a transaction before sending, which allows for error correction prior to computer processing.

- *Dual intensity:* Two different levels of brightness allow one level for data, the other for a form. A background image (such as an invoice) appears while data is being entered onto it. This is especially useful for standardized forms.

the family may be extended with weight and width variants.

13. Boldface: A heavyweight version of a type design, in which all strokes are thickened. The contrast between thick and thin strokes is often greater in bold than in roman.

14. Italic: Type that slants, and is generally lighter in weight than the roman within the same family. Almost always very different in design from the roman.

15. Sans serif: A type design without serifs, often with nearly uniform line weight.

Figure 7.5 The illustrated guide to type.

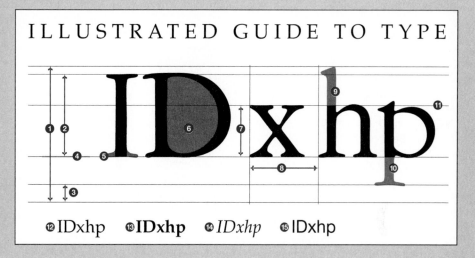

■ *Screen-splitting:* The device may monitor multiple tasks simultaneously, or it may be linked to separate computers with each machine communicating simultaneously with the device. The operator can even split the screen into sections, assigning each computer a separate area.

■ *Windowing:* A box appears on top of existing data. Special instructions dictate window size, location, background color, and foreground color. Windows are especially useful for error messages and user prompting. When the window is no longer needed, it disappears and the data it covered reappear. Some devices allow stacking of windows, each overlaid on top of the other.

■ *Graphics boxes:* These allow the user to draw lines, curves, squares, or other figures on the screen. Boxes look like special forms: payroll time cards, accounts receivable invoices, and so on.

■ *Hard-copy:* A printer attached to the device can produce a permanent record of displayed data. This option benefits programmers working away from the office or users who need to give receipts to customers.

Screens grow more popular every day, and their costs run from as low as $395 for a device with little memory to over $10,000 for one with a full range of video, color, levels of intensity, forms, screen-splitting, windows, graphics, and hard-copy capabilities.

Craig Smally had just finished designing the reports for Lumberjohn's new inventory system. He thought they were terrific. After two weeks of hard, uninterrupted work, during which Craig studied five installed inventory systems, he felt his design would provide users (store managers) with the sort of information that would make their jobs easier. Eager to show off his dream reports, he called a meeting for the company's seven managers.

To prepare for the presentation, Craig photocopied samples of all the reports, bound them in handsome folders, and placed them on the table in front of each manager.

Happy-go-lucky Joe Ferguson, manager of the downtown store, arrived with a burst of laughter, then sat down and began leafing through the folder. Before long, the smile on his face turned to a frown. What was the matter? wondered Craig. Maybe Joe just had an axe to grind against computers and analysts. Craig knew that some of Lumberjohn's managers thought the data processing people had forgotten that they serve the organization rather than vice versa, but he chalked those feelings up to jealousy.

Oh, it was true that when Lumberjohn first installed terminals in each store, data processing had made a mistake by not properly preparing the managers for the new devices, but the resulting resentment seemed way out of proportion to the problem. So the managers had trouble figuring out which buttons to push. Big deal. When new technology arrives, everyone has to learn new tricks. Why complain about the adjustment you must make every time you use a terminal, when that terminal is saving you valuable time? Craig reasoned.

Craig remembered how Ferguson had steamed over the memo that notified managers about Lumberjohn's new computer a few years back. Ferguson, it seemed, just couldn't get used to new ideas.

When the other six store managers arrived, the data processing manager began the meeting by praising his department and the job that Craig had done on the inventory system. But the more he talked, the more the managers' smiles turned to frowns.

Joe Ferguson, finding safety in numbers, finally cleared his throat and interrupted Craig's boss, complaining that data processing was telling users what they would get rather than asking them what they needed. He attacked the new computer, the terminals, and Craig's beloved inventory system. To prove his point, Joe ripped out one of Craig's reports.

"It doesn't take a mathematical genius to calculate that this report would result in 1,000 pages of printouts each day, seven days a week, for just one store! This one report would weigh almost ten pounds—a year's worth, over 3,600 pounds. I'd have to rent storage space and hire a new clerk just to deal with that ton-and-a-half of garbage!" Joe fumed. "I'm sick and tired of the tail wagging the dog around here."

The data processing manager grew defensive. "My people know what's best; we're the experts. We understand computers."

Joe threw his folder on the table. "Yeah? Well, I understand the lumber business, which pays your fancy salaries!"

Joe continued in a milder tone. "Look, when a customer comes into my store with a problem or a complaint about a product they bought from Lumberjohn's, the customer is always right. We exist to serve our customers. If we treat them fairly, they'll continue to be our customers; if we don't, they'll go elsewhere. Why doesn't that apply to you guys?"

Craig felt terrible. But he learned an important lesson: computers serve the users, and the users' needs require careful consideration. Before the meeting broke up, he promised to meet separately with each store manager so he could build their suggestions into his revised designs.

The primary advantage of personal computers or terminals is compactness, which allows them to sit conveniently near users. Disadvantages include the lack of a printed document. In addition to the cost of the device itself, an organization must calculate the costs associated with running a personal computer or terminal environment, such as extra hardware, security, and software. Maintenance costs for screen devices can be as low as $15/month.

FORMATTING REPORTS

Most reports, regardless of the medium used to generate them, contain all or some of the following design elements: report heading, page heading, control heading, column heading, detail line, control footing, page footing, and report footing. These design elements serve two purposes: they make the report easier to read and they provide a way to control the content of the report. Consult Figure 7.6 as you read the following.

A **report heading** states the title of the report itself. It appears only at the beginning of the report, sometimes on a separate or title page. It may include pertinent information such as the name of the company, the author of the document, and the date.

Page headings immediately follow the report heading. This category includes page numbers, date and or time of day the report was printed, and the name or number of the program generating the report.

Control headings are captions and titles that separate one group of data from another. Reports that do not contain distinct groups of data will not have control headings.

Column headings are captions that appear over vertical units of data. They identify what the data underneath represents.

A **detail line** displays the data for a single transaction. If the report will circulate outside the organization, someone will probably edit the data appearing in the detail line so that it looks more attractive. Editing a numeric field may involve inserting commas between digits or printing currency symbols and decimal points. For example, 1234.90 might be edited to read $1,234.90.

A **control footing** is the final part of a control heading and it usually provides data totals. Certain reports do not require control footings, nor do control headings require control footings, and vice versa.

Figure 7.6 A report summarizes captured data in a clear and organized fashion. Note the variety of design elements.

```
 Monday, Dec. 13, 1993          FLEET FEET                    Page:  1
 8:05 A.M.                   Accounts Receivable
                                Aging Report
                                                                    Page heading

 Customer  Contact     Telephone    Current  31-60  61-90  Over-90    Total
 Number    Person      Number                 Days   Days   Days      Owed
                                                                    Column heading

 000003    Webb, S     2136653456    948.23   0.00   0.00  500.00  1448.23
 000234    Lee, J      2137892341     23.89   0.00   0.00    0.00    23.89
 000512    Jacobs, T   2136789000    125.00  25.00   0.00    0.00   150.00
    *          *            *           *       *      *       *        *
    *          *            *           *       *      *       *        *  Detail lines
    *          *            *           *       *      *       *        *

 Grand Totals  ------------>     8907.23 456.00 245.00 1200.89 10809.12

                                                                    Report footing
```

Page footings appear at the bottom of every page. They may indicate page numbers or page totals.

A **report footing** occurs once, at the end of the report, indicating that the report has ended. It often lists overall report totals or some message to the user that this is the final page of the report. A final message is helpful because it indicates to the user that they have received all the pages in the report.

After thoroughly considering a report's eventual use, an analyst will decide which of the design elements to include in it. For example, since upper-level management makes decisions for the company as a whole, its reports might omit detail lines. Departmental managers, who need to monitor the activities of several subdepartments, may need reports with multiple levels of control headings or footings, while certain office workers especially need detail lines. Whatever the situation, an analyst should strive to tailor a report to the user's needs.

TYPES OF REPORTS

Most organizations find that they use a mixture of five types of reports: query, detail, summary, exception, and periodic. Although most conform to basic formatting requirements, each has a unique purpose and audience that dictate special format considerations.

Because **query reports** allow users to get immediate answers to their questions, they are quite common on PCs or terminals. For example, if an accounts receivable clerk wants to know whether a customer has sent a check for a particular invoice, the clerk can enter the customer and invoice numbers at a personal computer or terminal and obtain that particular account's status at once (Figure 7.7). Query reports can also provide users with summaries and lists of transactions.

A **detail report** displays all the pertinent facts about a situation. A vendor listing shows specific facts about each of Fleet Feet's vendors on separate lines and illustrates a detail report (Figure 7.8). Such reports create a handy information file for manual reference.

A **summary report** tabulates overall results, thus facilitating historical review or comparison of current to past activity. For instance, a summary report might compare last year's and this year's sales to date, thus helping management detect trends or review a salesperson's performance. Figure 7.9 reveals an increase in equipment sales this year over last year.

An **exception report** allows a user to specify criteria for isolating a certain set of data. The user in Figure 7.10 asked for a list of items that sold more units this year than last year. Note that only equipment and Nike Air Pegasus sales appear because other items, such as running shoes and clothing, did not meet the specified criteria. Inventory systems often employ exception reports to help users detect whenever stock levels fall below the reorder point. Such reports can save substantial clerical time and ensure accurate reordering.

A **periodic report** gives users data at regular time intervals. At the end of a month, an accounts receivable system might produce an aging report listing all customers whose balances have risen above a certain amount and whose payments have fallen more than 60 days past due.

Figure 7.7 Query reports require a terminal with on-line access to data.

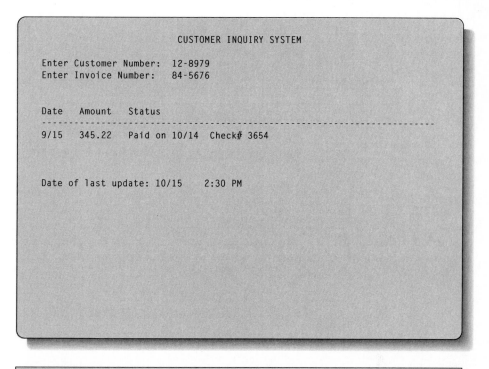

```
                          CUSTOMER INQUIRY SYSTEM

     Enter Customer Number:   12-8979
     Enter Invoice Number:    84-5676


     Date    Amount   Status
     - - - - - - - - - - - - - - - - - - - - - - - - - - - - - - - - - - - - - - - - - -
     9/15    345.22   Paid on 10/14   Check# 3654



     Date of last update: 10/15     2:30 PM
```

Figure 7.8 Detail reports display all relevant data about each transaction. Such reports can provide backup information for future reference.

```
Oct. 24, 1993                    FLEET FEET                      Page:  1
                            Numeric Vendor List

Vendor      Name               Type    Terms          Disc    Purchases
Number      Address                                               YTD

0-000000-1  Acme Office Supply  Cap    2% 10 Net 30   2.00    1,594.00
            123 Elm Street
            Scarborough, NY 10510

0-123456-7  Star Electronics    Mtl    Net 30         0.00    8,212.90
            423 Bancroft
            Berkeley, CA 94704

4-223344-5  Utah Telephone      Adv    5% 5 Net 10    5.00    9,895.00
            490 Broadway
            Orem, UT 84057
```

DESIGNING REPORTS FOR PRINTERS

The analyst must design every different page that the system will produce. After intense discussion with the user, the analyst draws the layout on a spacing chart, then submits the results for design review and user approval. Painstaking planning of reports helps smooth subsequent design stages, especially the next stage: database or record layout.

Figure 7.9 A summary report stresses overall totals rather than details, thus permitting a global review of data.

```
FLEET FEET
Sales by Category                                                As of: 10/24/93
========  ========        ========  ========  ========  ========
Category  Description        This      Last      This      Last
   #                        Month     Month      Year      Year
--------  -----------     --------  --------  --------  --------
1712-22   Running Shoes   $2,336.88 $2,113.55 $34,477.98 $45,607.88
5234-23   Clothing         3,856.77  2,406.66  44,712.44  56,896.07
8659-51   Equipment        2,555.91  3,550.22  89,377.12  81,652.90
```

Figure 7.10 Exception reports can help users determine trends. The reader saves time by viewing only pertinent data.

```
FLEET FEET
Sales Analysis: Items selling more this year      As of: 10/26/93
========  ========        ========  ========  ========  ========
Category  Description        This      Last      This      Last
   #                        Month     Month      Year      Year
--------  -----------     --------  --------  --------  --------
8659-51   Equipment        2,555.91  3,550.22  89,377.12  81,652.90
9000-34   Nike Air Pegasus   795.56    455.12   8,234.11   2,656.78
```

Good report design strives primarily to fulfill users' needs. After consulting users at length, the analyst creates a rough sketch of the proposed system's output. In many cases, analysts themselves design the format; in others, the user may already have developed a good report format—perhaps an old manual one—that the analyst need merely adapt to the new system; and in yet other cases, the analyst or user may borrow a format from another organization, vendor, or even the user's competitors.

With the rough sketch in hand, the analyst uses a printer layout sheet to draw the report to the user's specifications. As Figure 7.11 shows, each horizontal line of boxes represents one printed line, with line numbers on the left. The numbers across the top identify columns and specify where the data will eventually appear on the report.

When drawing the report format, Figure 7.12, the analyst uses special notations to describe pertinent facts about the design:

- A heavy line indicates the outline of the form
- Dashes represent perforations
- X's indicate placement of alphanumeric data on the form
- 9's indicate placement of numeric data on the form
- Preprinted information, such as corporate logos, are hand-written or drawn on the form
- Editing (comma and dollar symbol insertion for numeric data items, Z's for zero suppression, and so on) provides guidelines for the eventual appearance of the data

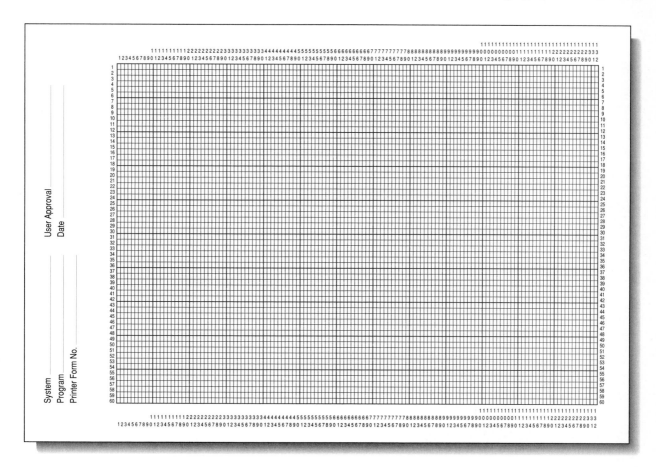

Figure 7.11 A printer layout sheet, or spacing chart, enables analysts to design reports.

Any layout design displays three types of information: preprinted, constant, and variable. Preprinted information includes the words that the organization or the form supplier will print on the paper beforehand. Constant information consists of headings that the computer will print or display on every form, while variable information will change for each printed line. For example, in Figure 7.12, we see a page heading, column headings identifying totals and sales data, and detail lines with specific sales category data.

Having completed the layout, the analyst submits it to the user for approval. If approved, the form will become part of the evolving system. If not, the analyst notes desired changes, redraws the form, and submits the revisions for approval. Several meetings and subsequent revisions may take place before analyst and user agree on a final design.

Many technologies exist for printing, varying from an adaptation of typewriters to photocopy machines. Each has different capabilities and print quality. Determining which type of printer to use requires the analyst to make trade-offs among speed, quality, cost, color, noise level, and graphics capabilities.

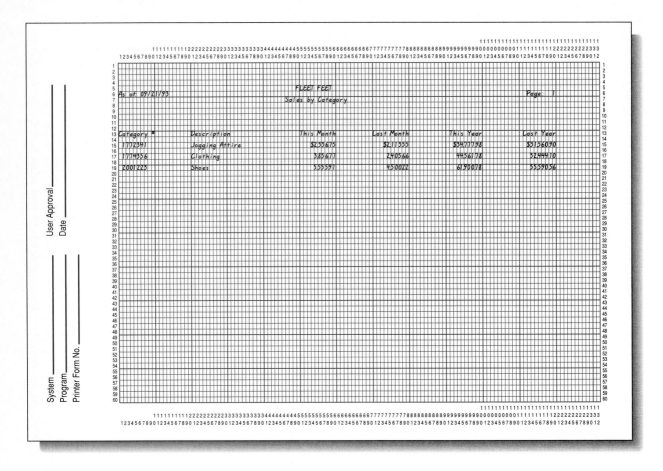

Figure 7.12 A spacing chart filled in with the format of the report.

DESIGNING REPORTS FOR SCREENS

The procedure for designing screen reports parallels that for printed reports. After consulting with the user, the analyst uses a screen spacing chart to sketch the proposed screen layout (Figure 7.13). The screen spacing chart in Figure 7.14 looks like a printer spacing chart, except that it depicts a screen with 80 characters across and 24 lines down (25 if the design includes a status line). Because a screen possesses certain highlighting capabilities (i.e., it can provide visual cues such as reverse, blinking, secured, or bright video), the analyst must define such requirements on the screen spacing chart. As you can see in Figure 7.15, highlighting notes would appear beside each field.

Personal computers and terminals can accommodate input, output, or both. Though analysts design input *after* output, they must consider some input requirements at this point. For example, a user may wish to use a personal computer to inquire about the balance of

Figure 7.13 Screen spacing or layout charts look like those for printers, but allow for 24 lines and 80 columns. They may also allow for a 25th, or status, line.

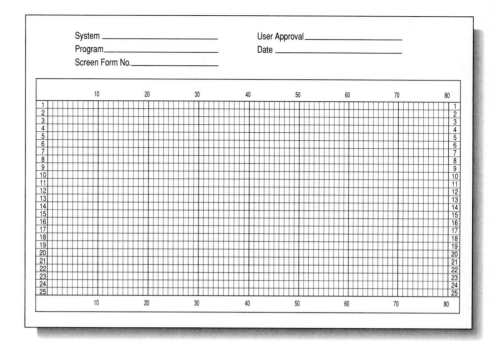

Figure 7.14 The analyst sketches the proposed design in the boxes on the form.

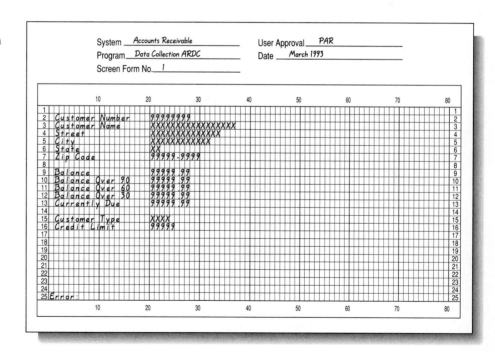

Figure 7.15 Highlighting notes enable the analyst to graphically specify screen requirements.

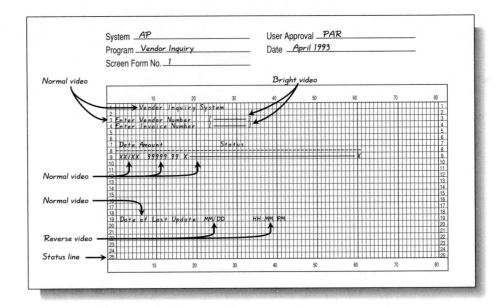

a customer's account. A screen design must prompt the user for the customer's account number before it can display the appropriate data. In other words, the device will ask for a little input before it produces any output.

If the analyst selects dual intensity, the screen design may show a form behind the data. The form may appear in normal video; the data, in bright video. The background form should resemble the form that the system will actually use.

Today's screens offer the analyst a variety of choices in screen clarity or resolution. Related to the number of dots of light or **pixels** (picture elements), screens are defined by the number of horizontal by vertical pixels. Some common resolutions are CGA (640 by 200), EGA (640 by 400), VGA (640 by 480), and super VGA or XGA (up to 1024 by 768).

An analyst should avoid placing too much data on one screen. Too often, beginning analysts attempt to squeeze everything onto a single screen, rather than resorting to two or more screens for a large report. If the report requires more than one screen, the status line notifies users which one they are seeing at a given time. Some personal computers and terminals offer sufficient memory to store both screens, thus allowing the user to easily flip from one to the other. Devices with insufficient memory require the user to store data collected on the first screen before entering data on the second.

PROTOTYPING REPORTS WITH CASE

An alternative to sketching the screen or printed report on a form has become increasingly popular. This technique, called prototyping, can speed up the output design process. With the word processor, prototyping, or CASE software, the analyst and the user can draw a simulation of the proposed screen or printed report.

GUIDELINES FOR SCREEN DESIGNS

Personal computers and terminals are everywhere. So many different kinds of people are using them that analysts must pay special attention to users' needs and capabilities when designing them. Clear, concise, and accessible screen design not only helps users enter data accurately, it also creates an atmosphere of friendliness and ease that can help users feel comfortable with and easily master the system.

Because users are not computer experts, the analyst should focus his or her attention on simplicity. Despite the complexity of the system and its programs, users' screens do not need to be ornate or confusing. The following basic rules can help screen designers achieve good results:

1. Since users are accustomed to reading printed media this way, screens should read from top to bottom and from left to right.

2. Because users feel more comfortable with words than numbers, screens should receive titles rather than their code numbers. Tops of columns, total lines, and footings should display captions.

3. Concise action verbs such as ENTER, SELECT, and CANCEL should direct operations. Extraneous pleasantries such as PLEASE, WHICH DO YOU WISH? only clutter up the screen.

4. Use a standardized vocabulary throughout (don't use "INPUT" in one place and "ENTER" in another).

5. Date display should reflect cultural conventions (e.g., dates in MM/DD/YY format for North Americans, while using YY/MM/DD format for the military or Europeans).

6. Always use the same termination key, both for terminating a screen and for returning to menus. (Don't use "Q"—quit—in one place, "Esc"—escape—in another, and "S"—stop—somewhere else.)

7. Test data to ensure correctness of entries. For example, provide for "Y" or "N" alternatives in a given situation, but do not allow any other possibilities.

8. Blinking video can attract the user's attention to crucial data.

9. Contrasting levels of intensity or colors provide visual clues for differentiating prompts from data. About 8% of all males have some color vision deficiency, commonly red-green. Never use color as the only method in which a system conveys important information.

10. If an application demands displays of more than one screen of data during an operation, users should have to stop a given display at its end and enter a command before being allowed to continue to the next screen.

11. Provide a clear pathway between multiple screens that depend on one another.

12. Have neutral observers review and test all designs.

13. Train users thoroughly on the various screens. Their training should focus on their daily interactions with the system.

14. Training manuals should describe both normal and abnormal operational procedures. Write them in the users' vocabulary, avoiding computer jargon whenever possible.

15. Keep screens free of technical jargon and computer buzzwords.

16. Help facilities can aid operators when they don't know what to do. Make them available at a single keystroke. Allow users to turn on and off the help facility when they want to. Pick a key to request the help facility.

17. Data and captions on the screen should:
    a. Align vertically
    b. Group in a logical fashion
    c. Have space between them
    d. Not require the operator to see a reference manual

18. If you do not know where to put the cursor, set it at row 24, column 80—the lower right-hand corner.

Basically, these rules reflect common sense and force analysts to keep users' needs and abilities at the top of the priority list when designing screens. No matter how sophisticated users may grow as they master their systems, they will still appreciate the simplest and easiest-to-use screens.

The design then goes through a repetitious process to review the design, fix the mistakes, review the design, and fix the mistakes until the user and the analyst are both happy with the report design. The iterative review and fix does allow the user to see their reports long before they get carved in stone during the implementation phase, and it thoroughly ties the analyst's efforts to the user's needs.

A system's reports and screens are often the most time-consuming part of the entire system development life cycle. With CASE software, prototyped screens can show menus, help messages, and video attributes such as reverse, blinking, or half-intensity—as well as the overall layout of the screen or report.

CASE software also uses a consistency checker that ensures the data shown on the report or screen are available from the system. The user thus gets the "look and feel" of the system during the design phase, not the development phase. The results are reports and screens that meet the real needs of the users; the system users want is the system users get.

To its credit, prototyping forces users to think through and fully specify their exact requirements before the analyst wastes valuable time on faulty designs.

CONTROL OF SYSTEM OUTPUTS

A **control system** is a set of procedures that ensures proper and approved system operation. You probably already know some controls: verification and validation during data collection and entry, time stamping (dating), page numbering of printed reports, and basic security provisions such as passwords, read and write access to files, and file backups. Floppy disks use several methods to make them read only and prevent accidental writing to the diskettes. A 5-1/4" floppy has a write-protect notch and uses a write-protect tab, while a 3-1/2" floppy has a hole and uses a sliding plastic protector to lock it.

A control system helps prevent computer-related fraud and can reduce errors. Once an error enters a computerized system, it takes a lot of time and money to remove it. If a department store mistakenly charges one customer's purchase to another customer's account, the store must eventually perform two operations to correct the error. First, it must remove the amount from the wrongly charged account, then it must transfer that amount to the correct account. In addition to wasting time and money, the department store risks losing the future business of both customers. A good control system can minimize such costly errors.

An Overview of Control Systems

The analyst can build controls into a system at the input, processing, or output stages. Controls are costly because they not only demand extra effort by the data entry staff during input, but they consume valuable computer time during processing or output. In the long run, however, the costs of good controls outweigh the costs of errors.

Control systems vary in complexity and sophistication, depending on an organization's needs and size. A retailer with few customers needs only a simple control system, whereas a large organization (such as Sears Roebuck and Company) with geographically dispersed offices and diverse points of data entry needs an elaborate one.

Whether simple or complex, control systems fall into two broad categories: machine and manual. A computer-based (machine) control system uses the computer to help detect errors. To test a date, for example, the computer can determine whether a month number falls between 1 and 12, a day number between 1 and 31, and a year number within an acceptable range. With a manual control system, someone visually checks for errors. For example, the data-entry operator might enter the customer's account number from a source document and have the computer display it along with the customer's name on the terminal screen, while the operator visually compares the screen display with the source document.

Quite often, combinations of machine and manual control systems make sense. Before entering data from a source document, an operator can hand-total specific fields; the operator then enters the data; the computer can total the same fields during the entry process that the operator hand-totaled; and at the end of the data entry process, the operator enters the hand-calculated total so that the computer can compare the two. If the two totals match, processing proceeds. If not, the operator must locate the error and correct it.

After an organization has collected, verified, validated, and processed its data, it wants to produce output, usually in the form of reports. The generation and distribution of reports demand additional controls. Simple output controls such as page numbers, dates, and times identify reports (refer back to Figure 7.6) and headings listing the name of the system that generated the report facilitate recognition and distribution.

Totaling and Crossfooting

Figure 7.16 Crossfooting and totaling control procedures form part of the report format and design.

Analysts may also build **totaling** and **crossfooting** controls into their report designs (Figure 7.16). Totaling adds all numbers in a specific column, while crossfooting adds all column totals, after which the computer can compare column totals to the final total. In Figure 7.16, total gross sales of $15,760.53 minus returns of $191.79 equal net sales of $15,568.74. Although some reports—such as an inventory report of merchandise on hand—do not lend themselves to totaling and crossfooting, in most cases, analysts should design this control element into their report layout.

```
FLEET FEET                       Sales by Salesperson              As of: 12/06/93
Sacramento, CA                                                         Page:   1
========================  ========================  ============  ===========  ============
Salesperson               Salesperson                      Sales      Returns     Net Sales
Number                    Name                              ($)          ($)          ($)         Crossfooting
========================  ========================  ============  ===========  ============

        83                Lardener, Scott              1,622.56        10.06      1,612.50
        95                Gloyd, Dorothy               5,001.23       100.21      4,901.02
       107                Lundberg, Carol              2,305.00         0.00      2,305.00
       235                Fulweiler, Jeff                345.90        45.80        300.10
       363                Olshefsky, Celia             6,009.45         9.00      6,000.45
       679                Chastain, Charlene              25.89         1.22         24.67
       890                Hyatt, Jan                     450.50        25.50        425.00

Totals for November                                 $15,760.53    $(191.79)    $15,568.74      Totals
```

Auditing

Another system control technique, **auditing**, traces a transaction through the system, verifying that intended processing actually occurred at correct times and locations. Accountants frequently speak of an "audit trail," which allows them to review their work for mistakes. Special software or the system itself can create such an audit trail. As a system processes a record, it can print out a copy of the old master record, the transaction record, and the new master record, noting the transaction's effect on the system (Figure 7.17).

Database management and inquiry software systems greatly assist in verifying that a system has made all desired changes. Almost every major database management system lets users inquire about stored data. By entering simple English-like commands, the user or auditor can examine database contents to ascertain accuracy (Figure 7.18). For example, a user wishing to locate a record enters "SELECT," which orders the database manager to search for the desired record. If it finds the record, the system reports so immediately. If not, the system reports that fact.

Security Checks

Since a company possesses a lot of sensitive and confidential data, its management wants to know that its systems are secure from unauthorized use or tampering. To provide assurances about security, analysts can employ a variety of effective control procedures. The obvious first step involves locking sensitive or negotiable documents, such as personnel files or blank payroll checks, in a vault somewhere other than the computer center.

Second, all documents should bear identification numbers along their top or bottom edge. Such numbers facilitate easy identification of a given form, including the system from which it came and whether it has fallen into unauthorized hands. Furthermore, printing dates on all documents helps users distinguish old from newly revised versions of forms.

Figure 7.17 Audit trails follow each transaction through the system, showing its effect on the files. Blanks indicate that the item either did not exist on the old master file or will not appear on the new one.

```
=================================================================
FLEET FEET              Customer Update Report        As of: 12/06/93
Sacramento, CA                                        Page:  1
Note: Letters on the right for transaction records mean: A-New customer,
C-Change in customer data, and D-Deletion of customer from file.
=================================================================
Change of Address
Old Master:           12332 Cindy Dunn       445 Sutter St.
New Master:           12332 Cindy Dunn       122 J St.
Transaction:          12332                   122 J St.            C

Deletion of a Customer
Old Master            47789 Tom Dobeck        5671 Hwy. 1
New Master:
Transaction:          47789                                       D

New Customer
Old Master:
New Master:           70956 Rebecca Siemers   1 Greenback Blvd.
Transaction:          70956 Rebecca Siemers   1 Greenback Blvd.   A
```

Figure 7.18 The SQL database manager allows novice users to interrogate the database. The user enters an English-like command, "SELECT," and the system displays its findings.

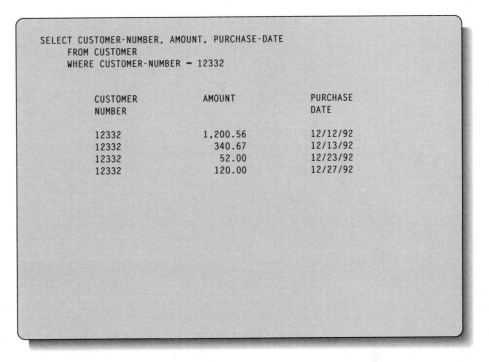

```
SELECT CUSTOMER-NUMBER, AMOUNT, PURCHASE-DATE
    FROM CUSTOMER
    WHERE CUSTOMER-NUMBER = 12332


        CUSTOMER            AMOUNT          PURCHASE
        NUMBER                              DATE

        12332             1,200.56          12/12/92
        12332               340.67          12/13/92
        12332                52.00          12/23/92
        12332               120.00          12/27/92
```

As a third means of control, the analyst should issue explicit directions about proper disposal of reports. If reports hold confidential data, such as gross pay or personal telephone numbers, they should be shredded rather than tossed in the trash or recycled. Analysts should be aware of the legal aspects of form control: Social Security numbers, driver's license numbers, gross income, telephone numbers, and so on, are privileged and private information; federal law prescribes the protection of such data.

Finally, forms suppliers should serially number valuable forms, such as blank checks or stock certificates, so users can record beginning and ending numbers before and after use. This also enables users to extract and destroy forms ruined during printer adjustment. One individual usually takes responsibility for maintaining form security, recording serial numbers, and reporting violations to an appropriate manager.

BUILDING A DATA DICTIONARY

Our look at report and screen designs has led us to draw or sketch a sample of what is wanted. In addition to the mock-up of the report, the analyst further defined the design with a data dictionary. A data dictionary helps state such requirements clearly and efficiently.

For an example of a data dictionary, refer back to the exception report in Figure 7.10 which lists sales statistics for items that sold better this year than last at Fleet Feet. By carefully considering each element of this report, we can quickly create its data dictionary:

SALES BY CATEGORY = Category Number +
Description +
Sales This Month +
Sales Last Month +
Sales This Year +
Sales Last Year +
Report Date

Category Number = General Ledger Account Number +
Store Number

Report Date = Month Number +
Day Number +
Year Number

Leveling stops when a term or field becomes self-defining. Three fields form the report date: month, day, and year numbers. Since none of these fields requires further clarification, leveling stops here.

The analyst can complete the data dictionary by adding further details about each item in the report. For example:

Field Name	Type	Length	Editing or Comments
Category Number	Numeric	6 char.	Edited XXXX-XX
Description	Alphanumeric	20 char.	Left-justified
Sales This Month	Numeric	10 digits	Edited $$$,$$9.99
Sales Last Month	Numeric	10 digits	Edited $$$,$$9.99
Sales This Year	Numeric	10 digits	Edited $$$,$$9.99
Sales Last Year	Numeric	10 digits	Edited $$$,$$9.99

Note that we have not yet dealt with the processing required to produce the final report. That processing description comes later, during the design of program modules.

In addition to defining and naming fields, the analyst must describe all other report properties:

- *Order:* sequence of data (for example, ascending by category number)
- *Totals:* items or categories to be totaled over a specific length of time (for example, sales totals for a specific month)
- *Frequency:* the schedule for generating reports (for example, monthly)

- *Volume:* the estimated number of pages or number of screens required by the report
- *Paper type* (applicable only to printed reports): the weight and bond of paper desired
- *Distribution:* who receives the report
- *Security:* the report's audience (internal, external, confidential, public) and method of disposal

The analyst can collect all such descriptive information in a single document, perhaps a standardized one developed by the computer services department. The collection becomes a part of the system documentation and is used by the analyst during database design.

SUMMARY

During output design, the second design step, the analyst establishes report formats and defines the data elements required to generate all reports. Users become deeply involved in this step, assisting the analyst in the designs and approving the results.

At this point, the analyst also selects the medium that will produce reports. Printers are most popular, followed by terminals that can display data stored on disks or tapes.

Some printers produce 10 characters per second, while others can generate thousands of lines per minute; some cost a few hundred dollars, others almost a million; some form characters with dot-matrix patterns, others with laser printing; and some offer graphics capabilities, while others do not.

CRTs, like printers, offer a wide range of characteristics, including color displays and special function keys. Data displayed on a CRT can blink and it can appear in varying intensities or even in inverse video.

The four basic types of reports are detail reports (which show all the pertinent facts about a transaction), summary reports (which show only overall totals or positions), exception reports (which list items that meet a predetermined condition), and query reports (which allow users to ask the computer for specific facts).

Regardless of the type, all reports share common elements. Headings show titles and footings provide overall results. Breaks indicate subtotals or temporary halts. Detail lines show all the pertinent facts about a single transaction.

System output controls protect reports from tampering or unauthorized use. Reports should be identified by form number, page counts, and date printed. Controls also involve the disposal of sensitive documents.

After designing all report formats, the analyst lists the data elements for each report in a data dictionary. In addition to noting the data required by each report, the analyst decides on the order of the data, how often to print the report, how many pages it will contain, what type of paper to use, and to whom the report should be distributed.

KEY TERMS

Internal reports
External reports
Turnaround document
Bond
Status line
Report heading
Page heading
Control heading
Column heading
Detail line
Control footing
Page footing

Report footing
Query report
Detail report
Summary report
Exception report
Periodic report
Pixels
Control system
Totaling
Crossfooting
Auditing

QUESTIONS FOR REVIEW AND DISCUSSION

Review Questions

1. What are the characteristics of a printer?

2. What are the five types of reports?

3. What are two basic report formats?

4. What should be included in the data dictionary for Figure 7.6?

5. What are three design criteria for CRT reports?

6. What criteria are there for specifying a printer?

7. What are three control techniques for monitoring system outputs?

8. Why do we prototype a system's reports and screen designs?

Discussion Questions

1. List the outputs from output design and what happens to each of them.

2. What does "signing off" mean to the user and analyst?

3. Explain the five characteristics of a well-designed report.

Application Questions

1. If you needed to print over 100,000 pages per month, which kind of line printer would be the best choice? The worst choice?

2. Is a check a turnaround document?

Research Questions

1. Obtain the names and addresses of four forms suppliers other than those given in this chapter. Where would you look for them?

2. From a supplier's product literature, obtain the cost difference between two-part and four-part carbon paper.

Designing the Reports for View Video's System

You'll remember that analyst Frank Pisciotta's feasibility study merely identified a few needed reports by title and function. Now Frank must discuss each report with its users, Mark and Cindy Stensaas, before he actually designs report formats and determines what data the system will need to print or display them.

In addition to the customer lists, View Video's tape check-out/check-in system should produce many other reports (Figure 7.19). Frank sketches on a printer spacing chart those reports he feels should be produced by View Video's 24 pin dot-matrix printer. For screen-based reports, he uses a screen layout form.

After Cindy and Mark okay the sketches, Frank writes a data dictionary for each report. Each report's data dictionary definition includes control factors: the report's order, any subtotals, final totals, or counts, plus the report's frequency, expected length, type of paper, distribution, and security. As usual, Frank includes dates, titles, and page numbers on all reports.

Customer lists (alphabetic or numeric) allow clerks and management to locate customer data when they know only the customer's name or phone number. These lists aid users in many ways: verifying a doubtful phone number, checking amounts owed by a particular customer, or counting how many customers owe money. Figures 7.20 and 7.21 show View Video's alphabetic customer list and its accompanying data dictionary.

Figure 7.19 Many reports will flow from View Video's system. After discussions with users, appraisal of new user needs, and examination of other video systems, the analyst may expand the number of reports beyond those proposed in the feasibility study.

Report Name	Function or Brief Description of Report
1. Alphabetic Customer List	An alphabetic listing of each customer, addresses, tapes rented, last activity date, and income generated.
2. Numeric Customer List	A numeric listing like the alphabetic one, except that it is in order by customer vendor.
3. Alphabetic Tape List	Detail list of every videotape owned, showing title, rating, income generated, times rented, and date last rented.
4. Numeric Tape List	A numeric list like the alphabetic one, except that it is in order by tape identification number.
5. Tapes by Category	An alphabetical list of tapes by category.
6. Overdue Accounts	A report showing the amounts of money each customer owes; customers not owing money are not included.
7. Daily Rentals	A chronological list of who rented what tape.
8. Customer Selection	Lists all the tapes that meet a specific set of criteria.
9. Viewing History	List of each customer and viewing habits by category and rating.

Figure 7.20 The alphabetic customer list contains vital data about every customer renting tapes.

```
Date:  10/03/93          View Video, The Movie People                Page:   1
                              CUSTOMER LIST
                             by Customer Name
==============    ==============    ========  ======  ========  ========  ==========
                                                                Date
Customer          Telephone         Customer  Tapes   Date      Last        Income
Name              Number            Balance   Rented  Joined    Rental    Generated
==============    ==============    ========  ======  ========  ========  ==========
Hakimoto, George  (415)333-8111       50.00      35  03/05/92  12/23/92      80.00
Laki, Michelle    (415)333-4182       25.00      14  11/21/93  12/19/93      28.00
Lanning, Floyd    (415)652-1140                   2  06/19/93  06/23/93       5.00
Macauley, Tammi   (415)122-5463        5.00     142  01/23/92  11/29/93     388.00
Nancebo, David    (415)122-6782                  56  01/12/93  12/12/93     150.00
       *                 *              *        *       *         *           *
       *                 *              *        *       *         *           *
       *                 *              *        *       *         *           *
                                    ========  ======                      ==========
Grand Totals                        1,230.00   9,267                       15,789.00
```

Figure 7.21 The data dictionary for View Video's alphabetic customer report.

System: Video Cassette Tape Rental
Name of Report: Alphabetic Customer List
Analyst: Frank Pisciotta
Date: 10/03/93

Element Name	Length	Data Type	Format
Customer name	32	Alphanumeric	none
Customer number	10	Alphanumeric	phone number
Customer balance	8	Numeric	money
Tapes rented	6	Numeric	9,999
Date joined	6	Numeric	date
Last rental	6	Numeric	date
Income earned	7	Numeric	money

Order of report:	Ascending by customer name.
Subtotals:	None.
Final totals:	Balance, tapes rented, income earned.
Counts:	None.
Frequency:	On demand.
Length:	One page for every 50 customers.
Type of paper:	8 1/2 by 11 white with perforations.
Distribution:	To owners and store manager.
Security:	Internal use only. Telephone numbers and financial data present on report.

The numeric list of customers displays the same data as its alphabetical equivalent, except that it is sorted into order by the customer's telephone number. This report allows the user to look up a customer if his or her telephone number is known.

The list of tapes by tape title is an important detail report for View Video, Figures 7.22 and 7.23. It gives the title of the tape, the tape number, year made, category, and rating. This report provides a ready reference to all tapes and can help customers determine if their favorite tape is owned by View Video.

Figure 7.22 The list of tapes by tape title.

```
Date: 10/03/93          View Video, The Movie People           Page:   1
                                TAPE LIST
                             by Tape Title

============================   ======   ======   ======   ======
                               Movie              Year
                               Number   Category  Made     Rating
Title                        
============================   ======   ======   ======   ======
Absence of Malice              1013     Drama     1982     PG
Agnes of God                   2144     Drama     1985     PG13
Airplane II                    3145     Comedy    1982     PG
All of Me                      3214     Comedy    1984     PG
Amadeus                        6123     Drama     1984     PG
American Dreamer               4589     Comedy    1984     PG
American Flyers                0123     Drama     1985     PG13
Annie                          3123     Musical   1982     PG
Any Which Way You Can          6456     Comedy    1980     PG
Arthur                         1234     Comedy    1981     PG
Author! Author!                4321     Comedy    1982     PG
```

Figure 7.23 The data dictionary for the list of tapes by title.

System:	Video Cassette Tape Rental
Name of Report:	Alphabetic Tape List
Analyst:	Frank Pisciotta
Date:	10/03/93

Element Name	Length	Data Type	Format
Title	28	Alphanumeric	
Movie number	4	Numeric	
Category	8	Alphanumeric	
Year made	4	Numeric	YYYY
Rating	4	Alphanumeric	

Order of report:	Ascending by title.
Subtotals:	None.
Final totals:	None.
Counts:	None.
Frequency:	On demand.
Length:	One page for every 50 titles.
Type of paper:	8 1/2 by 11 white with perforations.
Distribution:	To owners, store manager, out front.
Security:	None.

Another detail report and version of the previous report is the list of tapes by tape number. This report is useful when looking up a tape's title if the user knows the tape's number.

The new system will also produce a detailed report listing all the tapes grouped by category, Figures 7.24 and 7.25. This report enables customers to review all titles that belong to a category. For example, parents may want to know all the children's titles and this report can assist them in their search.

Figure 7.24 The titles by category shows all the tapes that belong to a single category group, regardless of rating or year made, and that start with the letter A.

```
Date: 10/03/93            View Video, The Movie People              Page:  1
                                CATEGORY LIST
Category: Comedy
=========   ====================================   ==========   ============
Movie                                              Year
Number      Title                                  Made         Rating
=========   ====================================   ==========   ============
   1234     Arthur                                 1981         PG
   3145     Airplane II                            1982         PG
   3214     All of Me                              1984         PG
   4321     Author! Author!                        1982         PG
   4589     American Dreamer                       1984         PG
   6456     Any Which Way You Can                  1980         PG

Category: Drama
=========   ====================================   ==========   ============
Movie                                              Year
Number      Title                                  Made         Rating
=========   ====================================   ==========   ============
   0123     American Flyers                        1985         PG13
   1013     Absence of Malice                      1982         PG
   2144     Agnes of God                           1985         PG13
   6123     Amadeus                                1984         PG

Category: Musical
=========   ====================================   ==========   ============
Movie                                              Year
Number      Title                                  Made         Rating
=========   ====================================   ==========   ============
   3123     Annie                                  1982         PG
```

Figure 7.25 The data dictionary for tapes by category.

System:	Video Cassette Tape Rental
Name of Report:	Tapes by Category
Analyst:	Frank Pisciotta
Date:	10/03/93

Element Name	Length	Data Type	Format
Movie number	4	Numeric	
Title	28	Alphanumeric	
Category	8	Alphanumeric	
Year made	4	Numeric	YYYY
Rating	4	Alphanumeric	

Order of report:	Ascending by category and by movie number.
Subtotals:	None.
Final totals:	None.
Counts:	None.
Frequency:	On demand.
Length:	One page for every 50 titles.
Type of paper:	8 1/2 by 11 white with perforations.
Distribution:	To owners, store manager, out front.
Security:	None.

The overdue customers report, Figures 7.26 and 7.27, lists all customers that owe money. It includes the customer's number (which corresponds to their telephone number), their name, and other valuable facts about the customer. Management may use this report to call these customers and remind them of their overdue accounts.

Another detail report, daily rentals, lists all rentals made during a particular day, Figures 7.28 and 7.29. It includes all pertinent data about the rental, including the customer, tape title, and clerk making the rental. Management may want to use the report to spot any trends in clerk check-outs or multiple copies of a tape rented.

Figure 7.26 The overdue accounts list shows those customers who owe View Video money.

```
Date:  10/05/93          View Video, The Movie People          Page:  1
                              OVERDUE ACCOUNTS
                               by Phone Number

==================    =====================    =========
                                                         Date
                                                         Last            Income
Phone                 Customer                 Customer  Rental          Generated
Number                Name                     Balance
==================    =====================    =========  =========     =========
(415)122-5463         Macauley, Tammi              5.00  11/29/93           388.00
(415)333-4182         Laki, Michelle              25.00  12/19/93            28.00
(415)333-8111         Hakimoto, George            50.00  12/23/92            80.00
         *                     *                     *         *                 *
         *                     *                     *         *                 *
         *                     *                     *         *                 *
                                                   =========                 =========
Grand Totals                                      1,230.00                 15,789.00
```

Figure 7.27 The data dictionary for View Video's overdue customer list.

System:	**Video Cassette Tape Rental**
Name of Report:	**Overdue Accounts**
Analyst:	**Frank Pisciotta**
Date:	**10/05/93**

Element Name	Length	Data Type	Format
Customer number	10	Alphanumeric	phone number
Customer name	32	Alphanumeric	none
Customer balance	8	Numeric	money
Last rental	6	Numeric	date
Income earned	7	Numeric	money

Order of report:	Ascending by customer phone number.
Subtotals:	None.
Final totals:	Balance, income earned.
Counts:	None.
Frequency:	On demand.
Length:	One page for every 50 customers.
Type of paper:	8 1/2 by 11 white with perforations.
Distribution:	To owners and store manager.
Security:	Internal use only. Telephone numbers and financial data present on report.

Figure 7.28 The daily report shows the tapes rented during a specific day.

```
Date:  10/05/93        View Video, The Movie People         Page:  1
                            DAILY RENTALS

======================  ====================  ===================  ============
Phone                   Customer              Title                Clerk
Number                  Name                                       Initials
======================  ====================  ===================  ============

(415)122-5463           Macauley, Tammi       Batman               KRE
(415)333-4182           Laki, Michelle        Dances with Wolves   REW
(415)333-8111           Hakimoto, George      Star Wars            NHY
(415)345-1243           Wurst, Kurt           Fantasia             KRE
(415)878-0666           Brohm, Kathy          Back to the Future   REW
(415)878-2312           Clark, Sue            Harold and Maude     KRE
(415)878-2331           Amato, Marjorie       E.T.                 NHY
(415)878-5678           Ruhakala, Ben         Dirty Dancing        KRE

Movies Rented on 12/20/93:  1,233
```

Figure 7.29 The data dictionary for View Video's daily rentals list.

System:	**Video Cassette Tape Rental**	
Name of Report:	**Daily Rentals**	
Analyst:	**Frank Pisciotta**	
Date:	**10/05/93**	

Element Name	Length	Data Type	Format
Customer number	10	Alphanumeric	phone number
Customer name	32	Alphanumeric	
Title	28	Alphanumeric	
Clerk's initials	3	Alphanumeric	uppercase

Order of report:	Ascending by customer phone number.
Subtotals:	None.
Final totals:	None.
Counts:	Number of movies rented.
Frequency:	Daily.
Length:	One page for every 50 rentals.
Type of paper:	8 1/2 by 11 white with perforations.
Distribution:	To owners and store manager.
Security:	Internal use only. Telephone numbers and rental data present on report.

The last report Frank designs is one that Mark and Cindy Stensaas see as a great aid to customers, Figures 7.30 and 7.31. All too often, customers have a difficult time finding a tape to rent and this query report would help to solve their problem. Mark and Cindy envision having a personal computer in the store that would be available to customers. The software would prompt customers for their choices in rating, year made, category, and movie stars. In the example shown in Figure 7.30, the customer prefers a movie with a PG rating or category, made during the 1980s, and having Goldie Hawn as the star. The computer then

Figure 7.30 A query report that allows customers to pick various viewing criteria. The computer then responds with a list of tapes meeting that criteria.

```
Date: 10/05/93           View Video, The Movie People              Page:  1
Time: 3:52 P.M.

SELECTION CRITERIA                        MOVIE STARS SELECTED
    Rating:    PG                              Hawn, Goldie
    Category:  No preference.
    Date Made: 1980s
========================  =======  ===========  ====================  =====
                                                                      Date
                                                                      Made
Title                     Rating   Category     Movie Stars
========================  =======  ===========  ====================  =====
Best Friends                PG     Comedy       Reynolds, Burt        1982
                                                Hawn, Goldie
                                                Silver, Ron
                                                Tandy, Jessica

Overboard                   PG     Comedy       Hawn, Goldie          1987
                                                Russell, Kurt

Protocol                    PG     Comedy       Hawn, Goldie          1984
                                                Sarandon, Chris
                                                De Young, Cliff

Seems Like Old Times        PG     Comedy       Hawn, Goldie          1980
                                                Chase, Chevy
                                                Grodin, Charles
                                                Guillaume, Robert

Swing Shift                 PG     Drama        Hawn, Goldie          1984
                                                Russell, Kurt
                                                Harris, Ed
                                                Ward, Fred
                                                Lahti, Christine
```

Figure 7.31 Data dictionary for View Video's customer inquiry.

System:	**Video Cassette Tape Rental**
Name of Report:	**Customer Selection**
Analyst:	**Frank Pisciotta**
Date:	**10/05/93**

Element Name	Length	Data Type	Format
Title	28	Alphanumeric	
Rating	4	Alphanumeric	
Category	8	Alphanumeric	
Movie star name	14	Alphanumeric	
Year made	4	Numeric	YYYY

Order of report:	Ascending by title.
Subtotals:	None.
Final totals:	None.
Counts:	None.
Frequency:	On customer demand.
Length:	One page for every 10 titles.
Type of paper:	8 1/2 by 11 white with perforations.
Distribution:	To customer.
Security:	None.

searches the database, looking for all tapes that meet this criteria, and finds six titles. When it finds a match, it reports all the facts about the title—including other stars that shared the bill with Goldie Hawn.

Frank's last report shows viewing history and patterns by customer, Figures 7.32 and 7.33. This report tells the Stensaases the types of videos each customer rents and can aid them when buying and marketing future videotapes.

Frank has now defined all the reports that the users need, including the formats and data elements required to generate them. The output design phase for View Video concludes and the next step in the systems process, input design, begins.

Working With View Video

In most cases, users will discover other reports that they want after using the system and seeing the value of the data and how it can help them operate their business better. This is an event that all software developers need to anticipate instead of resisting. It is especially the case when software is sold as a package and the owners have their own ideas of what is missing from the system.

Case Study Exercises

Design the following new reports. Show their formats and write the data dictionaries.

1. A viewing history by category.
2. A list showing each clerk and the number of tapes they checked out.
3. All customers that have rented more than 40 titles and joined within the last year.
4. An analysis of each title, showing data about the title and the number of times the title was rented.
5. An analysis report showing the number of tapes checked out each day in the prior month.

Figure 7.32 The viewing history shows viewer preferences.

```
Date: 10/05/93          View Video, The Movie People              Page:  1
                            VIEWING HISTORY
                            by Phone Number

==================  ==================  ============================  =============
                                                Year Made                 Rating
Phone               Customer
Number              Name                1990 1980 1970 1960 Before    NR  G  PG  R
==================  ==================  ============================  =============
(415)122-5463       Macauley, Tammi      10    3    8    2     0       5  7  12  19
(415)122-6782       Nancebo, David        5    2    1    0     0       0  8   0   0
(415)333-4182       Laki, Michelle       12    0    0    0     0       0  0   0  12
(415)333-8111       Hakimoto, George      0    0    0    0    15      15  0   0   0
(415)652-1140       Lanning, Floyd        4    0    0   15    14      14 10   0   5
```

Figure 7.33 The data dictionary for View Video's viewing history list.

System: Video Cassette Tape Rental
Name of Report: Viewing History
Analyst: Frank Pisciotta
Date: 10/05/93

Element Name	Length	Data Type	Format
Customer number	10	Alphanumeric	phone number
Customer name	32	Alphanumeric	none
Year made	4	Numeric	YYYY
Rating	4	Alphanumeric	

Order of report:	Ascending by customer phone number.
Subtotals:	None.
Final totals:	None.
Counts:	Breakout rentals and count by year made and rating.
Frequency:	On demand.
Length:	One page for every 50 customers.
Type of paper:	8 1/2 by 11 white with perforations.
Distribution:	To owners and store manager.
Security:	Internal use only. Telephone numbers and viewing data present on report.

CHAPTER 8

Input Design and Data Collection Screens

GOALS

After reading this chapter, you should be able to do the following:

- List devices commonly used for data entry
- Distinguish between verification and validation
- Identify at least five types of validation
- Define a check digit and describe appropriate applications
- Discuss the advantages and disadvantages of terminals
- Design a data collection screen

PREVIEW

Once our reports are designed—along with their accompanying data dictionary definitions—the analyst can undertake input design (Figure 8.1). During this stage of the design process, the analyst, consulting the data flow diagrams developed during analysis, determines the source and method for collecting data and getting them into the system. Depending on the nature of the application, an organization may enter data into its computer via on-line terminals, personal computers, or other devices, such as bar code readers, cash registers, optical scanners, or hand-held wands.

Earlier, we discussed the importance of data output: reports enable users to make decisions, control activities, and monitor individual and organizational performance. The quality of these reports depends, of course, on the accuracy, timeliness, and completeness of the input used to generate them. Faulty input results in faulty reports (Garbage In, Garbage Out: GIGO). For example, incorrect input, such as an erroneous account number, may generate a wrong customer billing. Likewise, late input may delay customer billing and hurt the company's positive cash flow; incomplete input, such as a missing account number, can cause costly repetition of a task. As with output design, the analyst must pay strict attention to user needs and abilities, for they—the users, not the analyst—are ultimately responsible for the organization's data collection.

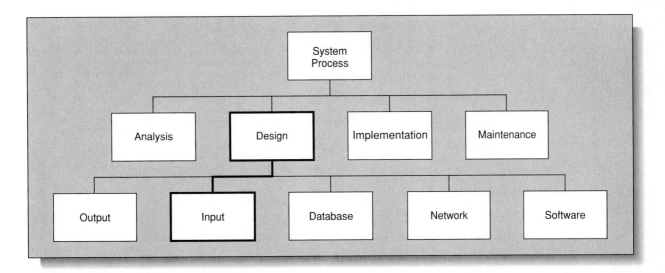

Figure 8.1 Input design determines the format and validation criteria for data entering the system. It is the second task of the systems design phase of the systems process.

Analysts must struggle to keep up-to-date on all the new hardware devices that vendors offer. The latest piece of fancy hardware may offer unprecedented features, but the cost may outweigh its worth. For example, while a new optical page reader that costs over $250,000 and can read hundreds of pages a minute may work fine for the Internal Revenue Service, it would not make sense for a small department store. As the saying goes, "You want the right tool for the right job."

METHODS OF DATA ENTRY

Data enters an organization's information system from both internal and external sources. Internal data, from employees and management, might include time cards or vouchers (Figure 8.2). Some internal data, such as a customer's previous year's total purchases, exists within the system itself—stored, perhaps, on the computer's tape or disk. External data comes from people and agencies outside the organization and might include credit applications, invoices, orders, payments, or packing slips (Figure 8.3).

Most organizations collect their original data on a form, such as a time card or a purchase order, called a **source document**. However, the organization must eventually transcribe all or part of the data from the source document into a form accessible to its computers.

After pinpointing the data sources, an analyst determines the method of data entry: manual or direct. **Manual entry** requires that someone use a terminal or personal computer to key the data. Each of these devices, as their names imply, stores data on different media: terminals on a central disk and personal computers on floppy or hard disks.

Direct entry machines read data directly into the computer. The cash registers at many supermarkets, for example, can automatically read the Universal Product Code (UPC) stamped on merchandise as a clerk passes each item over a glass-covered scanner surface.

Figure 8.2 Internal data comes from company personnel. If the data arrives in hand-written form, as it often does, someone must reformat it before entering it into the computer.

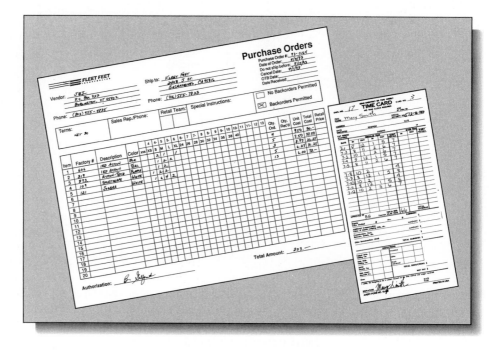

Figure 8.3 Data from customers include credit application forms, payments, and charges. Such data usually arrive in hand-written form and must be organized into a format acceptable to the computer.

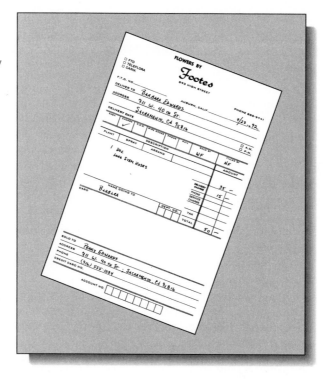

In some stores, the computer actually announces each price. Although this type of equipment now costs more than manual entry devices, it reads data more quickly and reliably, and it involves less processing time. If these features reduce errors by a factor of two or three with increased speed, an organization might recover the extra cost in less than a year. The UPC readers also offer by-product data about purchasing patterns by product, time of day, and day of week, and may even associate the purchase back to the customer.

CONTROLLING DATA ENTRY

Potential for error exists with any type of data-entry device. Stock clerks may write the wrong price on a sales tag and key entry personnel may strike wrong keys, posting inaccurate data into an account or inadvertently posting correct data to a wrong account. Similarly, direct entry machines occasionally malfunction, and they cannot read smeared UPC codes.

Consequently, any system must include a way to detect data-entry errors, because once an error enters a system, it has a way of compounding problems. In an inventory system, for example, an item incorrectly shown as being in stock will result in the company's not reordering on time, thus losing a future sale and customer goodwill. It will also take additional, unanticipated amounts of time and money to find and correct the error. During input design, analysts should devote a lot of time to designing methods that detect and eliminate errors at every step of the input process.

Input devices offer various types of error-detection techniques. Some require rekeying of data; others use the power of the computer itself. We call these two different error-detection controls verification and validation.

Verification

With **verification**, a person checks data by rekeying it. Verification of data is usually done by a different person from the one making the original entry. The rationale for using two different people is that the same person tends to make similar errors and will therefore typically make the same mistake at the same place on the second pass through the data.

Such re-examination or rekeying can greatly reduce errors. For instance, if an operator of an input device normally incurs a one-half of a percent error rate during initial data entry, the chances of making an identical mistake a second time declines to only a 25 in a million chance ($0.005 \times 0.005 = 0.000025$, or 25 millionths). Clearly, double entry can almost eliminate typographical mistakes.

Validation

Because **validation** employs the intelligence of the terminal or computer itself, it offers much more sophisticated error-detection techniques. For example, a computer can reject data that does not fall within a predetermined range. If a program sets the lower and upper boundaries for a certain drug's dosage at 10 and 30 mg (milligrams), the computer can automatically reject 7.5 or 42 mg, flashing an error message on the terminal screen.

A system can employ any of nine common validation tests:

1. *Class test:* Determines whether the data is numeric or alphabetic (for example, a drug number should contain only digits 0 to 9).

2. *Sign test:* Determines whether the algebraic value of the numeric data is less than, greater than, or equal to zero (for example, a drug number is larger than zero).

3. *Reasonableness test:* Rejects data that should never occur (for example, the program would reject more than six grams of streptomycin because that dosage would kill the patient).

4. *Sequence test:* Ensures that data appears in a certain order (for example, the key field of the current record must have a value greater than that of the previous record).

5. *Range test:* Determines that the data falls between an upper and lower limit (for example, a month number must fall between one and twelve).

6. *Presence test:* Ensures that the data are present to allow further processing (for example, the operator must enter a drug number before entering anything else).

7. *Code test:* The data value must match one of several acceptable predetermined values (the year number must be 91, 92, or 93).

8. *Combination test:* A certain group of data elements must all be present at the same time and must be correct (drug number, dosage, date, and patient number must all appear).

9. *Date test:* Date requires a two-digit month number, a two-digit day number, and a two-digit year number (such as 12/02/93).

Let's assume that Albany Veterinary Clinic enters four data elements for each prescribed drug.

Data Element	Validation Requirements
Drug Number	Numeric (0001000–0099999)
Dosage	Alphanumeric (mg)
Date Administered	MM/DD/YY where MM: 1–12; DD: 1–28, 30, or 31; YY: 91, 92, 93
Patient Number	AA-DDDD where AA: Alphabetic (initials of last and first name) and DDDD: Numeric (a sequential number)

If an operator enters

2345, 10mg, 20/13/92, AS-356S

validation should identify two errors: an incorrect month number (20 exceeds the acceptable 12) and an improper right half of the patient number (no letter should appear there). The terminal might notify the operator of the error by flashing that data element, as well as printing an error message on the terminal screen. The operator could then correct the entries to read:

2345, 10mg, 02/13/92, AS-3568

Batch or Control Totals

In addition to verification and validation, an analyst may specify **batch** or **control totals**. This control technique has the computer compare the accumulation of all the individual amounts in the group with a predetermined total. If the two totals match, the assumption is that the data are error free. If the two totals differ, an error has occurred.

Suppose Albany Veterinary Clinic decides to employ batch totals to help detect errors on payments received from customers. If so, a clerk would hand-total all checks received each day, then enter all the customers' numbers and amounts received at a CRT. After posting the last check, the clerk would enter the hand-calculated total to see whether it agrees with the machine-calculated one. If so, the system could go on to use that data to update the accounts. If not, the computer could print a list of all the entries, which the clerk could visually check for discrepancies. Having located the error, the clerk can then return to the terminal, re-enter the correct data, and compare the totals again.

Check Digits, Transpositions, and Slides

Check digits, transpositions, and slides can help detect data-entry errors. With these three detection schemes, the computer performs mathematical calculations on the data. Easy to program, these schemes can quickly find certain kinds of errors.

The **check digit** scheme is a formula-based error-detection routine. After an operator enters a data element, the computer analyzes it to ascertain that it follows a specific pattern or adheres to a specific formula.

If you look on the back cover of this book, you will see an ISBN (International Standard Book Number). Each copyrighted or public domain literary work has a unique ISBN, which is tested by using the check digit scheme. Suppose the ISBN for a particular textbook is 0-534-00615-9. The following formula can test for the accuracy of that number:

1. Multiply successive digits (starting with the second digit) by 9, 8, 7, 6, 5, 4, 3, and 2, respectively.
2. Add the products.
3. Divide by 11.
4. The result should be a number with no remainder.

Applying this formula to our hypothetical text, we get

$$9 \times 5 + 8 \times 3 + 7 \times 4 + 6 \times 0 + 5 \times 0 + 4 \times 6 + 3 \times 1 + 2 \times 5 + 1 \times 9 =$$
$$45 + 24 + 28 + 0 + 0 + 24 + 3 + 10 + 9 =$$
$$143$$

After we divide by eleven ($143 \div 11 = 13$), we get a result with no remainder. If we had gotten a remainder, the number is wrong because all ISBN numbers follow this formula without exception. Many check digit schemes exist, but this one is the one most frequently employed by software writers.

Check digits are everywhere; for example, in bank account numbers, airplane tickets, and credit card numbers. In all cases, the check digit is appended to the real number (in our example, the ISBN) to make the formula work properly. A major criticism of the U.S. Social Security number system is that it does not incorporate a check digit. Errors on SSNs are quite common and expensive to correct.

Check digit testing works beautifully for account numbers or other data that identify a customer, vendor, employee, or subscriber with a unique assigned identifier. Check digits are not useful with amounts of money, customer names, or units of measure, which may vary and therefore do not lend themselves to a predetermined formula.

Another formula-based method for detecting errors involves transposed numbers. A **transposition** occurs when the order of the numbers entered is changed. For example, suppose one data-entry operator enters 357, while another operator enters 537 (a transposition). When such a transposition error occurs, the difference between the two numbers will always be evenly divisible by 9. In this case, the difference between the two numbers (537 − 357 = 180) can indeed be evenly divided by 9 (180 ÷ 9 = 20). Such a calculation does identify a transposition error, but it does not reveal the correct data entry. Someone would have to go back and check for correctness. This method of detecting transposition errors only works on numeric fields, and it must take place during verification of two data entries.

A third formula-based error detection method is the **slide**. Slides detect an incorrectly placed decimal point. Suppose a data-entry operator initially enters 71.25 for the cost of veterinary services, then during verification discovers that the number 71.25 has been entered as the payment. If we can evenly divide the difference between the two numbers by 9, with no remainder, a slide has probably occurred. As with transposition errors, slide error detection only applies to numeric fields and can take place only during verification.

Visual Verification and Computer-Assisted Validation

The most complex checking procedure combines visual verification and computer-assisted validation. Let's assume that Albany Veterinarian Clinic uses a terminal. Analyst Marcie Crowley could program the computer to display a patient's name immediately after the clerk enters his or her account number, thereby allowing a quick visual check for accuracy. If the terminal display matches the name on the source document, the data entry continues. If not, the clerk starts over.

We could repeat the process for the drug number: enter the drug number, have the computer look it up and display its name on the terminal screen, then visually verify it for accuracy. If the number is correct, the computer processes the entire transaction; if not, the clerk must re-enter the data.

DATA-ENTRY HARDWARE

Before designing source documents, the analyst must select the most appropriate data-entry hardware. If an organization must process a large volume of data, an optical scanner may work because it provides direct input to the computer, allows for continuous file updating, and processes data at low cost (despite its high purchase price).

WORKING WITH PEOPLE
ERGONOMICS AND DESIGN

Gaylen Lewis, responsible for entering data into the payroll system at Haley's Department Store, worried that he would not be able to input all the necessary data for this week's paychecks. The Christmas holidays, overtime, vacations, and a number of new part-time workers made an already difficult job almost impossible. As if that were not enough to ruin Gaylen's day, the computer's response time had grown intolerably slow because so many other systems (such as inventory) require year-end processing.

Gaylen thought about how much more efficiently the payroll system could run if he could organize the company's time cards in a way that would make data entry easier. Eventually, he shared his frustrations with his supervisor, Joyce Isheim. To Gaylen's delight, Joyce acted on the payroll problem immediately. Meeting with the payroll supervisor, Greg Goodwin, Joyce and Gaylen found him equally frustrated with the payroll system. "After all," he said, "from our employees' point-of-view, paychecks are Haley's most important output. You can't cash an end-of-the-year inventory report." Within a few days, Gaylen found himself working directly with Greg on the problem.

First, they examined Haley's time cards. Gaylen pointed out how hard it was to locate desired data: the first data entry, employee number, appeared at the bottom of the time card; the second entry, date, appeared at the top. In addition, total hours came from the bottom of the card, sick time from the middle, and employee vacation again from the bottom. Greg agreed that searching the card for data wasted a lot of time. Why not design it to read from top to bottom and left to right in the order of data entry? "I can't believe we've lived with this for so many years," Greg mused. Gaylen nodded. "It wasn't such a problem when we only had 50 employees."

Since Greg had been involved with the design of the original system, he reasonably estimated that it would take about seven hours to repair the program, test it, and retrain Gaylen to use it. Wanting to avoid late paychecks at all costs, Greg helped Gaylen finish the current period's data entry, then stayed after work to repair the system.

The following morning, Greg and Gaylen reviewed the new design. Gaylen expressed his enthusiasm. "The new time cards should be a breeze to enter. But," he went on, "I've got another big headache."

He explained that Haley's growth had also outpaced his department's physical space. "Just last week, a data-entry operator ruined a whole stack of time cards when they fell into the sink beside the coffee pot." Although Greg felt sympathetic about the need for improved working space, that problem wouldn't be so easy to solve. They experimented by moving the terminal into an adjoining typist's room with a large desk, ample lighting, and a comfortable chair. Though a more suitable environment, the noise level from the typing distracted Gaylen. Finally, Greg suggested that Gaylen and Joyce redesign the original workspace, using ideas from companies specializing in ergonomics, the science of creating work spaces with people in mind.

Gaylen and Joyce found new chairs from a defunct division of the company that allowed for variable heights and widths and also tilted to fit each worker comfortably. The adjustable desks they ordered contained ample compartments for time cards, and they accommodated the terminals nicely. And special lighting could be positioned so that it didn't reflect off a terminal screen.

After the new equipment was installed, Gaylen looked forward to coming into work each day. Joyce summed it up for everyone. "In this computer age, it's so easy for management to forget that *people* still do the work."

In another situation, an application may require a turnaround document—perhaps a two-part printed page designed to tear into pieces, one a check and the other a receipt. Here the analyst may find a terminal more appropriate because of low volume or because handwritten data are re-keyed into the system.

To illustrate the trade-offs an analyst faces when choosing input hardware, let's consider the case of Albany Veterinary Clinic, which has been using its computer for billing and payroll. It now wants to improve the efficiency of its recordkeeping and ensure accurate reporting of the handling of controlled drugs to state agencies. To add the drug inventory application to the system, Albany hires a consulting analyst, Marcie Crowley, who, while reviewing Albany's data-entry needs, isolates four data elements: drug number, dosage, date administered, and patient number.

During discussions with the clinic's eight veterinarians, Marcie learns that Albany averages 240 prescriptions daily. Based on this information alone, she might recommend buying a personal computer, terminal, or optical scanner for this new application. However, as we will soon see, a number of other factors will affect her decision.

Terminals

Because terminals allow so many users to access an organization's computerized systems directly, they are popular as data-entry devices. Equally important, in a terminal-oriented environment, users can locate and correct errors *before* the computer actually processes the data.

Terminals fall into two categories: dumb and intelligent. Dumb terminals cost the least and offer the fewest options or features. Most have little or no memory, limited cursor capabilities, and some lack even the ten-key numeric pads found on adding machines. You will most likely find these terminals where users demand minimum cost and possess the least computer sophistication.

An intelligent terminal, on the other hand, contains a built-in microprocessor and memory, a keyboard that includes a numeric ten-key pad, and a variety of useful cursor capabilities. This type of terminal, though it costs more, offers the analyst a wide variety of design options. You will most likely find them in high-volume data-entry situations in which users possess some degree of computer sophistication.

For Albany's needs, the analyst might consider two quite different terminals: the Hewlett Packard 740 (a dumb terminal) and a DEC VT-340 (an intelligent terminal made by Digital Equipment Corporation), either of which could do this job nicely.

Like most intelligent terminals, the DEC VT-340 has several built-in features: function keys that the clinic can pre-program to store dates and the codes identifying frequently prescribed drugs, automatic cursor positioning to speed data entry, and flashing and reverse video to highlight input errors. This smart terminal, which costs around $1,000, could provide quick and efficient data entry.

However, Albany could save substantial money by purchasing an HP 740, which costs under $400. Though the HP 740 is a dumb terminal, the analyst could design the input screen to compensate for the terminal's lack of memory and error-detection techniques. We'll examine this more thoroughly when we discuss designing input screens. For now, just remember that a dumb terminal such as the HP 740 lacks sophistication and creates a more time-consuming data-entry process, which may, in the long run, offset its low initial cost.

Albany might decide to buy the VT-340 for two reasons: Marcie Crowley's prediction that Albany would soon recoup the cost difference between the HP 740 and the VT-340 through faster and more reliable data entry, and because Albany already owns three older model DEC VT-220s, which means that users would not have to learn how to operate a completely new type of terminal.

Personal Computers

Personal computers also make good data-entry devices. Like a terminal, a personal computer can provide a direct link to a mini- or mainframe computer. The special software needed to connect the two typically permits the PC to perform three functions: emulation, uploading of data, and downloading of data.

When operating in **emulation** mode, the personal computer behaves like a terminal (if the user wishes it to do so), but it still retains its ability to behave as a stand-alone personal computer. Walker, Richard, and Quinn (Seattle, Washington) sells emulation software that makes the IBM PC and PS/2 personal computers—as well as various clones of those machines—emulate a wide range of real terminals from IBM, DEC, Data General, and Hewlett Packard. Other software and hardware vendors have accomplished the same effect for the Apple Macintosh. The cost of the extra software and a possible emulation card can run $200 or more.

Uploading of data allows a person using a personal computer to transmit his or her data to the host computer. The user may have written a report or developed a spreadsheet on a personal computer and want to transfer the data to a minicomputer or mainframe. A user entering data prior to connection to the mini- or mainframe computer can save a lot of money by not requiring central system upgrades over a user who consumes a lot of direct time with the large machine.

Downloading is the converse of uploading. Here, the mini- or mainframe computer sends a stream of data to the personal computer for storage on a disk. Later, the user of the personal computer can view the data at his or her convenience and transfer it to a word processor, spreadsheet, or database manager. This last approach has won many converts among users who want to work on their data at their leisure or at home.

Emulation, uploading, and downloading features will propel personal computers past terminals in many applications, especially as the price differential between the two devices continues to fall.

Optical Character Readers

All key entry input devices suffer a common drawback: a human operator must key in all data. Direct entry devices such as optical readers (Figure 8.4) can read data directly to the computer, disk, or tape from paper media by scanning a code printed on the product (such as a can of peas), on a tag attached to the product (such as a sport coat), or on an ordinary sheet of paper (such as IRS 1040 forms). Optical readers cost from a few hundred dollars (for hand-held wands) to hundreds of thousands of dollars (for machines capable of reading three thousand sheets of hand-written data an hour).

Since optical readers can connect directly to computers, they effectively bypass the keyboarding step of terminal devices. Error-detection techniques parallel those for smart terminals. The user programs the reader to know where data are found on each source document and then formulates specific error-checking rules, such as testing for numeric value and checking for specific values (A through F) or the presence of a value (an amount of money).

Figure 8.4 This optical reader can "read" data from the bar code on the side of a product. The code identifies the product, but does not contain the price of the product.

If the machine detects an error, it rejects that item or page and does not permit it to travel in the same direction as an acceptable source document. One pitfall of optical readers occurs with hand-written data or with data entered in the wrong field. In these instances, the optical reader will reject even accurate data.

At Albany Clinic, Marcie Crowley tests a hand-held wand reader as a potential data-entry device for the new drug system. Capable of reading the bar codes printed on prescription bottle labels and relatively inexpensive (less than $200), the wand could be linked to a personal computer or to an intelligent terminal. That would work fine for drugs administered on the premises, but in a field situation, someone without a wand would later have to enter data manually. And, of course, in neither case could the wand automatically read a patient's name, the dosage, or date.

To avoid having to buy two data-entry systems (one requiring the keying of all data entry on their VT-340, the other using a hand-held wand with their VT-340), Marcie sticks with her earlier decision to key all data on a VT-340 terminal.

OTHER DESIGN CONSIDERATIONS

In addition to the basic considerations about trade-offs among hardware and software solutions to data-entry problems, the contemporary analyst can choose from an array of state-of-the-art features that enhance computerization. Among them are color, sound, various types of menus, and function keys.

GUIDELINES FOR DESIGNING DATA-ENTRY SCREENS

Users communicate with application systems running on their computers almost exclusively through the terminal or CRT. Considering the tens of millions of people now interacting with computers on a daily basis, it behooves the analyst to design display screens as clearly, concisely, and user-friendly as possible. The following guidelines can help analysts design such a screen:

1. Keep screens simple, having them serve a single logical user activity.
2. Make screens easy to read. They should:
   a. Look alike, using a common format
   b. Have a title, date, time, and page number
   c. Similar data should appear in the same place on all screens
   d. Read top down, left to right
   e. Have captions that are centered over the area that they represent
3. At the beginning of an application, position the cursor to the first field in which users will enter data. When errors occur, the cursor should move to the first field in error.
4. Color is a positive design element. Use the same color for similar elements, such as red for errors, green for money, and so on. More than four colors on a screen can overpower the user.
5. Use lowercase letters for prompts. They are less formidable and make the screen "user easy."
6. Design the data format to follow cultural conventions. For example, a screen can display properly positioned hyphens for Social Security or phone numbers.
7. Insert asterisks (or some other character) to tell the operator exactly what is required in all mandatory fields.
8. Use a consistent method for signaling error messages.
9. Place the error message in the same place on every screen.
10. To ensure that users see all errors, the screen should return intact with the erroneous data. Highlight each field with blinking text or audio signals.
11. Use vertical and/or horizontal lines to divide the screen into quadrants.
12. Liberally employ menu screens, which casual users greatly appreciate. Build an accelerated screen for advanced users.
13. Include passwords to tighten security.
14. Write directions in a clear, logical step-by-step fashion according to the way that users normally do their jobs.
15. Keep directions free of technical jargon and computer buzzwords.
16. Try to ensure that data asked for once are not requested again, except for verification.
17. Adopt a convention for Yes/No responses. For example, use "Y" for yes; anything else is automatically no.
18. When telling a user to press a specific key, put a special character around it. For example, [Y] means to enter a Y.
19. Minimize the operator's need to use the clear key and vertical or horizontal tab keys; this speeds data entry and reduces the likelihood of errors.
20. Test your screen with some real data for an hour. Then have a real data-entry operator test your screen. Watch how he or she operates the screen. These two tests may lead to some screen redesign.
21. It is OK not to use the whole screen.

Prompt, friendly, clear, and understandable input screens boost user productivity, while slow and confusing ones increase frustration over computerization of any task.

Color

Used judiciously so that the screen is not cluttered with arcade-like effects, color can communicate important messages to users. The effective use of color involves three variables: hue, saturation, and brightness. Hue refers to the actual tint, such as red, green, blue, yellow, or others. Saturation refers to the purity of the color. For example, pure red is a highly saturated color, while pink is a diluted form of red. Brightness determines how light or dark the color appears (i.e., how much white the color contains). Most people find their eyes drawn first to bright, highly saturated colors.

In most designs, a given color conveys a certain meaning. For example, green usually signals an acceptable condition, orange and red indicate mistakes, and yellow can caution or warn the user. Regardless of the meanings an analyst assigns to colors, once defined, those meanings should remain consistent throughout the design.

Most manufacturers of color screens claim that they can display 256 or more colors. However, specialists report that the human mind can follow only 4 to 7 colors at one time. Consequently, when designing screens with color, the analyst should remember the slogan "Less is more."

Though not widespread, some color vision deficiency does occur in eight percent of all European and American males and one-half percent of females. When it does afflict a user, the analyst can compensate with other types of signals, such as sound, printed messages, shades of gray, cross hatching, stripes, blinking text, or allowing the user to customize the colors.

Poor background lighting washes out color and makes screens harder to see. Therefore, the analyst should become familiar with the environmental conditions under which the application will operate before picking color as a design element.

Color printers do not achieve the quality of CRTs. Over the next few years, this problem will probably disappear, but until then, color printers will benefit few applications.

Color can greatly enhance windows or boxes. When designing these, however, it usually makes sense to keep the outlines of the windows or boxes black.

Marcie Crowley, Albany's analyst, considers colors and how to employ them in collection screens for the clinic. She chooses to use cyan (light blue) as the background color for screens with black letters. Error messages will appear on the bottom (25th line) in red.

Sound

Sound synthesis mechanisms can also provide a useful design element. Sound works especially well for applications in which the user should be aware of an operation's status or should be alerted that something unexpected may happen. For example, a sound signal can prompt an inattentive user that the operation has concluded or whether it demands the user's attention.

As with color, overuse of sound can cause more annoyance than benefit. Loud or harsh sounds can intimidate users, and tunes or jingles can sound silly or trite. It usually makes sense to test the sound on the user for a long period of time before fully committing to it. In any case, each sound signal should be sufficiently unique for users to distinguish between different sounds, and are adjustable from loud to nonexistent.

In many instances, the best effect is achieved by coupling subtle sound with a visual cue on the screen. This works especially well when the user has turned the sound off, has moved out of hearing range, or suffers from a hearing deficiency.

Marcie decides that sound does not seem to have merit in the clinical setting. The background noises of dogs, cats, birds, and other animals make the clinic a noisy place and any other noises emitted by a terminal or personal computer would probably confuse the operator.

Menus

Most contemporary software employs **menus** from which users can select the functions they wish to perform (Figure 8.5). Once the user types the number to the left of a desired menu selection, presses a unique letter for that entry, and uses a mouse or the arrow keys, the menu disappears and the application begins. In some cases, a selection from a main menu will cause a submenu to appear on the screen. The submenu, which depends hierarchically on the selection from the main menu, should only contain items related to the main menu. Additional layers of sub-submenus and sub-sub-submenus, though possible, add significantly more layers of complexity to an application. Remember, users want to be able to grasp what's happening without undue thought or delay.

Pull-down and pop-up menus offer two viable alternatives to the traditional vertical list menus. **Pull-down menus** present the main choices as words across the top line or the bottom line of the screen. Most commercial personal computer software uses pull down menus.

The **pop-up menu** does not appear in the menu bar (across the top of the screen), but on the screen when needed. Pop-ups work best for setting a value or choosing from a list of related items such as baud rates, which disk drive to activate, and which font style to use for a laser printer or desktop publishing application.

Pop-ups behave like the other two types of menus, allowing the user to move around within the menu box and make a selection. A box usually surrounds a pop-up, which also contains the current or default value and lists of the values of related items.

With the wide variety, number, and skill levels of the staff in the Albany Veterinary Clinic, Marcie Crowley decides that a main menu bar with pull-down options is the best for all users—especially since many of them are familiar with other popular products such as Windows, Lotus 1-2-3, and other personal computer programs.

Figure 8.5 Manzanita Software's BusinessWorks PC's Main Menu uses layered or hierarchical menus with numbers to the left of each selection. Entering a "3" on the Main Menu brings up BusinessWorks PC's Accounts Receivable System.

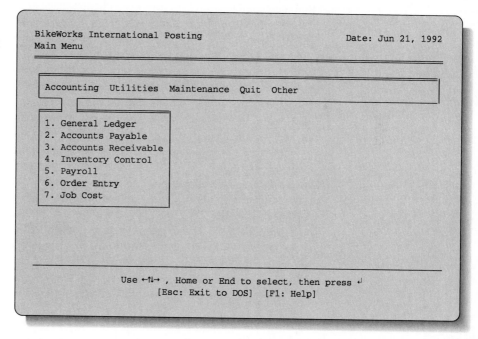

Keyboards and Function Keys

Function keys, special keys on a keyboard that can perform a special operation, have become commonplace features on most personal computers and terminals (Figure 8.6). They fall into one of two categories: dedicated and nondedicated. Dedicated keys do not allow users to define their functions, but perform a built-in operation, whereas nondedicated ones do permit users to define their use via an application program.

Typical dedicated function keys include:

Del	Removes the character to the left of the cursor
Home	Moves the cursor to the upper-left corner of a screen or to the top selection in a menu box
End	The opposite of Home; moves the cursor to the lower-left corner of the screen or to the last selection in a menu box
PgUp	Scrolls the screen or menu box forward one logical grouping
PgDn	Scrolls the screen or menu box backward one logical grouping
Esc	The escape key usually takes the user backward one level

The nondedicated function keys, usually labeled F1 through F10 (although some personal computers and terminals go as far as F15), are adjustable by the application. This means that as the user performs the application, the purposes of the function keys change, with their new purposes appearing at the bottom or top of the screen. Lotus 1-2-3 nicely exemplifies the changing nature of nondedicated function keys. Pressing the slash key while operating within the worksheet brings up the command menu. Once the user selects an option, such as File, the function keys change purpose again. Now they will perform Retrieve, Save, Combine, Xtract, Erase, List, Import, and Directory. Alternatively, the alphabetic keys R, S, C, X, E, L, I, and D will perform these same actions.

Most data collection applications require function keys dedicated to "Help" and "Quit." A "Help" key assists the user by providing explanations during the operation of the application. One software vendor uses the question mark key (?) for this purpose. Whenever a user needs help, he or she can merely press the question mark key, which causes a window to pop up with a brief explanation. Further help can be obtained by typing two question marks (??), which calls up a more detailed explanation in the pop-up window.

Data collection design also invariably includes a "Quit" key, which allows users to stop a process in progress. A West Coast software vendor sells new key caps just for this purpose with its software packages. After installing the software, the user can remove the existing left square bracket ([) key cap and replace it with a "DONE" key cap. To exit the application, the user can then simply press the "DONE" key.

Besides function keys, certain keystrokes need to perform a common function across all applications. Microsoft's Presentation Manager and Windows, Hewlett Packard's NewWave, and Apple Macintosh software provide for such consistency among applications. Because of this integrated keystroke definition across applications, users can move quickly from one application to another without having to relearn how the keyboard functions.

Figure 8.6 Common keystrokes and their results as advocated by Microsoft ® Windows ™ SDK, version 3.1.

8.6a These navigation keys move the cursor at various increments around the screen.

Key	Unmodified Key Moves Cursor To...	CTRL+Key Moves Cursor To...
HOME	Beginning of line. (Leftmost position in current line.)	Beginning of data. (Top left position in current field or document.)
END	End of line. (Rightmost position occupied by data in current line.)	End of data. (Bottom right position occupied by data in current field or document.)
PAGE UP	Screen up. (Previous screen same horizontal position.)	Screen left/beginning. (Top of window; or, moves left one screen.)
PAGE DOWN	Screen down. (Next screen, same horizontal position.)	Screen right/end. (Bottom of window; or, moves right one screen.)
LEFT ARROW	Left one unit.	Left one (larger) unit.
RIGHT ARROW	Right one unit.	Right one (larger) unit.
UP ARROW	Up one unit/line.	Up one (larger) unit.
DOWN ARROW	Down one unit/line.	Down one (larger) unit.
TAB	Dialogs: Next field; may move left to right or top to bottom at designer's discretion; after last field, wraps to first. (SHIFT+ TAB moves in the reverse order.)	(Not defined.)

8.6b Toggle keys control how the keyboard functions. They turn a particular function on or off each time it is pressed and released.

Key	Function
INS	Toggles between Insert mode (new text characters push old one to right) and Overtype mode (new text characters overwrite old ones).
CAPS LOCK	Pressing alphabetic key yields uppercase.
NUM LOCK	Numeric keypad keys yield numbers rather than direction.
SCROLL LOCK	Navigation keys scroll data without moving cursor; existing selections are preserved.
F8	Toggles Extend mode. In this mode, selection behaves as if the SHIFT key is locked down for all direction keys and mouse actions.
SHIFT+F8	Toggles Add mode, which allows disjoint selection through the keyboard. In Add mode, navigation keys move the focus without affecting existing selections, and pressing the SPACEBAR toggles the selection state of an item.

(continued)

End-users can thus focus on completing tasks, rather than spending a lot of time learning individual applications. Consistency in key usage is paramount for speeding data entry and learning how to use the software.

Marcie agrees that software should have a standard function key definition. She reviews the common ones found on popular personal computer software systems and decides to follow their lead.

Figure 8.6 *(continued)*
8.6c These two editing keys cause deletions of characters on the screen.

Key	Recommended Function
DEL	If there is a selection: Deletes entire selection.
	If there is an insertion point and no selection: Deletes character to *right* of insertion point.
BACKSPACE	If there is a selection: Deletes entire selection.
	If there is an insertion point and no selection: Deletes character to *left* of insertion point.

8.6d Modifier keys switch the mode of a key and are used in combination with another function key.

Key	Typical Functions
SHIFT	With alphanumeric keys, yields uppercase or the character inscribed on the top half of the key.
	With mouse click or navigation keys, extends or shrinks the contiguous selection range.
	With function keys, alters meaning of action (for example, F1 brings up the Help application window, pressing SHIFT+F1 enters Help mode.)
CTRL	With mouse click, selects or deselects an item without affecting previous selections.
	With alphabetic keys, yields shortcuts.
	With navigation keys, typically moves cursor by a larger unit than the unmodified key.
ALT	With alphabetic key, navigates to the menu or control marked with that key as a mnemonic.

Icons

Made famous by the Apple Macintosh computer but developed by Xerox, an **icon** is a small "picture" representation of an object (Figure 8.7). Icons can represent garbage cans, printers, an in-tray, programs, a document, tools, and so on. End-users move these different icons to perform operations on the data. For example, to erase a file on a Macintosh, the user drags the icon representing the file on top of the garbage can icon and then selects "Empty the Trash" from a pull-down menu. Microsoft's Windows now puts icons on the PC/PS-2 and compatibles.

Mouse

A **mouse** is a palm-sized pointing and drawing device that uses hand motion as a means of moving the cursor on the monitor. Once the pointing device, usually an arrow, is placed at the proper selection, the user presses a button on the mouse to make the selection.

As with keyboards and function keys, systems employ mice and their buttons differently. Standardization of button clicks is a must if users are to have a common structure among applications (Figure 8.8). Mice come in one-, two-, and three-button varieties, with

Figure 8.7 Icons can take the form of wastebaskets, hard disks, floppy disks, printers, word processing documents, filing cabinets, or whatever is appropriate.

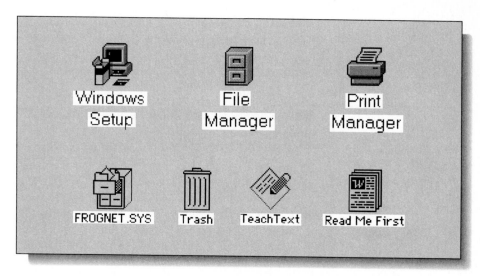

Figure 8.8 Common mouse operations advocated by the Microsoft ® Windows ™ SDK, version 3.1.

Operation	Definition	Common Usage (Using Button 1)
Point	Move pointer ("hot spot") to desired screen location.	Navigates; prepares for selection or for operation of control.
Press	Press and hold the button.	Identifies object to be selected.
Click	Press and release button without moving mouse.	Selects insertion point or item; operates control; activates inactive window or control.
Double-click	Press and release button twice within specified interval, without moving mouse.	Shortcut for common operations; for example, activates icon, opens file, selects word.
Drag	Press button and hold while moving mouse.	Identifies range of objects; moves or resizes items.
Double-drag	Press button twice and hold while moving mouse.	Identifies selection by larger unit (for example, words).

the three-button mouse having the most flexibility. In most software systems, there are keystroke equivalents to mouse point-and-click selections.

Since most of Albany Veterinary Clinic's use of the system will require keying data values for pet owners, animal problems, drug prescriptions, and billing issues, Marcie Crowley decides against a mouse or icon interface. Before making the decision herself, she observes how the staff works and asks them for ideas and opinions. Their reasoning is based on the amount of time that users spend talking on the telephone and holding pets, drugs, and related paperwork. They usually won't have a free hand or the space to roll a mouse around their hectic workplace.

INTERFACE DESIGN

An **interface** is a shared boundary between two objects. We see interfaces every day in our kitchens, automobiles, and home computers. The interface for a microwave oven consists of the set of buttons, dials, and displays that we use to operate the microwave. Likewise, our Video Cassette Recorders (VCRs) have an interface that allows us to program them to record or play back various programs. Some interfaces are better than others and the VCR is one of the worst; millions of people have given up on learning to program them. Why? Because the interface is too complex for many users to operate. As a consequence, you can now "program" a newer VCR simply by entering a number; recording that selected program is automatic.

The data-entry screen defines the interface between the user and the application software system. A good interface invites use, while a poor one can create havoc. The difficulty in designing an interface is the set of trade-offs the designer must make: color or monochrome, text or icon, mouse or keyboard—the list is endless.

Besides the technical issues, there are many "soft" reasons why interface design is so hard. Chief among these are the political ones. Users have seen interfaces in other software systems, know them, and want their software to look, act, and feel like another system. Microsoft, Borland, WordPerfect, Lotus, Apple, IBM, Hewlett Packard, and others have spent millions of dollars on their interfaces. A conscientious analyst can learn many lessons from these interface designs.

If an analyst is working in an environment where the interface is in place, few choices exist. The new software must use the same design as the existing ones. All function keys, visual effects, and color usage should match, with only a few rare deviations.

However, in a new application, the analyst can define the interface. Regardless of technical or "soft" issues, all interfaces must obey some common themes. At the top of the list is formation of a team of users, programmers, and the analyst. The team should start with a review of the application and the functions that the various screens should perform, and then look at other interfaces. From here, the group can spot the elements they like and don't like, including color, flow between functions, sizes of objects, use of capital versus lowercase characters, and so on.

Part of any interface design is keeping the user informed as to what is taking place behind the scenes. In a few situations, the software will require extra time, such as when searching a file to find a requested record or when processing a complex calculation. In these cases, the interface should tell the user—perhaps with a pop-up window—that it is "working" or that it is reading the 100th record of 500 that it needs to search. The Macintosh has a rotating object (such as a clock) so the user knows at a glance that the computer is performing an action.

12 STEPS TO BETTER MENUS

Menu. The very word beckons, promising a tantalizing array of options. Whatever guise, menus remain the primary means of presenting program choices and the principal way users make things happen in software. Just as there are fine restaurants and fly-in-the-soup beaneries, menu systems vary from four-star to the kind that send you groping for the antacid.

Virtually all PC applications are erected on a network of menus that serve as the primary interface. Menus simplify learning and using the software, and make the most of screen real estate, always a precious commodity. The 12 rules of thumb that follow will help you create effective menuing systems.

1. Make the most of the screen: Menus fall into two major categories: vertical list and horizontal list.

2. Maintain program context: Whatever its form, a menu system should provide the clearest form of context, telling the user where they are in the application, how they got there, and how to retrace to the exit.

3. Build a reliable escape hatch: Mix horizontal topline menus with vertical list menus to achieve efficient use of screen space and easy readability of vertical lists.

4. Limit menu length: Restrict menus to between 4 and 10 options. Psychologists suggest that comprehension and recall are easiest with 7 or fewer items.

5. Nest menus judiciously: Where menuing systems are concerned, broad and shallow is better than narrow and deep. Just as a single hallway that opens onto nine rooms represents a more efficient design than a hallway that branches to three rooms, each of which leads to another and then another still, so menu layouts should be simple and direct.

6. Arrange selections by importance: Place more frequently used options at the top or left side of the menu and less frequently used ones to the bottom or right.

7. Keep names short, sweet, and active: Effective naming of menu options is like good writing; concise and to the point. Two or three words per name is plenty. Use uppercase sparingly; initial capital and lowercase letters are easier to read than all uppercase.

8. Flag submenu options: Users should need only a quick look to distinguish a selection that leads to a submenu from one that triggers a program action. Arrowheads or trailing ellipses are ways to mark submenu selections.

9. Encourage one keystroke selection: Choosing a selection is simply a matter of one keystroke, either a single letter or digit. First letter selections are rapidly committed to memory.

10. Distinguish options visually: Use color, underlining, or capitalization of the first (and sometimes the second) letter to set apart options. Menus for all applications require work and rework until every shred of ambiguity is eradicated.

11. Make movement simple and consistent: Pick a way to navigate in the menus, do not change it, and use the same way in all menus.

12. Implement lateral movement: Allow movement from one menu at the same level to another at the same level. This is especially valuable for users who can't quite find the command they want.

In designing menus, don't expect to get it right the first time. Look for users whose expertise and sophistication match those of your intended audience. Talk to the users and solicit their views. Set aside time for testing and retesting. Plan for change.

With a list of interface definitions ready, the analyst can begin the task of constructing a prototype by using a package such as ObjectVision from Borland. Once the prototype is completed, users can "test-drive" the interface and modify it as they see weak spots. Once all members of the team are in agreement, the analyst is ready to design real screens with the interface in mind.

DESIGNING DATA-ENTRY SCREENS

Personal computers and terminals can place data at users' fingertips, allowing them to call up specific data and make timely decisions based on that data. A personal computer or terminal-oriented system also shifts responsibility for data entry and accuracy from data-entry operators, who may not comprehend the meaning of the data they enter, to users, who are familiar with the data. This heightens user awareness of the importance of accurate and timely data entry, as well as of the entire data processing function.

To make users comfortable with this shift in responsibility, analysts have had to work hard at building user-friendly systems that noncomputer professionals can easily understand and use. Most users want a data input process that explains itself, prompts appropriate responses, facilitates easy error detection, and pulls them smoothly through crucial steps of the application.

When designing data entry for a personal computer or terminal-oriented system, the analyst "paints" the screen with identifying words and formats. Most screens permit 24 horizontal lines, 80 vertical rows, and 2 intensities or brightnesses. The lower intensity might display directions for data entry, while the brighter one can show actual data being input. If input data survive verification and validation, the personal computer or terminal might lower the intensity of that data to reflect that they have passed required tests. Data items that fail might remain brighter or even flash an error signal.

Data collection screens display headings that define their purposes. Often, the analyst will use the heading for the system's title, date, time of day, and even the last time the program was used (Figure 8.9). By varying light intensities and employing flashing error messages, the analyst can avoid specifying a separate place on the screen for such information. In this case, whenever the user positions the cursor to an incorrect field, an error message will automatically occur.

For example, Albany Clinic's new system allows operators to enter only a drug number, then the computer automatically retrieves and fills in the drug's name for visual verification. After operators assure themselves of the proper drug, they can proceed to enter the dosage, date, and other pertinent data.

The screen design in Figure 8.9 also illustrates the types of prompts that help data-entry operators enter data correctly. For example, "MMDDYY" requires double-digit month (01 to 12), day (01 to 31), and year (00 to 99) entries. "Y or N" facilitates error detection and "QUIT to stop" permits the user to exit from the system.

Some applications require multiple screens. For example, when the Albany Clinic treats a new cat or dog, data about that patient and its owner are so extensive that it will take two screens to capture and verify all of it. The first screen may gather data about the owner of the animal (owner name, address, and phone number), while the second may capture data about the animal itself (animal name, type, weight, and allergies). Sometimes, analysts feel compelled to gather all data on one screen for efficiency's sake. However, it usually makes more sense to follow the rule that each screen should have a single purpose and restrict itself to logically related data.

Figure 8.9 A data-entry screen for Albany Veterinary Clinic. Note the headings, prompts, and spaces for the operator responses.

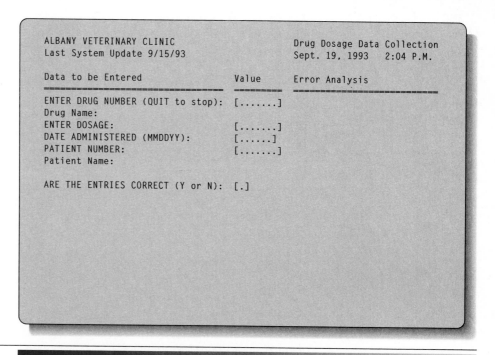

```
ALBANY VETERINARY CLINIC                    Drug Dosage Data Collection
Last System Update 9/15/93                  Sept. 19, 1993   2:04 P.M.

Data to be Entered              Value      Error Analysis
═══════════════════════════     ═══════    ═══════════════════════════
ENTER DRUG NUMBER (QUIT to stop):  [.......]
Drug Name:
ENTER DOSAGE:                      [.......]
DATE ADMINISTERED (MMDDYY):        [......]
PATIENT NUMBER:                    [.......]
Patient Name:

ARE THE ENTRIES CORRECT (Y or N):  [.]
```

USING CASE TO DESIGN SCREENS

Our look at CASE software has revealed many capabilities. We started with diagrams of system data flows, refining them to increasing levels of detail. We constructed entity-relationship diagrams to model the files that a system requires and the inter-relationships those files would have with each other. We developed reports that the system needed with relative ease, detailing what we wanted to see, plus headings, totals, and the order of the presented information. While not technically a part of the CASE software, database managers assisted us in defining the physical files from the logical ones we built in the E-R diagrams.

Through all these processes, the CASE software has kept track of the system's facts in its data dictionary—really the heart of the CASE software. While many more stages in the systems process remain, CASE can help the analyst with data collection and screen designs. The dictionary has a complete record of all data stores (files) and attributes (fields) and can report these to you in a variety of reports (Figure 8.10). The dictionary list tells everything the CASE software knows about the Donor-SSN, including in which data stores (files) it is found, which reports use the data element, and which screens collect or display the Donor-SSN.

Part of the input design phase for a CASE software system is completing the attribute data dictionary for every attribute that the system uses. Ideally, the definition of attributes takes place during database design, but sometimes attributes are skipped since a complete definition can wait until input design.

Figure 8.10 A typical report of a data element from a CASE software system. While every CASE product is different, most capture specific facts about data elements or attributes.

ATTRIBUTE DATA DICTIONARY

System :	Blood Bank Donations
Attribute name :	Donor-SSN
Analyst :	Neal Hundt
Date :	12/03/93

Long name:	Donor Social Security Number
Type:	Numeric
Length:	9
Edit mask:	999-99-9999
Heading:	Donor SSN
Prompt:	Donor SSN
Help message:	This is the unique donor identification number
Default:	Required entry
Error message:	Enter a numeric value
Sign:	None
Initial value:	None
Right justified:	Yes
Left justified:	No
Units:	None
Blank if zero:	No
Maximum value:	999999999
Minimum value:	000000001
Data stores:	Donor, Donation
Reports:	Alphabetic donor list, Numeric donor list, Available donations
Screens:	Donors, Donations

Once all attribute definitions are finished, the CASE software is prepared to assist the analyst in the actual screen design. Usually an interactive process, the CASE software asks the analyst for the following information:

- Title of the screen
- Use of current date or time on the screen
- Colors that the analyst wishes to use
- Definitions of function keys specific for the screen
- Boxes or lines that the analyst wants drawn on the screen
- Names of the attributes to collect
- Attribute locations on the screen (rows and columns)
- Data store in which to hold the collected data
- Location and color of error messages
- Pop-up windows for help messages
- Pop-up windows for selecting codes or displaying other context-sensitive data

While answering all the questions, the CASE software builds a model or prototype of the screen for the analyst to see and test-drive. The analyst can thus modify the design interactively, moving headings, prompts, and error message locations around so that the best screen design is achieved.

The CASE software "knows" the prompts to use, validation checks to make, and text for error messages to display. It also selects editing formats for the attribute, length of the attribute, and data type of the attribute. All of these facts are stored in the data dictionary repository and the CASE system calls upon the repository when building the screen.

With these facts at its disposal, including the newly completed screen design, why can't the CASE software "write" the program for the screen itself? In fact, some CASE systems can! The software produces a source program, perhaps in COBOL or C, that programmers can modify, compile, and run.

DOCUMENTING SCREEN LAYOUTS

The popularity of screens, personal computers, and terminals has stimulated software vendors to develop special systems that aid analysts and programmers in building screens. For instance, one vendor permits the screen designer to develop the desired screen, define all validation requirements for each field (including all state abbreviation, ZIP, and telephone area codes), compile the design (just like a programmer compiles a COBOL, C, Ada, or Pascal program), and incorporate the screen as an operational program. The resulting screen, when used, can directly update a database or build a batch disk file for later processing. If errors are detected during the batch update, the faulty records remain in the file, with appropriate error messages. The next time a user performs data entry, all error records are available for correction.

Screen designs depict the exact appearance of the screen:

- Name and title of the screen
- List of all the entries required by the operator
- Positions on the screen where data are entered
- Location of all error messages
- Rules governing how the screen collection will end

After designing the screens, the analyst submits them—along with screen documentation—for management approval (Figure 8.11). Screen documentation parallels screen design and stipulates:

- Name and title of the screen
- Date and name of the analyst designing the screen
- Purpose of the screen
- List of data elements, their type, number of characters, and validation requirements
- Processing requirements that indicate the computer's response to data entries

Figure 8.11 Documentation of the screen design for Albany Veterinary Clinic's drug administration. Screen designs and documentation require user and management approval.

```
System :  Drug Inventory                        Date: 9/24/93
Analyst:  Marcie Crowley                        Comments:
Purpose:  Collection of Drug Administration
================================================================

Data Element            Type              Length    Validation Requirements

Drug Number             Alphanumeric        7       Presence, not blank
Dosage                  Alphanumeric        7       Presence, not blank
Date                    Numeric             6       MMDDYY where:
                                                         MM: 1-12
                                                         DD: 1-28, 30, or 31
                                                         YY: 91, 92, 93
                                                    Presence
Patient Number          Alphanumeric        7       Presence, not blank
Correct Entry           Alphanumeric        1       Presence, "Y", "y",
                                                         "N", "n"
_____

Processing Requirements:
   1. All fields will be validated.
   2. When drug number is entered, the computer will display its name.
   3. When the patient number is entered, the computer will display the patient's name.
   4. When "Quit" is entered as a drug number, the system will terminate.
   5. Validation will be performed when operator enters a "Y." An "N" entry will cause this transaction
      to be discarded.
```

Management and users can now review the analyst's work to decide whether or not the design meets their needs. If not, the analyst will need to revise the screen design and documentation and resubmit it for approval.

Screen layouts and accompanying documentation join report layouts and documentation, as well as the database schema, as part of the design phase of the systems process.

DESIGNING FLEET FEET'S DATA COLLECTION SCREENS

Peggy Adams-Russell, Fleet Feet's analyst, starts her input design with her level 1 data flow diagram for the accounts payable system (Figure 8.12). She concentrates on inward and outward data flows, to/from process circles, and to/from sources or sinks of data. These arrows all require some type of data-entry screen, either to collect specific facts or a menu to select various options.

Her examination reveals that the data flows generally fall into one of four broad categories: invoices, checks, vendors, and reports. Peggy decides that these categories will make excellent options for her accounts payable main menu bar (Figure 8.13). Her menu follows her other designs, having the title and date on top, the system name on the second line, and the body of the data collection screen and directions on function key usage at the bottom of the screen.

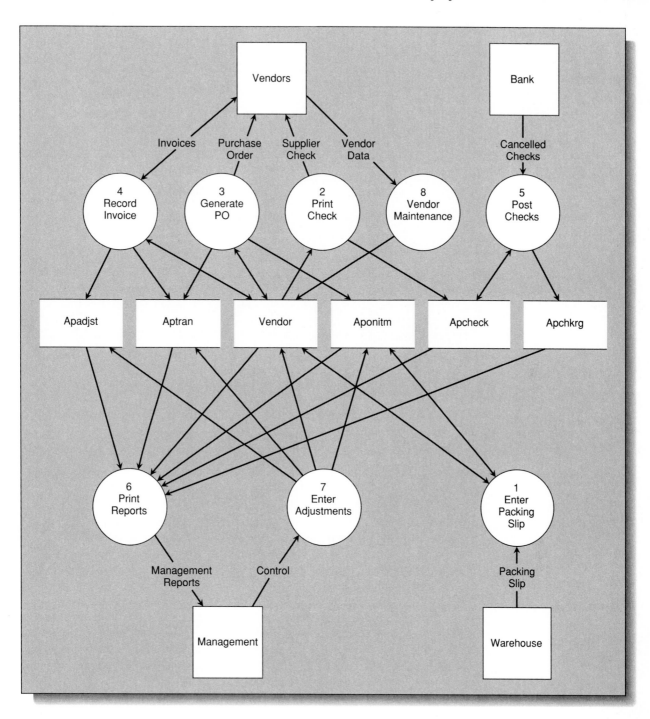

Figure 8.12 Fleet Feet's
level 1 data flow diagram.

Figure 8.13 Fleet Feet's accounts payable menu bar, showing the suboptions for the invoices selection.

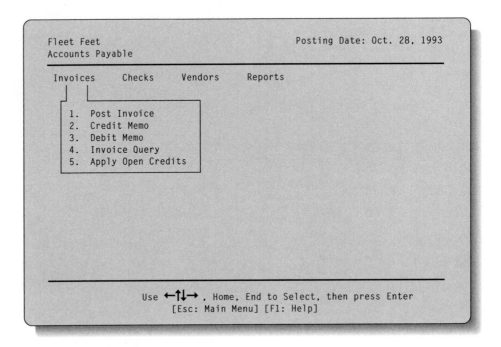

```
Fleet Feet                                    Posting Date: Oct. 28, 1993
Accounts Payable
_____

  Invoices     Checks     Vendors     Reports
 ┌─┐    ┌─┐
 │ └────┘ │
 │  1.  Post Invoice      │
 │  2.  Credit Memo       │
 │  3.  Debit Memo        │
 │  4.  Invoice Query     │
 │  5.  Apply Open Credits│
 └────────────────────────┘




_____
          Use ←↑↓→ , Home, End to Select, then press Enter
                   [Esc: Main Menu] [F1: Help]
```

For each of the four categories, Peggy decides on the subtasks within that option:

1. *Invoices:* Record an invoice, credit memo, debit memo, query by invoice number, or apply open credits.
2. *Checks:* Print supplier checks, reconcile cancelled checks, void checks, query by check number, and enter hand-written checks.
3. *Vendors:* Maintain vendors (add, change, or delete), query by vendor number, print purchase order, enter packing slip, and vendor invoice query.
4. *Reports:* Detail, aging, open invoice, cash flow, cash requirements, vendor list, check register, payment selection, vendor history, and a purchases journal.

She lists the subtasks in each category by priority, with the most often used task first and the least used option last.

Peggy starts with the first and most often used data collection screen, "Post invoice" (Figure 8.14). She collects a few sample invoices and notices that the data collected needs to focus on the vendor, the invoice, and the accounting or general ledger distribution. Peggy makes her data collection screen mimic a real invoice and divides the screen into three parts. She uses left and right brackets, shown as [and], to outline the position of the cursor during the actual data-entry process.

With consistency in mind, Peggy makes this screen look like all the others, displaying Fleet Feet's name and the date on the first line. She places the option within accounts

Figure 8.14 Posting an invoice is the most common task within the accounts payable system.

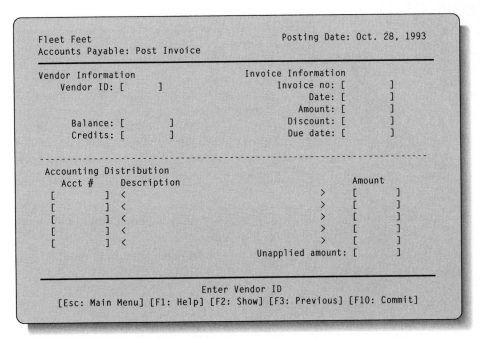

```
 Fleet Feet                              Posting Date: Oct. 28, 1993
 Accounts Payable: Post Invoice

 Vendor Information                   Invoice Information
     Vendor ID: [        ]              Invoice no: [        ]
                                               Date: [        ]
                                             Amount: [        ]
       Balance: [        ]             Discount: [        ]
       Credits: [        ]             Due date: [        ]


 - - - - - - - - - - - - - - - - - - - - - - - - - - - - - - - - - - - - - -

 Accounting Distribution
     Acct #      Description                              Amount
     [        ] <                              >    [        ]
     [        ] <                              >    [        ]
     [        ] <                              >    [        ]
     [        ] <                              >    [        ]
     [        ] <                              >    [        ]
                                    Unapplied amount: [        ]


 _____

                            Enter Vendor ID
        [Esc: Main Menu] [F1: Help] [F2: Show] [F3: Previous] [F10: Commit]
```

payable on the second line. The body of the screen is next. She uses line 23 to display a hint message ("Enter Vendor ID"). Function key usage is shown on the last line.

The flow of the entries will start with the vendor information first. The user enters the vendor identification number and the computer provides the vendor's name and address underneath, along with other factual data. The user then enters the invoice information, number, amount, date, and so on.

The last part is distributing the amount of the invoice to several general ledger accounts. After the user enters the account number, the computer will provide the account's name. The user enters that amount, which is debited to that account. To credit an account, Peggy decides to have the user enter a negative amount. After each accounting entry, the unapplied amount at the bottom of the screen is adjusted. When it reaches zero, the transaction is balanced. The user then presses F10 to commit (save) the invoice.

After committing the invoice, the data part of the screen is erased and the system returns the cursor to the vendor identification field. When users are finished recording invoices, they press the Esc (escape) key and are returned to the main menu.

In actual practice, Peggy would have to design many screens and write the associated documentation for each. The final product of her design would consist of perhaps 50 to 60 pages of screen designs and documentation.

Peggy will want to review each design with Fleet Feet staff before progressing any further. She does not wait until she is completely done with her design, but shows each to specific users so that she can adjust them during the design activities. Peggy knows that user input is valuable and will result in a better system—one that the user will like and use to solve their accounts payable problem.

SUMMARY

Input design is the fourth stage in the design phase of the systems process. During this stage, the analyst consults users to determine the most efficient way to collect necessary data and control their accuracy.

Today's trend toward on-line applications lends itself to a personal computer or terminal-oriented environment. Analysts must pay special attention to user-friendly systems that permit quick error detection. Verification, validation, control totals, check digits, transposition, slides, visual verification, and computer-aided validation help catch, control, and reduce errors.

Catching errors on input is far less costly than trying to repair the errors once the data are stored on disk or tape. Our reports (screen or printed) are only as good as the data we collect; if the data we collect are in error, our reports will have errors.

Validation is the most widely used means of error detection. With validation, the computer itself checks the data for accuracy with a series of tests: class, sign, reasonableness, sequence, range, presence, code, combination, or date. If the entry fails the appropriate test, the user is notified and must repair the data.

The hardware to collect data can vary from dumb or smart terminals to personal computers and optical readers. Personal computers, terminals, and on-line systems are rapidly replacing older batch processing devices such as punched cards for data collection. Optical readers work best when an organization faces massive amounts of data.

Screen design can also include such design elements as color, sound, various types of menus, function keys, and icons. Each of these design elements can make an application easier for users to learn, operate, and understand.

The result of input design is a description of how the organization will collect its data. This description takes two forms: screen or key-entry input form design, and details about error-detection methods (validation or verification). This description joins report designs as part of the system's documentation.

KEY TERMS

Source document
Manual entry
Direct entry
Verification
Validation
Batch total
Control total
Check digit
Transposition
Slide

Emulation
Uploading
Downloading
Menus
Pull-down menu
Pop-up menu
Function keys
Icon
Mouse
Interface

QUESTIONS FOR REVIEW AND DISCUSSION

Review Questions

1. What are some devices commonly used for data entry?
2. What is the difference between verification and validation?
3. What are some of the types of validations?
4. Why do we use a check digit?
5. What would a data collection screen designed to gather a person's Social Security number, name, address, date of birth, and telephone number look like?
6. What are the advantages of terminals over personal computers?

Discussion Questions

1. List the inputs to the input design phase.
2. List the outputs from the input design phase.
3. How much data can be displayed on a terminal screen? How many rows and columns are there?
4. Make a list of three types of menus.
5. List five design elements for screens.

Application Questions

1. Make a chart of the number of days in each month that would be necessary to properly validate a date entered in MMDDYY form. Don't forget to consider leap years in your chart.
2. Does an optical reader require verification of the data it reads?
3. What validation methods could be performed on a field that holds an amount of money?

Research Questions

1. Is it necessary to design a screen for View Video to print an alphabetic list of customers? Why?
2. Explain when it is necessary to have a second level screen menu.
3. Does the database schema have any effect on input design?

CASE STUDY
View Video's Input Design

Frank Pisciotta's input design for View Video's tape check-out/check-in system is finally finished. He is ready to turn now to input design.

When Mark and Cindy Stensaas started their business, they decided to buy (rather than write from scratch) the fundamental accounting software they needed. After a long search, the two owners agreed on BusinessWorks PC, written by Manzanita Software Systems of Roseville, California. They picked modules for payroll, general ledger, accounts receivable, and accounts payable. This software has operated satisfactorily for other organizations for over five years, so Frank chooses to pattern the new videotape system after Manzanita's system.

Before designing any screens for data collection purposes, Frank decides on a few preliminary design criteria. Every screen will display the title of the organization on the first line, along with the date. The second line will identify which screen the user is operating. On line 22, Frank decides to draw a horizontal line to separate the data-entry area from the explanation area. Under the horizontal line, Frank will tell the user the purpose of the specific entry, such as "Enter the customer's telephone number." Line 24 is dedicated to identifying function keys and their purposes at this point in the program. Frank decides that he will always use function key F1 for help, F2 for a check-out, F3 for a return, and the escape (Esc) key to back the user out of their current function. Since check-outs and returns (F2 and F3) are the most often used functions of the software, Frank decides to make them "hot keys." At any time, a user can press F2 or F3 and go directly into a check-out or return without having to first back out of other menus and then navigate into the check-out or return menus. Whatever function the user was performing prior to pressing F2 or F3 is abandoned.

A second design criteria is color. Frank decides after watching BusinessWorks PC, Microsoft Windows, Lotus 1-2-3, and a few other software systems that he will follow their lead on color. His screens will have a black background with cyan (light blue) letters for the organization's title, date, horizontal lines, and explanations. Menus will have a blue background with white line "boxes." Words appearing inside the boxes will be in white. Frank decides not to let users change the colors; it only adds complexity to the software and with many clerks using the system, he wants them to see a consistent color interface.

The first component of his design is a main menu (Figure 8.15), which will lead users easily through the application, asking what they want to do and steering them to the desired function. Franks sees six main activities: check-out, returns (check-in), customers, movies, reports, and utilities.

The main menu is accompanied by a detailed data dictionary description of its use (Figure 8.16). This documentation, coupled with the screen designs themselves, will serve two purposes: it will tell programmers what to do, and it will form the basis of an instructional packet that Frank can give to users who will operate the new system.

Since check-out and returns are the most frequently used tasks, Frank decides to design them next. When a user activates "Check-out" from the main menu, the clerk must enter appropriate data for the check-out (Figure 8.17). Accompanying the screen design is a data dictionary definition of the screen that explains how it operates and the edits that the system will have to make on the data entered (Figure 8.18). One of View Video's rules is that a customer can check out more than one tape at a time. Frank provides for this business rule via a function key. The computer will return the customer's name and videotape's title for

Figure 8.15 Frank Pisciotta's design for the main menu screen and directions for its use. The videotape system consists of many functions. The user selects desired functions by choosing the appropriate entry from the main selector menu.

```
View Video                                    Dec. 11, 1993
Main Menu                                     Initials: KRW

Check-out   Return   cuStomers   Movies   rEports   Utilities


            _____
              Use ←↑↓→, Home, End to Select, then press Enter
         [Esc: Exits to DOS]  [F1: Help]  [F2: Check-out]  [F3: Return]
```

Figure 8.16 Each screen design is accompanied by a data dictionary describing it. These descriptions have two functions: they establish objectives for the programmer and they show users how to enter data.

System: Video Cassette Tape Rental
Name of Screen: Main Menu
Analyst: Frank Pisciotta
Date: 11/03/93

To access the main menu at the DOS prompt, type VTMM and press the "Enter" or "Return" key on the keyboard.

Before any selection is allowed, the user must enter his or her three-letter initials. The system must check to make sure they are legal and if not, place an error message on the screen in a pop-up window in red. For example, KRW are the initials for Kathleen Rae Williams.

Select from the entries shown using arrow keys, Home, End, or enter the capitalized letter of a selection. For example, pressing U activates the Utilities selection, R is used for returns, and E is for rEports.

Function keys
Esc	Returns to DOS
F1	Provides context-sensitive help on the field where cursor is positioned
F2	Calls up the Check-out function
F3	Calls up the Return a tape function

the clerk to verify visually. The clerk can then ask the customer for his or her name and see if they are associated with that customer account. The clerk can bring up other names in a pop-up window with a function key.

The return screen (Figures 8.19 and 8.20) looks almost identical to the check-out screen. In this case, the clerk must still enter the customer's telephone number and the date returned, or accept the default date. The clerk then scans the bar codes on the tape and the

Figure 8.17 The screen
for a cassette check-out.

```
View Video                                          Dec. 11, 1993
Check-out                                           Initials: KRW
_____


         Customer phone: [                ]
           Date rented: [121193]
               Tape-id: [      ]

         Customer name: [                            ]
            Tape title: [                            ]
                        [                            ]
                        [                            ]
                        [                            ]
                        [                            ]
                        [                            ]



         _____

                    Enter the customer's telephone number
          [Esc: Main Menu] [F1: Help] [F3: Return] [F4: Names] [F5: Next tape]
```

Figure 8.18
Documentation of the
"Check-out" screen.
Documentation can also
include examples of how
users should enter the data.

System :	Video Cassette Tape Rental
Name of Screen:	Check-out
Analyst :	Frank Pisciotta
Date :	11/03/93

Enter the data as follows:

Customer phone	10 numeric digits. Will redisplay with dashes. Example: 9167823456 becomes 916-782-3456
Date rented	MMDDYY format. System will set today's date as the default; if different, enter correct date. System to validate date entered as legal.
Tape id	Scan the bar-coded tape number or enter it manually. System to validate that it is legal. Example: 0245
Customer name	System to look up the customer's name, using the customer phone.
Tape title	System to look up the tape's title, using the tape id.

Function keys

Esc	Returns to main menu.
F1	Provides context-sensitive help on the field where cursor is positioned.
F3	Abandons transaction and calls up the Return a tape function.
F4	Shows names of other customers using this telephone number. Places names in pop-up menu.
F5	Allows entry of a second, third, fourth, fifth, or sixth tape for this customer. Clerk does not have to re-enter the customer's phone or date rented. Shows successive tape titles under first title.

Figure 8.19 The screen for a tape return.

```
View Video                                      Dec. 11, 1993
Return                                          Initials: KRW
_____

          Customer phone: [              ]
          Date returned: [121193]
               Tape-id: [     ]

          Customer name: [                        ]
            Tape title: [                        ]
                        [                        ]
                        [                        ]
                        [                        ]
                        [                        ]
                        [                        ]


_____
                 Enter the customer's telephone number
        [Esc: Main Menu] [F1: Help] [F2: Check-out] [F5: Next tape]
```

Figure 8.20 Documentation of the "Return" screen.

System:	Video Cassette Tape Rental
Name of Screen:	Return
Analyst:	Frank Pisciotta
Date:	11/03/93

Enter the data as follows:

Customer phone	10 numeric digits. Will redisplay with dashes. Example: 9167823456 becomes 916-782-3456
Date returned	MMDDYY format. System will set today's date as the default; if different, enter correct date. System to validate date entered as legal.
Tape id	Scan the bar-coded tape number or enter it manually. System to validate that it is legal. Example: 0245
Customer name	System to look up the customer's name, using the customer phone.
Tape title	System to look up the tape's title, using the tape id.
Function keys	
Esc	Returns to main menu.
F1	Provides context-sensitive help on the field where cursor is positioned.
F2	Abandons transaction and calls up the Check-out a tape function.
F5	Allows entry of a second, third, fourth, fifth, or sixth tape for this customer. Clerk does not have to re-enter the customer's phone or date returned. Shows successive tape titles under first title.

system returns the tape's title. If the customer is returning multiple tapes, the same function key (F5) allows this procedure to repeat.

The customer function allows a clerk to enter a new customer, change an existing customer's data, or delete a customer from the system (Figure 8.21). With this routine, the system captures new or revised data about each customer: phone number and name, address, date joined, and date last rented. As usual, Frank documents this screen design with an explanation of its purpose and function in user-friendly language (Figure 8.22). The deletion of a customer requires the user to enter the customer's telephone number and visually verify that this is the correct customer. If all tapes are returned and the customer owes no money, the deletion is deemed acceptable and the data are removed from the database.

The movie function adds new movies to the collection, allows the user to change data about a specific movie, or removes a movie from the collection (Figure 8.23). This screen is quite complex because it must also allow for entering the names of the stars in the movie, as well as the data about a movie. Another pop-up window is necessary for the star's data (Figure 8.24). Activated by a function key, the window allows entry of names when the user is entering a new movie and shows the names when changing a movie. When a movie is deleted, the star references are also removed, as well as the data about all check-outs and returns.

Next comes report generation. During the output design stage of the systems life cycle, all reports were defined and described. This function gives the user access to the reports (Figure 8.25, page 290). The user picks the specific report that he or she wants and then is allowed to "filter" the report by giving a specific subgroup and not having all the data shown (Figure 8.26, page 290). It may take a long time, perhaps over an hour, to print a lengthy report, so the print program will run while the user does something else. Such simultaneous running, which exploits the computer's full potential, also typifies interactive on-line systems.

Figure 8.21 The screen to add a new customer, change an existing customer, or drop a customer.

```
┌─────────────────────────────────────────────────────────────────┐
│  View Video                                      Dec. 11, 1993    │
│  Customer                                        Initials: KRW    │
│  ───────────────────────────────────────────────────────────────│
│           Customer phone: [              ]                        │
│                                                                   │
│               Last name: [                  ]                     │
│              First name: [                  ]                     │
│           Middle initial: [ ]                                     │
│           Street address: [                          ]            │
│                    City: [            ]                           │
│                   State: [   ]                                    │
│                ZIP code: [     ]                                  │
│             Date joined: [       ]                                │
│         Income generated: [         ]                             │
│         Date last rented: [       ]                               │
│            Tapes rented: [     ]                                  │
│              Credit card: [              ]                        │
│         Credit card type: [       ]                               │
│         Expiration date : [       ]                               │
│  ───────────────────────────────────────────────────────────────│
│              Enter the customer's telephone number                │
│  [Esc: Main Menu] [F1: Help] [F2: Check-out] [F3: Return] [F6: Borrower] │
└─────────────────────────────────────────────────────────────────┘
```

Figure 8.22
Documentation of the
"Customer" screen.

System:	Video Cassette Tape Rental
Name of Screen:	Customer
Analyst:	Frank Pisciotta
Date:	11/03/93

Enter the data as follows:

Customer phone
10 numeric digits. Will redisplay with dashes.
Example: 9167823456 becomes 916-782-3456

Last name
Customer's last name must have an entry.
Example: Pisciotta

First name
Customer's first name must have an entry.
Example: Frank

Middle initial
Optional entry; a single letter; system will uppercase it; no period necessary.
Example: R

Street address
Mailing address up to 30 characters.
Example: 950 South Lincoln Way

City
City name.
Example: Moorhead

State
Two-letter abbreviation. System to check for a match.
Example: MN

ZIP code
Up to 9-digit postal service ZIP code.
Example: 56560

Date joined
MMDDYY format. System will set today's date as the default; if different,
enter correct date. System to validate date entered as legal.
Example: 100391

Income generated
Default is 0.00.

Date last rented
MMDDYY format. System will set today's date as the default; if different,
enter correct date. System to validate date entered as legal.
Example: 120391

Tapes rented
Number of tapes rented since the customer joined. Default is 0.
Example: 15

Credit card
Actual credit card number.
Example: 0123-456-7890

Credit card type
Issuer of the credit card.
Example: VISA

Expiration date
Date the credit card is due to expire.
Example: 0994

Function keys

    Esc — Returns to main menu.

    F1 — Provides context-sensitive help on the field where cursor is positioned.

    F2 — Abandons transaction and calls up the Check-out a tape function.

    F3 — Abandons transaction and calls up the Return a tape function.

    F6 — Allows entry of a second, third, fourth, fifth, or sixth borrower for this customer. Uses a pop-up window in lower-right corner.

    F10 — Deletes the customer after entering the customer's telephone number and visually verifying that this is the correct customer. Makes sure all tapes checked out are returned before completing the deletion. Removes all check-out and return records if the deletion is acceptable. Not shown on screen; user must know this function.

Figure 8.23 The screen to add a new movie, change an existing movie, or drop a movie.

```
┌─────────────────────────────────────────────────────────────────────┐
│  View Video                                         Dec. 11, 1993     │
│  Movies                                             Initials: KRW     │
│  ─────────────────────────────────────────────────────────────────  │
│                                                                       │
│                    Tape id: [     ]                                   │
│                                                                       │
│                      Title: [                    ]                    │
│                  Year made: [    ]                                    │
│                     Rating: [    ]                                    │
│                   Category: [         ]                               │
│             Date purchased: [      ]                                  │
│            Date last rented: [      ]                                 │
│               Times rented: [     ]                                   │
│                       Cost: [       ]                                 │
│           Income generated: [        ]                               │
│                Tape status: [ ]                                       │
│                                                                       │
│                                                                       │
│  ─────────────────────────────────────────────────────────────────  │
│                    Enter the videotape's id number                    │
│     [Esc: Main Menu] [F1: Help] [F2: Check-out] [F3: Return] [F6: Stars] │
└─────────────────────────────────────────────────────────────────────┘
```

The main menu's sixth capability performs operational utility functions (Figure 8.27). Each utility function operates in a pop-up window on the left of the utility submenu. Frank defines each function in his screen data dictionary definition (Figure 8.28). During the process phase of design activities, Frank will completely describe each function in far greater detail.

Frank's collection of screen designs, validation specifications, and processing requirements provides the basis for screen documentation. The programmer will consult this documentation frequently while developing necessary programs during the development stage of the systems process. Well-designed screens make the programmer's job much easier, as well as less costly.

Working With View Video

Frank Pisciotta's screens are not 100% complete. He has not yet built the screens for the Utilities and Movies functions. Drawing and documenting screens is a very important task. Getting the screen right early saves valuable programming time, retains consistency between screens, and helps guarantee accuracy for the user during the important data-entry phase.

Case Study Exercises

Draw screen designs for the following:
1. Entering, changing, and deleting clerk's initials.
2. Entering, changing, and deleting the Store table.
3. Setting the security for a clerk.
4. The completion of the rental rate screen.
5. A modification to reports that would add two new reports.
6. Entering the movie star's name in the Movies data collection screen.

Figure 8.24
Documentation of the
"Movies" screen.

System: Video Cassette Tape Rental
Name of Screen: Movies
Analyst: Frank Pisciotta
Date: 11/03/93

Enter the data as follows:

Tape id
5 numeric digits.
Example: 12345

Title
Tape's title.
Example: Star Wars

Year made
Four-digit year number.
Example: 1980

Rating
As set by the film maker.
Example: R

Category
From one of the common categories.
Example: Sci Fi

Date purchased
MMDDYY format. System will set today's date as the default; if different, enter correct date. System to validate date entered as legal.
Example: 100391

Date last rented
MMDDYY format. System will set today's date as the default; if different, enter correct date. System to validate date entered as legal.
Example: 120391

Times rented
The number of times customers have rented this tape. For a new tape, it is set to zero.
Example: 41

Cost
Amount paid for the tape; do not enter $ or commas.
Example: 35.00

Income generated
Amount of money the tape has made.
Example: 120.50

Tape status
A code describing the current disposition of the tape.
Example: R for rented, L for lost, D for damaged

Function keys

Esc
Returns to main menu.

F1
Provides context-sensitive help on the field where cursor is positioned.

F2
Abandons transaction and calls up the Check-out a tape function.

F3
Abandons transaction and calls up the Return a tape function.

F6
Allows entry of up to six stars for this tape. Uses a pop-up window in lower-right corner. Type the star's name as the prompts request.

F9
Shows codes and their meanings.

F10
Deletes the tape after entering the tape's identification number and visually verifying that this is the correct tape. Makes sure the tape is returned before completing the deletion. Removes all check-out and return records if the deletion is acceptable.

Figure 8.25 Screen for report generation. Frank Pisciotta's design calls for a submenu for each report coming off the main menu screen. The user selects desired functions by choosing the appropriate entry from the submenu.

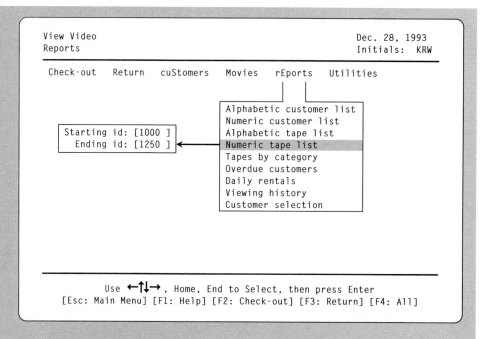

Figure 8.26 Documentation of the "Reports" screen.

System: Video Cassette Tape Rental
Name of Screen: Reports
Analyst: Frank Pisciotta
Date: 11/03/93

Use the up or down arrow keys, Home, or End keys to select the report desired. When selected, press the "Enter" or "Return" key. A second window pops up, allowing the user to select a specific group of items to include in the report. If all the items are desired, press the F4 key.

Enter the data as follows:

Starting id   5 numeric digits.
                     Example: 1000

Ending id   Up to 5 numeric digits.
                     Example: 1250

Function keys

  Esc   Returns to main menu.

  F1   Provides context-sensitive help on the field where cursor is positioned.

  F2   Abandons transaction and calls up the Check-out a tape function.

  F3   Abandons transaction and calls up the Return a tape function.

  F4   Prints all data in the selected category.

Figure 8.27 Screen for
Utility functions. Frank
Pisciotta's design calls for
another submenu for utility
routine. The user selects
the desired utility function
by choosing the
appropriate entry from the
submenu.

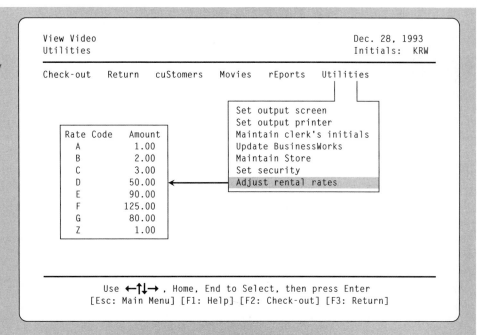

```
View Video                                         Dec. 28, 1993
Utilities                                          Initials:  KRW


 Check-out    Return    cuStomers    Movies    rEports    Utilities
                                                             │
                                      ┌──────────────────────────────┐
                                      │ Set output screen            │
                                      │ Set output printer           │
          ┌────────────────────┐     │ Maintain clerk's initials    │
          │ Rate Code   Amount  │     │ Update BusinessWorks         │
          │    A          1.00  │     │ Maintain Store               │
          │    B          2.00  │     │ Set security                 │
          │    C          3.00  │     │ Adjust rental rates          │
          │    D         50.00 ◄├─────┴──────────────────────────────┘
          │    E         90.00  │
          │    F        125.00  │
          │    G         80.00  │
          │    Z          1.00  │
          └────────────────────┘



 ───────────────────────────────────────────────────────────────
             Use ←↑↓→ , Home, End to Select, then press Enter
          [Esc: Main Menu] [F1: Help] [F2: Check-out] [F3: Return]
```

Figure 8.28
Documentation of the
"Utilities" screen.

System: Video Cassette Tape Rental
Name of Screen: Utilities
Analyst: Frank Pisciotta
Date: 11/03/93

Use the up or down arrow keys, Home, or End keys to select the utility function desired. When selected, press the "Enter" or "Return" key. A second window pops up, allowing the user to perform that specific utility function.

Utility function	Description.
Set output screen	Defines the output device for all reports as the screen.
Set output printer	Defines the output device for all reports as the printer.
Maintain clerk's initials	Allows the user to enter, change, or delete clerk's initials. Scrolling window on the left shows current names and initials.
Update BusinessWorks PC	Posts financial transactions to the BusinessWorks PC accounting package.
Maintain Store	Allows the user to enter or change the Store table. This table holds the store's name, address, city, state, ZIP code, and phone number.
Set security	Each operator/clerk has a security code and password associated with their initials. Codes less than 10 have all provisions. Codes greater than or equal to 10 can only perform check-out and return functions.
Adjust rental rates	Each title is associated with a rental code that has a daily price. The rates are settable by a user with a low-security code number.

Function keys	
Esc	Returns to main menu.
F1	Provides context-sensitive help on the field where cursor is positioned.
F2	Abandons transaction and calls up the Check-out a tape function.
F3	Abandons transaction and calls up the Return a tape function.

File Design

GOALS

After reading this chapter, you should be able to do the following:

- Explain the seven types of files
- Draw the binary patterns for a character in EBCDIC, ASCII, true binary, and packed decimal
- State the advantages and disadvantages of sequential, indexed, and direct file organizations
- Describe the difference between a compact disk (CD) and a typical hard disk
- Name the elements of a file design
- Design record layouts for disk or tape files
- List two control methods for files

PREVIEW

During the file design phase of systems design, the analyst selects the storage requirements best suited for storing the system's data (Figure 9.1). Many analysts thoroughly enjoy file design because it draws upon so many of their computer skills and knowledge. When undertaking this crucial phase of systems design, you must not only determine where and in what order to store data, but you must also figure out how to configure essential data and choose identifiers with which users can easily retrieve data. In this chapter, we will examine various methods for storing data on both disks and tapes.

To build the description of a file, the analyst uses the report formats, data collection screens, and data dictionaries developed earlier in the systems process. These descriptions ultimately form part of the system specifications, which is the overall documentation for a project.

AN INTRODUCTION TO FILE DESIGN

A computer system can store data different ways: in its internal memory or RAM (random access memory); on tape; or on a diskette, hard disk, or compact disk (CD). Each method offers trade-offs in speed, cost, and capacity.

Internal memory is ideal for the computer's software program, data, and operating system (the program that monitors, controls, and manages the activities and resources of a computer system). However, internal memory is expensive and has low capacity when compared to tapes or disks. Internal memory is also volatile, meaning that it loses data when the power is turned off.

Tape is inexpensive and has a high data capacity. However, accessing the data it holds can vary from a few thousandths of a second to several minutes.

Disks hold vast amounts of data (hundreds of millions of characters to billions of characters) and are relatively inexpensive (on a cost-per-character basis). Accessing a disk's data can take a few thousandths of a second, which is slower than accessing memory-stored data.

Compact disks have large-storage capacity (in excess of 700 million characters) at low cost. The trade-off is that they are "read only" (meaning that you cannot write to them, only read what is already on them) and they have slow access times, as long as half a second.

Storage Capabilities

Figure 9.1 Database or file design determines the content and format of files that will hold a system's data.

Internal machine memory is very fast (it can retrieve data in less than a millionth of a second), but expensive (about .005 cent per character). Tape or disk memory is slower (usually operating in thousandths of a second), but less expensive (about .0001 cent per character).

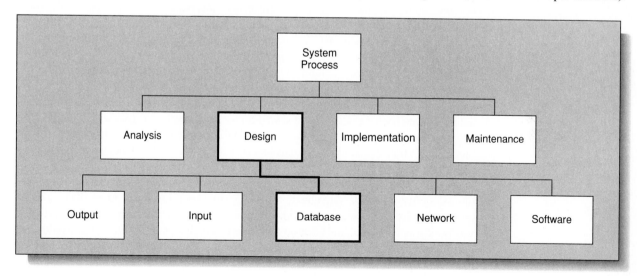

If Fleet Feet's accounts payable system requires 100 characters of data for each of 10,000 vendors, it will need a computer with one million internal memory locations ($100 \times 10,000 = 1,000,000$) to hold the entire file in memory. Although most modern computers can store this much data in internal memory, an organization that retrieves only one customer's record at a time will avoid relying only on internal memory. Although faster than the other media, large amounts of internal memory are prohibitively expensive. In such an application, an organization would use external tapes and disks, which can store massive amounts of data at a much lower cost. In fact, companies such as Sears or Exxon that have extremely large databases rely on both tapes and disks.

Currently, tape and disk manufacturers offer a wide variety of storage devices that can store a billion characters on a disk at a cost of only a few thousandths of a cent per character. On the other hand, disks run so much more slowly than internal memory, they would pose problems in an application that demanded speedy access to a huge database. Still, most disks retrieve data in under twenty milliseconds (twenty thousandths of a second).

Tapes, although even slower than disks, can also store billions of characters, and they cost even less per character. After weighing the trade-offs among these costs, speed, and volume factors, most organizations end up using disks, tapes, or both to create and maintain their files.

Types of Files

Regardless of an organization's size or the particular application in question, an analyst will set up seven kinds of files: master, transaction, back-up, temporary, suspense, table, and program.

A **master file** stores all the permanent or semipermanent data relevant to a particular application until that data grow obsolete. For an accounts receivable application, an organization such as Fleet Feet would store a customer's account number, name, address, telephone number, and various balances in a master file. In an accounts payable system, a master file would store data about each vendor. In a blood bank, the master file would keep data on donors.

Transaction, or **detail**, **files** hold data describing the details of a business activity. Continuing the example above, a detail file for a charge customer at Fleet Feet would include the customer's account number, the date of a particular purchase, the amount owed, and the identifying number of the store where the customer conducted the transaction. This data would remain on file until the customer settled the account. Fleet Feet might then delete the data from the detail file or save it for historical purposes. Likewise, in an AP system, the transaction file keeps data on a purchase (what, when, cost, and so on) and the blood bank's transaction file keeps data on every donation (such as who, when, blood type, and where donated).

Back-up files provide copies of other files, thus ensuring data replacement in the event of loss, theft, or sabotage. Organizations usually store back-up files on tapes (instead of the more expensive disks) somewhere other than where they store the originals. Then, if a file was accidentally deleted, the back-up files can restore the lost files. Fleet Feet would keep back-up copies of both their master and transaction files in a storage facility unconnected to the main processing center.

Organizations use **temporary**, or **scratch**, **files** to store data for a very short time. In this case, a program may extract certain data from a master or transaction file and store it on a reel of tape from which yet another program can then take the data and use it for a special report. Management may analyze this report for trends, then destroy it. In an accounts receivable system, a scratch file may contain a list of customers who have not paid their debts for over 90 days. Such a report could facilitate debt collection and help management improve cash flow. Since the data change with every payment customers make, the AR scratch file needs rebuilding every time it is needed.

Suspense files record errors detected by validation programs. Most batch processing systems authenticate data before they update other files. If they detect errors, they can store those errors in a file for later research and correction. Typical types of errors include missing data (such as the customer account number in an accounts receivable system), incorrect transaction dates (month numbers not between 1 and 12), improper abbreviations (incorrect 2-letter state), and so on. In some cases, the system collects erroneous data in an error or audit report that a clerical worker can review, correct, and then re-enter.

Table files hold infrequently changing tabular data such as payroll tax tables, discount codes, or error messages. These types of files typically provide the necessary data that a program needs for computation or reference. For instance, a West Coast software organization stores all of their software's error messages in a table file, so that whenever programs bump into errors, they can read the appropriate error message from the table file and display it to the user. In some cases, word processing software allows users to build such files with customized error messages that suit their exact needs.

Table files also assist changeover to another language, such as French or Italian. All that needs to change are the entries in the table file; the program does not need modification or recompiling. A translator simply rewrites the words in the new language.

Program files, as the name implies, store the source language programs needed to operate the system. Program files allow users to modify their programs without re-keying them, to transfer a program to another computer of the same or different manufacturerer, or to access specific portions of one program when writing a new program. Fleet Feet's program files are as valuable to the company as its data files.

DATA STORAGE METHODS

All files contain two types of data: alphanumeric and numeric. **Alphanumeric data** items include letters, special characters, and numbers. **Numeric data** items contain numbers only. A retail store's customer numbers may contain only numbers (such as the numeric 7148) or numbers and letters (the alphanumeric RE7148). A Social Security number with dashes (503-57-9888) is alphanumeric, whereas a Social Security number *without* dashes (503569888) is numeric. Generally, the distinction centers on how an organization uses the data: numeric data undergo calculation, alphanumeric do not.

A system designed to store massive amounts of data will require great efficiency because unless the computer can easily accommodate billions of characters, its disks or tapes will overflow. With the right storage method, a designer could compress a data file that might consume 240 million bytes to one that will consume only 128 million bytes—a savings of almost 50 percent!

Storing Alphanumeric Data

Several methods exist for storing alphanumeric data in either internal or external memory, but the two most popular are EBCDIC and ASCII. IBM computers commonly use the 8-bit binary coding system called **EBCDIC**. EBCDIC is an acronym for Extended Binary Coded Decimal Interchange Code. The 7- or 8-bit coding system found on many minicomputers and personal computers is called **ASCII.** ASCII is an acronym for American Standard Code for Information Interchange.

Both EBCDIC and ASCII utilize a binary format wherein groups of bits with specific on-off patterns represent particular characters. For example, the binary pattern 11010011 represents the capital letter L in EBCDIC, while 11001100 represents L in ASCII. The digit 5 in EBCDIC is 11110101; in ASCII, it is 00110101. Notice that the EBCDIC binary pattern for L is "larger" than its ASCII counterpart. Likewise, the EBCDIC 5 is larger than the ASCII 5. Also note that EBCDIC 5 is "larger" than EBCDIC L, whereas ASCII 5 is smaller than ASCII L.

While the differences between EBCDIC and ASCII may seem trivial, computers using EBCDIC will produce different results than computers handling ASCII. For example, when an EBCDIC-based system sorts alphanumeric data, it reads alphabetic fields before numeric ones (A–Z, 0–9); an ASCII-based system does the reverse (0–9, A–Z). The order of characters is called the **collating sequence**. Consider each data item in a master file of alphanumeric automobile license plate numbers:

219XLA, LEZ697, NCH763, 145AVE, 119XLA

The EBCDIC method would sort the plate numbers as:

LEZ697 NCH763 119XLA 145AVE 219XLA

The ASCII method would produce:

119XLA 145AVE 219XLA LEZ697 NCH763

The fact that ASCII and EBCDIC follow different collating sequences becomes a critical issue whenever an organization must exchange data between ASCII- and EBCDIC-based computers. To do so, data require resorting to fit the receiving system's format. Despite this translation problem, however, neither EBCDIC nor ASCII operates more efficiently than the other when storing data.

EBCDIC and ASCII are the right choice for storing customer or vendor names, addresses, cities, state abbreviations, country names, and any other value that is not numeric. The question that often arises is how to store fields such as telephone numbers, Social Security numbers, and so on. The answer is not hard and fast. If the field is numeric and not used for any calculational purpose, then pick either EBCDIC or ASCII—whichever is supported by the computer in use.

Storing Numeric Data

Computer memories, tapes, and disks can store numeric data in two ways: character or numeric. Character is used for display purposes to a screen or a printer, while numeric is for computation.

ASCII and EBCDIC can store numeric data digit-by-digit, in character form. For example, to store a 01778 ZIP code for a customer in a vendor master file, both would assign 8 bits to each of the five digits, resulting in a total of 40 bits.

ASCII:	00110000	00110001	00110111	00110111	00111000
EBCDIC:	11110000	11110001	11110111	11110111	11111000
	0	1	7	7	8

Although commonly called EBCDIC or ASCII, these formats sometimes go by the name of zoned decimal since the left half of each 8 bits (called a nibble) is always 0011 in ASCII and 1111 in EBCDIC.

Packed decimal stores two digits in each eight-bit byte and provides yet another popular method for storing numeric data. The last four bits on the right of every packed decimal number indicate sign; for example, 1110 means negative, 1111 means positive. If a number enters the system without a sign, packed decimal assumes that it is positive. The 01778 ZIP code would look like this in packed decimal.

Packed Decimal:	00000001	01110111	10001111
	0 1	7 7	8 +

Packed decimal (known as COMP-3 in COBOL) needs only 24 bits of memory (3 bytes) to store the same numeric data that requires 40 bits (5 bytes) in the character version of EBCDIC and ASCII. Though packed decimal offers considerable space savings, it applies only to numeric data items. Many organizations use it to compact numeric data fields such as account balances that are used for computational purposes.

Straight binary offers a fourth method for storing numeric data using the base 2 number system. As its name implies, this format stores data in its true binary form. Thus, 01778 code looks like this in straight binary:

Straight Binary: 00000110 11110010

Although straight binary requires the least amount of memory (only 16 bits, or 2 bytes, for a ZIP code), it can become cumbersome whenever decimals are involved because bits are not easily translatable into decimals. Straight binary usually works best in an application in which an organization must keep large tables of data on file, such as income tax tables or lists of airline departure and arrival times.

As you can see, all forms of these methods present advantages and disadvantages and involve trade-offs with the different amounts of memory. When selecting one for a given application, an analyst must weigh the hardware's coding system, the volume of data the system must handle, available tape or disk space, and an organization's speed requirements. If, for example, a company already has a computer that requires certain numeric data to appear in packed decimal form for computational purposes, it should plan to store all other numeric data items, such as account balances, that way. Were EBCDIC chosen for account balances in a packed decimal environment, the computer would have to convert the EBCDIC data into packed decimal, perform a necessary operation (such as altering the balance), then reconvert the new data into EBCDIC. Of course, for applications involving alphanumeric data, the organization would have no choice but to use ASCII or EBCDIC, since packed decimal applies only to numeric data.

MEDIA AND DATA STORAGE TECHNIQUES

In the 1960s, analysts could not routinely specify the still relatively rare and very expensive disks, so they most often relied on tape. Since then, however, disk prices have fallen. Because today's users frequently desire the personal computers or terminal-oriented systems that permit direct, immediate access to data, analysts have turned increasingly to specifying disks as the media for a file. Nevertheless, some applications still lend themselves better to tape.

Tapes and Sequential Files

When designing a file for tape, analysts must understand the medium's physical properties. Most tapes are one-half inch wide and they come on reels or cartridges that are 600, 1,200, or 2,400 feet long. Data may be stored on a tape in EBCDIC, ASCII, or any of the numeric methods by writing the binary patterns across the tape in tracks. Tape used in the late 1960s and early 1970s used seven tracks. Now tapes use nine tracks: eight for data and one for parity (a built-in error-detection technique). BPI (bits per inch) refers to the different densities available for recording data on tape: 800, 1600, or 6250 BPI, for example. A tape drive reads or writes the data on the tape at speeds varying from 37.5 to over 200 inches per second.

Tapes also read or write data record by record. In other words, users cannot randomly skip from one record to another at will, but must process data sequentially, record by record. This limitation dictates that tape-oriented systems must store data in a certain order, so that users can access the data according to unique record keys such as account, customer, part, and Social Security numbers. Files that store records according to a key field are called **sequential files**.

When designing systems that read or write data record by record, analysts allow for a space between each record or between groups of records. The term *blocking* refers to a grouping of two or more logical records to form a single physical record (Figure 9.2). The **blocking factor** equals the number of logical records included in each physical record. For example, a file with a blocking factor of 10 would contain 10 logical records between every physical record.

THE SUPER-MINI VERSUS THE PC

Bob Opie, job superintendent and estimator at Continental Hiller Construction Company, loved to fool around with electronic devices, so his coworkers weren't surprised when he brought a new 4-pound Compaq laptop personal computer into the office. But they did wonder why Bob thought he needed a portable Compaq when he had his own terminal for tapping into the company's sophisticated time-sharing system. Continental Hiller's Digital Equipment Corporation VAX ran state-of-the-art word processing software, boasted a database manager called RDB, supported many programming languages, had an electronic mail system, allowed programming C and COBOL and other languages, and offered a giant menu of canned programs that Bob could call up from his terminal. By contrast, Bob's Compaq had a dinky 24-pin dot-matrix printer, a 1.4MB floppy disk drive, a built-in modem with fax capability, and an 80MB hard disk. Its meager 6 MB memory paled next to the VAX's 32 MB and the disk drive of 80MB didn't compare with the 4,000MB available on the VAX.

Undismayed by his colleagues' smirks, Bob began explaining the software for the Compaq: Lotus 1-2-3 (an integrated spreadsheet, graphics, and database system), WordPerfect word processor, dBASE database manager, PFS: Plan scheduling and planning system, Crosstalk data communications software, and graphical Windows operating software. He surprised his colleagues by showing them how his Compaq could link to the VAX with the Crosstalk software so that it functioned just like a terminal at the office, home, or on a construction job site. Bob finished his explanation by saying that all this software was available on a wide variety of other makes and models of personal computers. And with his fax modem linked to his cellular telephone, he was put into instant communication with the company office, suppliers, and anyone else.

After only two weeks of experimenting, Bob began using his Compaq for job costing. Collecting data on the cost of concrete at the job site, Bob entered it on his Compaq using dBASE, then sent the data to the home office for review. Bob loved the fact that the Compaq did not need air conditioning or special power sources, and he enjoyed transmitting data directly from the site without having to return to this office or leave his car. In fact, many times he faxed responses via his cellular telephone from inside a building site where construction workers were pouring concrete!

As job superintendent, Bob routinely scheduled subcontractors, a very time-consuming task. Not only did that task require him to track a number of overlapping schedules and costs incurred on the job, it involved producing a weekly status report to management. Luckily, the Compaq planning software performed these functions, too. Now Bob can enter data on sequence and cost factors for each subcontractor, then project a schedule for each step of the job. Not only will PFS:Plan show the proper sequencing, but it will print a graph showing when all the subcontractors should start and stop work. It also allows for changes, which helps Bob adjust to inevitable delays. Most importantly, after using it, Bob discovered that delaying a crane scheduled for the job by two days could save Continental Hiller about $1,150!

Bob found Lotus 1-2-3 useful for tracking subcontractors and comparing Hiller's estimate of a particular phase of a job with actual costs incurred. Bob used rows in the Lotus 1-2-3 spreadsheet system for each subcontractor and columns for the various costs. The rightmost two columns displayed percentages and the difference between expected and actual costs. The bottom row totaled each column. Whenever contractors told Bob they'd finished their work, Bob inspected the work and authorized payment, entering the amount into the Compaq so Lotus 1-2-3 would automatically compute totals, percentages, and differences. He was the only Hiller estimator who could quote up-to-the-minute costs.

The Compaq's WordPerfect word processing put the icing on the cake. With it, Bob could write each week's report on the job site or at home, edit it, check the spelling, and send it to the office for management review via his modem. No longer did he get bogged down with the dictating, typing, editing, retyping, and duplicating that could only be done from the main office.

Most of Continental Hiller's staff grew to respect Bob's success. When Bob returned to the main office for a monthly project review, his boss asked him to train two other field engineers on new IBM Personal System/2s the company had decided to buy. He even handed Bob a check to cover the cost of Bob's Compaq and its software. "Sometimes," mused his boss, "Less *is* better."

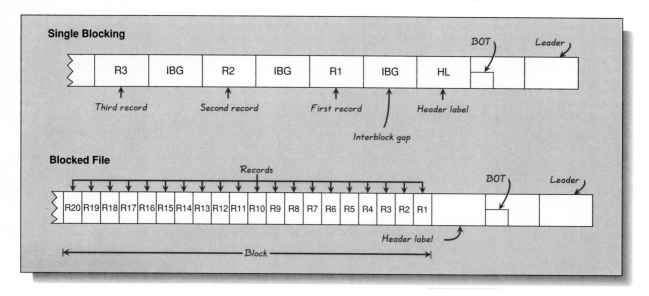

Figure 9.2 Blocking permits an organization to keep more records on a tape because it reduces the number of interblock gaps. In the top diagram, the tape holds only three records due to the space taken up by the interblock gaps. These gaps slow the speed of processing because the tape drive has to start and stop each time it encounters a gap. The bottom diagram is a blocked file with a blocking factor of 20, which means that 20 records can be read or written before the tape drive stops.

Physical records are separated by a space, called the interblock gap (IBG) or the inter-record gap (IRG). The space is about 3/10 of an inch (0.3") in length and is used by the tape drive to get it up to speed to read or write a physical record and then to stop when the physical record is completely read. Thus, when the blocking factor is 10, we have a physical record (with 10 logical records), an IRG, another physical record, another IRG, and the the process repeats itself for all the records in the file.

To further speed processing, the designer may specify buffering. A **buffer** is an extra section of memory set aside to allow the tape to read ahead of itself or to hold data for later writing to the tape drive. The advantage of buffering is that while the program is working on the last physical set of records, the tape drive can read the next set of records into memory for processing. Then when the program calls for the next set, there is no waiting for the tape drive to read and transfer the data to memory. The disadvantage is the extra cost of memory set aside for the buffer. Since the price of memory falls every year, buffering is becoming an attractive way to speed processing activity.

If Fleet Feet's inventory system stores data concerning 10,000 different items, and if it allows 200 characters to identify each one (product number, description, quantity on hand, location in the warehouse, number of items sold last year, reorder point, and so on), the tape must hold 2,000,000 characters (200 × 10,000 = 2,000,000). Using a 1600 BPI tape to store the data record by record would involve 200/1600, or 1/8 inch (0.125") of tape for each record. With a 0.3" interblock gap between each record, the total length of tape required to store such an inventory file would be:

$$
\begin{aligned}
\text{Length} &= \text{Number of records in file} \times (\text{Record length} + 0.3) \\
&= 10,000 \times (0.125 + .3) \\
&= 10,000 \times 0.425 \\
&= 4,250 \text{ inches}
\end{aligned}
$$

On the other hand, we might block these records into groups of ten, resulting in 1,000 groups with an interblock gap between each group. This approach would consume much less tape:

$$
\begin{aligned}
\text{Length} &= \text{Number of blocks} \times (\text{Length of each block} + 0.3) \\
&= 1,000 \times [(0.125 \times 10) + 0.3] \\
&= 1,000 \times (1.25 + 0.3) \\
&= 1,000 \times 1.55 \\
&= 1,550 \text{ inches}
\end{aligned}
$$

Analysts should take great pains to choose optimum blocking factors. An inappropriately small one will waste tape; an inappropriately large one may decrease the amount of memory available to hold all the data and the program that processes the data. Most computer manufacturers provide guidelines for blocking data on particular tape drives. These guidelines strive to balance the density of the tape, the speed of the device, and the record length to make optimum use of the tape drive.

For instance, one computer manufacturer suggests that the number of characters in the block should come as close to 32,767 as possible. For our 200-byte records, this would result in 163 records per block ($32,767 \div 200 = 163$).

Blocking not only saves tape, it also reduces computer processing time. When a tape drive reads a record from tape, it starts and stops the tape at each interblock gap, a process that takes about five thousandths of a second (0.005 second). Thus, our unblocked Fleet Feet inventory file would experience 10,000 starts and stops for a total of 100 seconds [$10,000 \times (0.005 \times 2)$]. However, if we blocked it by tens, the file would contain only 1,000 gaps, resulting in only 1,000 stops and starts for a total of only 10 seconds [$1,000 \times (0.005 \times 2)$].

If stock moves in and out of Fleet Feet's warehouse rapidly, an inventory system may need to record 3,000 changes per week, and it must post each change to the inventory master file. If the company has organized the tape inventory master by record key (product number), the system must sort randomly occurring changes before posting them to the master file. In other words, the computer must read 3,000 changes, sort them, and then write them out—in order, by product number (Figure 9.3). Thus, 6,000 records require processing just to rearrange 3,000 changes into the right order.

Once the system has sorted all changes by product number, it can match them to items in the master files and then record them in the correct place. To update a tape master file, the system reads each record from the master file, then changes or copies it without change, thereby creating a new version of it. To update a 10,000 item inventory file, the system would have to process 23,000 records: 10,000 reads, 3,000 changes, and 10,000 writes. Unaltered copies of all changes and of the original master file provide a back-up in case something happens to the new master file, such as inadvertent erasure.

Tapes and their sequential file processing offer several advantages. First, processing automatically creates back-up files. Second, using one tape usually costs less than using multiple disks, and the tape can completely store a great deal of data. Third, since tape devices have enjoyed popularity since the 1960s, many programmers feel comfortable with them and have developed clever strategies for handling all types of record lengths. Finally, an organization can readily send a reel of tape through the mail, which usually costs less than contemporary data communication links.

Figure 9.3 Data flow diagram illustrating inventory master file update. There must be a tape for the changes, master file before the update, and master file after the update.

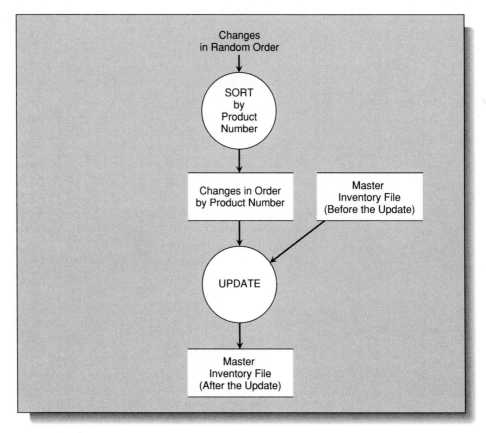

On the other hand, sequential storage prohibits direct retrieval of records. Tape drives are expensive (as much as $50,000) and a system that can process changes, maintain a master file, and create a new master file requires three independent drives. Often, a reel of tape stores only a single file, in which case an organization would need many reels to store all necessary records.

Though the medium of choice for storing files in the past, tape will continue giving way to disks, just as punched cards gave way to tape. Even so, tapes may continue to enjoy popularity as disk back-ups for quite awhile.

Disks

Disks resemble phonograph records in diameter and thickness. We classify disks as floppy (usually found on personal computers), hard, or compact (CD).

Disk capacity (the amount of data that the disk holds) varies from manufacturer to manufacturer, ranging from 1MB to over 1,000MB (MB stands for a million bytes; 1,000MB equals approximately 200,000 typed pages of text). Technically, 1MB is really 1,024,000 bytes since 1MB really represents 1,000 groups of 1,024 bytes. However, most computer users round off the extra 24,000 extra bytes and call it a million bytes for short.

Both floppy and hard disks can store data in EBCDIC, ASCII, packed decimal, or binary formats. Each approach involves writing binary patterns in concentric circles called tracks. Tracks are, in turn, subdivided into sectors that resemble arcs of a circle. Each sector has its own address.

A read-write head attached to an access mechanism moves across the disk to access data, enabling users to retrieve data from any sector on the disk by positioning the read-write mechanism over the desired track and waiting for the wanted sector to spin past the read-write head. This ability to pick and choose specific sectors, bypassing all those in between, gives a disk one of its most attractive features: direct access ability.

We can increase the amount of data available to the read-write mechanism by stacking ten or more disks together. Ten disks stacked together allow us to write data on the top and bottom of each disk, for a total of twenty surfaces. We call the arcs lying under the read-write access mechanism **cylinders**. Some disks contain as many as 500 cylinders.

To record, store, or retrieve data, the access mechanism moves to the proper cylinder. Unlike a phonograph's arm and needle that can only read what it finds, the disk's access mechanism can both read and write data from or to a sector, although not at the same time. With the arm in position, the drive waits for the correct sector to spin under the access mechanism so it can read the data on that sector. The cylinder concept speeds data access since the read and write access mechanism does not require repositioning to read or write another track, thus saving valuable access time.

If we stack ten disks with 500 cylinders and 40 sectors per track, each holding 500 characters, we achieve the following capacity:

$$
\begin{aligned}
\text{Capacity} &= \text{Number of surfaces} \times \text{Number of cylinders} \times \text{Number of sectors} \times \\
&\quad \text{Number of characters} \\
&= 20 \times 500 \times 40 \times 500 \\
&= 200{,}000{,}000 \text{ bytes (200MB)}
\end{aligned}
$$

In some cases, analysts may select disks for sequential processing. For example, a disk can emulate a tape if data appears sector by sector, cylinder by cylinder, as if the sectors were continuous. In certain applications, the higher cost of disks may be offset by the fact that computer operators will avoid the time-consuming task of mounting and dismounting tape reels during sequential file update.

Disks are not cheap: a 200MB disk drive can cost $1,000 or more. Prices, though, have dropped significantly and capacities and speeds have increased dramatically since the late 1960s. Personal computers have reduced the cost of small hard (20–50MB) disks to well under $250. Since disks offer both sequential and random access capabilities, their popularity will continue in the years ahead.

Indexed Files

As we have seen, the individually addressable sectors on disks, which accommodate both sequential and random record processing, provide a powerful advantage over tapes. To update a record, the program needs merely to locate the desired record, read it into memory, change it appropriately, then write the desired record back to the same sector from which it came. To locate a desired record, a system will usually use one of two techniques: indexed file organization or direct file organization.

Indexed file organization uses a "table" to locate records on a disk. For example, to find data on product number C-64744 in an inventory system, the program would scan the table of product numbers, locate the desired product number, and then read its sector location (Figure 9.4). The indexed file organization software maintains the table of product numbers and sector locations, and it "automatically" searches the table to find the desired product number. In addition, the indexed software can add and delete product numbers as new items come into stock or old ones go out of stock. From a programmer's perspective, the indexed software needs to know the product number and it returns the entire record (if there is one) or an error code if the requested record does not exist.

Let's apply this procedure to our 10,000 product inventory system, in which the analyst wishes to record 3,000 changes. Since indexed file organization allows random record retrieval, we do not need to instruct our computer to sort changes. Rather, indexed allows us to make updates to the master file by reading each of the 3,000 changes, searching the table, reading the records of the addresses indicated by the table, altering the records in memory, and writing the records back to their original locations. To record 3,000 changes, indexed processes 9,000 records, reading 3,000 transactions and 3,000 indexed masters, then writing 3,000 records back to the indexed file. That is quite different from the 23,000 total records processed for a sequential tape system, amounting to over 60% savings.

For large applications, indexed file organization can seem unwieldy, requiring enormous memory for vast numbers of records. However, an analyst can solve this problem by dividing the main table into a number of subtables. By combining the first entries of each subtable into a "master" table, indexed files can locate a specific record by scanning the new table to find the proper subtable, then reading the subtable into memory. It then scans the subtable until the record keys match, after which it reads the record, makes the changes, then writes the altered record back to the disk.

Indexed files require that a system store the record key stored in a "table." To add a record to the file, the system must store its key and location in an overflow table. Periodically, the software rebuilds the tables and files to keep processing efficient. Some computer manufacturers include software that routinely performs all the table operations, rebuilding tables whenever necessary.

Figure 9.4 The indexed table locates a record on a disk. Thus, product number C-64744 is found at sector 12228.

Product Number	Location or Sector Number
A - 01118	12224
A - 78561	12225
A - 98356	12226
B - 34655	12227
C - 64744	12228
G - 75190	12229
K - 89757	12230
M - 78900	12231
Z - 89222	12232

Indexed sequential software offers great flexibility because it allows both random and sequential access to data. That is, it can read a specific record (such as product number C-64744) or it can process the entire file in order (a list of items on hand in ascending order by product number). Disadvantages include the extra work involved in maintaining the tables, the amount of memory required to store tables, and the extra disk space for the index and overflow areas.

Direct-Access Files

If an application demands random access of records, **direct file** organization offers a viable alternative to indexed files. Direct files convert the record key into the relative record number with a **hashing algorithm**. A common hashing algorithm is called divide and remainder. First, we divide the record key by a prime number (a number only divisible by itself and 1) to determine the remainder. The prime number must be greater than the number of actual records, plus contain an allotment for future file expansion. For a 10,000 product inventory system, the analyst might pick the prime number 11,001, which allows for 1,001 additional expansion positions (about 10%). For product number C-64744, the division yields:

$$\text{Location} = \text{Remainder (Record key} \div \text{Prime number)}$$
$$= (64,744 \div 11,001)$$
$$= 5, \text{Remainder } 9,739$$

The remainder—9,739—identifies the record number, C-64744, in the file. With no table to search or maintain, the system can locate a record with a simple act of division.

If a record key contains alphabetic characters, as many part numbers, license plate numbers, and driver's license numbers do, we cannot divide until we have either stripped the record key of its nonnumeric characters or converted them to numbers. One hashing algorithm for alphanumeric conversion is the Soundex system. Soundex converts letters as follows:

B, F, P, V are assigned a 1
C, G, J, K, Q, S, X, Z are assigned a 2
D, T are assigned a 3
L is assigned a 4
M, N are assigned a 5
R is assigned a 6
A zero is assigned to A, E, H, I, O, U, W, Y

Thus, the word KLEGER becomes 240206 and BURNS becomes 10652 in the Soundex system. This system, however, functions like a one-way street: you can't convert numbers back into words.

Direct file processing works much like indexed: reading a change record, calculating the location, reading the disk, modifying the record, and then writing the new version back to the same location. As with indexed, 3,000 changes require 9,000 reads and writes.

The direct file hashing algorithm method easily supports applications demanding quick record retrieval, because locating and reading the desired record into memory usually requires a single access to the disk. It involves a simple calculation for finding the record number, permits both numeric or alphanumeric keys, and is easily implemented with a few COBOL, C, Ada, or Pascal instructions.

A disadvantage of direct files and their hashing algorithms comes from the fact that calculations on two different record keys may result in identical remainders or collisions (sometimes called crashes or synonyms). For example, product number C-64744 and F-42742 both yield a remainder of 9,739 when divided by 11,001. When collisions occur, an indicator is stored in the first record (C-64744) to warn a user of the crash. The indicator reveals where the record number F-42742 really resides. Given the opportunity for collisions, extra disk space is allotted for a record that would otherwise collide with another.

The random order of the file creates another disadvantage of direct files. Since records do not appear in any specific order, a sorting step must occur before listing or otherwise processing the file in sequence.

Also, when a direct file becomes full, which sometimes happens, a programmer must write a special one-time program to rebuild it with expansion space. With tape, on the other hand, as the file fills up, you can simply use additional tape.

Disk files do not generate automatic back-ups. To provide an extra copy of the file, we must copy it to another disk or reel of tape. Fortunately, many operating systems contain utility routines to copy disk files to tape.

The choice between sequential, direct, or indexed files revolves around users' needs. Users who want instant access to data will benefit from direct or indexed techniques. Users in batch environments, where data are collected for later processing (perhaps at night when demand for computer time has slackened) may not need to depart from the sequential access tradition.

Compact Disks and CD-ROMs

Imagine a single disk the size of a 5-1/4" floppy diskette (actually 12 centimeters in diameter) that stores the entire *Encyclopedia Britannica*, about 750MB! Data storage of this magnitude would normally occupy 6 feet of library shelf space. Yet this high-density storage is now available in the form of compact disks or CD-ROMs (compact disks, read-only memory).

Borrowed from the technology of audio disks, this storage media uses laser beams to record and read digital data on disks ranging in size from 4.7" to 14". A single CD laser disk holds the equivalent of 18 reels of tape. Today, most of these devices record data, but are not reusable. Reusable laser disks are available, but they are still expensive.

Laser disk technology involves focusing a thin laser beam, 1 micron wide (a thousandth of a millimeter) on the disk surface. A complex collection of mirrors and lenses etches a spot (called a pit) or no etching (called land) on the disk, which represents either a binary zero or one. To read data from CD-ROM, a low-power laser is focused on the disk and reflected light is sensed by a photodetector. The detector converts the binary zero or ones to the computer, which interprets it as text, graphics, or sound for use in the application.

CDs are profoundly affecting many computer applications, especially in the fields of training and entertainment. Apple Computer, Inc., uses CDs as a distribution media for all types of software (operating systems, utilities, examples of how to write code, and so on), manuals (that you can edit and print on your laser printer), marketing data, and short clips of full motion video (with sound) for multimedia systems.

Connecting CD-ROMs to computers with appropriate software allows the analyst access to exciting applications. Video images in the form of tutorials, documentaries, drama, art, and so on are recorded and instantly addressed. The user thus controls the outcome as he or she responds to questions or scenes that appear on the screen. This process provides custom-tailored instruction to the user, allowing them to control the end result that best satisfies their own needs, interests, and abilities.

CD-ROMs find their niche in applications that require the storage of images or historical data. They operate like other types of random-access disks, yet have super high storage capabilities, beyond that of regular disks. Analysts should consider this media when the system demands graphics, read access, ultra-high storage capacity, or compact size.

DESIGNING DISK OR TAPE FILES

When designing a file storage method for a computer system, the analyst must choose the proper storage medium (magnetic disk, CD, or tape) and the best method for accessing data from the file: sequential, direct (random), or indexed.

Having settled upon a medium and an access method, the analyst actually begins to design the file to match the requirements of all report formats and their accompanying data dictionaries. For example, our Fleet Feet inventory system requires a stock status report, which is a summary of inventory activity for the month, listing additions, deletions, quantities on hand, and quantities on order.

Considering the nature of the report, the analyst, Peggy Adams-Russell, might decide to construct two files: a master file containing data about each item (called ITEMS), and a transaction file for changing data about items (called ITEMDTLS for Item Details). Since indexed sequential provides both direct and sequential access to data, and since she anticipates the need for both types of access, Peggy might specify disks for this application.

An important part of the design revolves around selecting a field to serve as the unique identifier or the **record key**. For our Fleet Feet inventory system, we need a product number for each item the chain sells. Peggy discovers that every manufacturer has their own product number and she decides to use theirs rather than have a special Fleet Feet product number—prefixing each with a letter R for Reebock, N for Nike, and so on. A different product number would require Fleet Feet to cross-reference the manufacturer's product number when ordering stock, which would be time-consuming and prone to errors.

Most files require a record key. In an accounts receivable system, every customer needs an account or customer number. Just as citizens are identified by Social Security and driver's license numbers, in an accounts payable system, vendors are identified by vendor numbers.

When sketching the record formats, analysts use file layout sheets (Figure 9.5). Each sheet can hold several record formats and each should include information on the number of fields in the record, the data format and the size of each field, and the order in which fields will appear. Each record in our Fleet Feet inventory master file, ITEMS, will contain

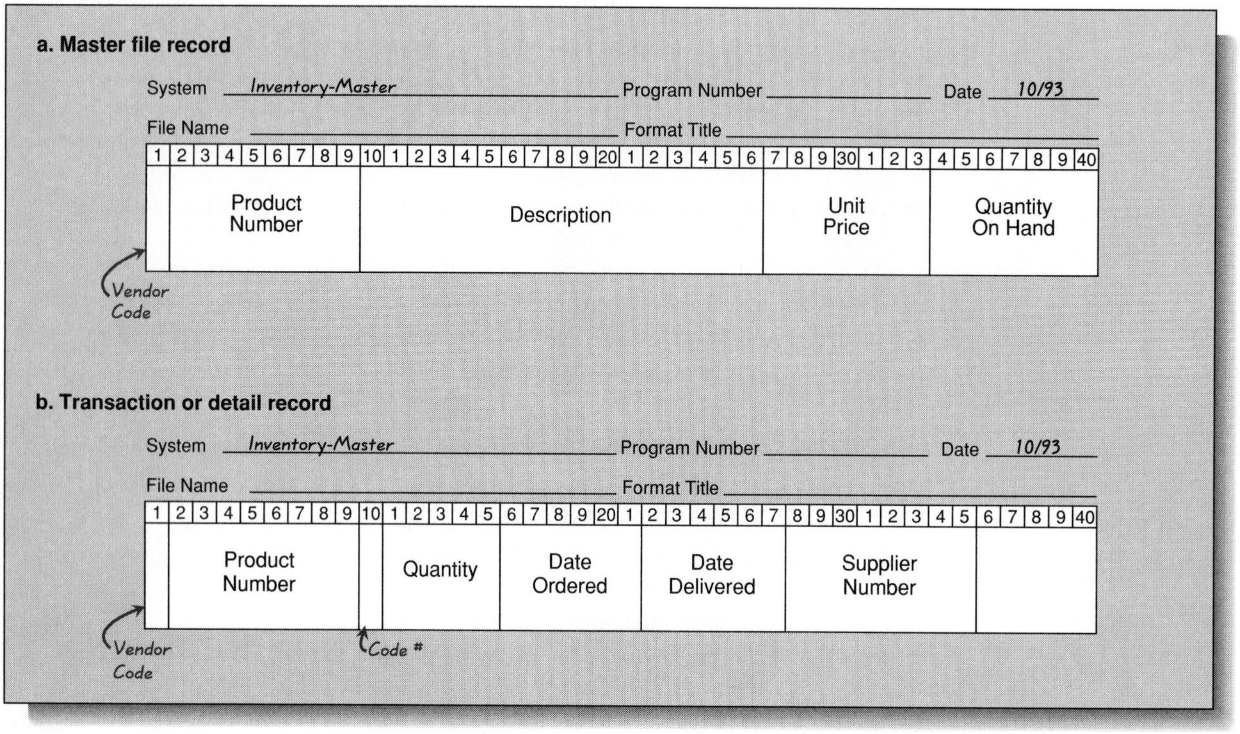

a. Master file record

System *Inventory-Master* ____ Program Number ____ Date *10/93*

File Name ____ Format Title ____

1 2 3 4 5 6 7 8 9 10 1 2 3 4 5 6 7 8 9 20	1 2 3 4 5 6 7 8 9 30 1 2 3	4 5 6 7 8 9 40

| Product Number | Description | Unit Price | Quantity On Hand |

Vendor Code

b. Transaction or detail record

System *Inventory-Master* ____ Program Number ____ Date *10/93*

File Name ____ Format Title ____

| Product Number | Quantity | Date Ordered | Date Delivered | Supplier Number | |

Vendor Code *Code #*

Figure 9.5 File or record layout sheets allow analysts to describe a record, name it, date the design, define the fields, and identify the person who designed it. This layout sheet shows both the inventory master and detail record. The analyst uses a vertical line to separate fields.

5 fields stored in indexed seqential format, allowing 42 characters of storage space per record, and the file will appear in sequential order. The ITEMSDTLS file will allow for 6 fields stored on disk, 42 characters per record, and a sequential order.

Placement or order of the fields is immaterial. However, tradition calls for the record key to appear first, with the next most important field second and continuing to the least important field last.

Data for both master and detail file records will come from two sources: customer orders that reduce inventory, and orders from suppliers for new items to replenish inventory (Figure 9.6). Both sources require validation to ensure the correctness of the data. Errors in dates (month number not 1–12), product numbers, or code numbers (1: reduction in inventory, 2: replenish inventory) should cause the system to reject that transaction. Data for correct orders should update the master file and alter the "quantity on hand" field, thus providing the organization with a precise record of quantities in the warehouse.

Suppose, on the other hand, that the analyst decides to employ the indexed seqential and disk storage technique for the master inventory file, but chooses to store the detail file on tape. The former application demands immediate user access, while the latter does not. In fact, the system will need transaction details only once for update purposes. As soon as it has updated the master file, it can set the detail file aside for later study or back-up purposes.

Having decided to store the transaction details on tape, the analyst must block the records to pack as much data as possible on each inch of tape, thereby minimizing processing time. If the manager of computing services mandates an even number of records

Figure 9.6 This data flow diagram shows the flow of data through an inventory system.

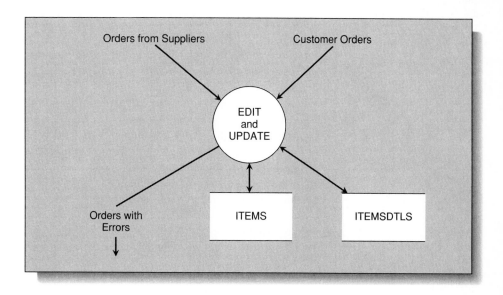

per block plus a total block length of around 2,000, the analyst would arrive at an appropriate blocking factor of 60. Since each detail record contains 42 characters, a factor of 60 results in 2,520 characters ($60 \times 42 = 2,520$) per block. On a 1600 BPI tape, such a block would consume 1.58 inches ($2,520 \div 1,600 = 1.58$).

Let's assume that during the analysis phase of the systems life cycle, the analyst noted that the organization typically processes 12,000 detail records each month. This fact would indicate a need for 200 blocks ($12,000 \div 60 = 200$). The length of the detail or physical file then equals 200 blocks of 1.58 inches plus 200×0.3 of an inch interblock gaps, or a total of 376 inches of tape [$200 \times (1.58 + 0.3) = 376$]. On a 2,400-foot reel of tape, the detail file would use 376 inches, or 31.3 feet (about 2 percent of the whole tape). Clearly, processing time is minimized.

After Peggy defines her files, she considers what indexes she will need. For the master ITEMS file, she will want quick access by product number. Similarly, for her ITEMDTLS file, she will want an index by product number to allow rapid retrieval of all the records about a specific product.

FIXED- VERSUS VARIABLE-LENGTH RECORDS

To update a sequential, indexed, or direct master file, a system must maintain a transaction file, which will permit the transfer of changes to the master file. In a transaction file, the length of each record and the number of fields in each record can be identical, or **fixed-length**. Despite this simplicity, fixed-length records do not aid logical associations between records in the transaction file and their master file counterparts.

Variable-length records provide a viable alternative to fixed-length records. With this approach, the lengths of records can vary, depending on the number of transaction records coupled to them (Figure 9.7a). For example, an accounts receivable system could store data about a customer (account number, address, telephone number, and various balances) in a fixed-length position of a record called the root area, while storing each charge or payment the customer makes in a variable portion. By appending payment data onto the master record, the system no longer requires a separate transaction file, and it will not have trouble associating the transaction data with the master file data because both reside in a single record.

The shortest variable-length record would consist of the root area alone, whereas the longest one might permit storage of hundreds or thousands of changes. The formula below calculates the length of a variable-length record:

Record length = Length of fixed segment + Length of a variable segment ×
Number of variable-length segments

For example, assume that each fixed-length segment equals 256 characters and that each variable-length segment takes 50 characters. We call the number of transactions (the number of variable-length segments) segment occurrences. If four customers generate 2-, 0-, 10-, and 30-segment occurrences respectively, their total record length would be:

Customer 1: Record length = 256 + (2 × 50) = 256 + 100 = 356
Customer 2: Record length = 256 + (0 × 50) = 256 + 0 = 256
Customer 3: Record length = 256 + (10 × 50 = 256 + 500 = 756
Customer 4: Record length = 256 + (30 × 50) = 256 + 1500 = 1756

The variable-length record concept can ease the data storage of a wide range of applications, including those in accounting, banking, schools, and libraries.

Applied to Fleet Feet's accounts payable system, the variable-length record approach would specify that the root segment should hold data about each vendor: vendor number, name, address, and amount owed (Figure 9.7b). The variable segment would hold the data about a purchase or payment authorization.

Some word processing programs use variable-length records to track each line that the user enters. Since each line has a different physical length, the word processor places a line length count at the front of each line or marks the end of the line with a special character, one that a user cannot enter from the keyboard.

Variable-length record files make the most sense in applications where subsequent transactions require association with corresponding master records, elongating the original record. However, since a system using this approach cannot update data in an on-line transaction environment (rewriting a longer record over the top of a shorter one), it usually works best for sequential files on tapes or on disks operating in a batch-processing environment. In most cases, variable-length records are avoided because of the problems that arise in maintaining accurate record lengths.

Figure 9.7 Use of variable-length records crosses a wide range of applications.

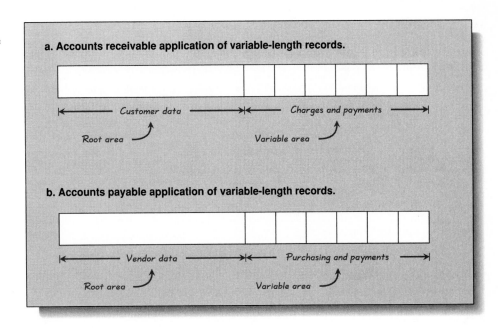

a. Accounts receivable application of variable-length records.

Customer data — *Charges and payments*

Root area *Variable area*

b. Accounts payable application of variable-length records.

Vendor data — *Purchasing and payments*

Root area *Variable area*

FILE AND PROCESSING CONTROLS

As with output controls, the analyst can choose from a wide variety of file and processing controls. Good controls will protect the integrity and accuracy of files and make them less prone to errors or tampering.

Record Counts

A **record count** simply tallies the number of records read from or written to a file and is one of the easiest control procedures. If the system does sequential file processing, its transaction file may undergo three basic types of updates: additions, changes, and deletions. The record count control procedure uses the following formula to monitor additions and deletions:

Number of records in the New Master File (NM) =
  Number of records in the Old Master File (OM) −
  Number of records dropped from the Old Master File (DR) +
  Number of records added to the New Master File (AD).

WORKING WITH TECHNOLOGY
GUIDELINES FOR DESIGNING FILES

All storage systems depend on what you wish to keep there. You would not think of renting a large garage to store income tax records, nor would you try to stuff a tennis racket into a shoebox. Similarly, a computer system's files depend on the user's requirements and upon hardware and software restrictions.

In the 1960s and 1970s, disks were expensive "shoeboxes" with high costs and low capacity, but many of today's disk drives hold vast volumes of data inexpensively. Applications can exploit increased disk capacity to put important data at users' fingertips.

Despite advancements in disk technology, tape still offers a viable alternative for storing data, especially if the application involves hundreds of billions of characters or does not require data at the users' immediate disposal via a personal computer or terminal. Tape is also seeing a revival of interest in the personal computer marketplace when used for backing up hard disk drives and as a medium for distributing software to customers.

Whether a system uses disks or tapes and searches sequentially, randomly, or indexed sequentially, certain "rules of thumb" govern file design:

1. From the data flow diagram (DFD), identify all data stores or files.

2. Assign each file or data store a unique and meaningful name.

3. Specify the type of file: master, detail, table, and so on.

4. List the data elements in the file, using a data dictionary format. Include in this definition the data type (alphanumeric or numeric), as well as the length of the data element. If the element is numeric, decide if you want to store it in ASCII or EBCDIC and zoned decimal or true binary.

5. Identify the access requirements for the file (ascending by vendor name, random by vendor number, ZIP code for bulk mailings, and so on).

6. Determine the media for storage: disk or tape.

7. Estimate the number of records in the file.

8. Calculate the frequency of record additions, deletions, or changes. This tells you the file's volatility. Highly volatile files belong on disk, while static files can reside on tape.

9. Define the order of the file's data elements. Place key fields first, followed by changeable data, with an expansion area for additions.

10. Pick a key field that uniquely identifies each record (e.g., a Social Security number, part number, customer number, or vendor number).

11. Consider adding a data element for date of last activity for some types of records (e.g., customer payments or purchases).

12. Calculate the record length by adding together the lengths of each data element in the file. A hardware specialist can tell you if the length will fit a particular storage device. Block the file accordingly.

13. Specify each file's need for back-up.

Because input-output operations consume much more computer time than internal operations, good file design—which smoothes data storage and retrieval—can reduce costs and improve performance. The fewer READ or WRITE operations the system makes, the faster it will run, and the less it will cost over time.

Abbreviated, this becomes:

$$NM = OM - DR + AD.$$

Since existing records with changes do not affect the total number of records in either the old or new master file, the formula need not account for them.

The analyst may specify record count controls in an update program and have the computer tabulate and print these four quantities. Then a control clerk can visually match these totals to a manual total to ensure accuracy.

Back-Up

Back-up, or file copying on a routine basis, offers another simple control procedure for both on-line and batch systems. The operations staff within most computer installations routinely copies the entire contents of their disks onto tape, creating a back-up in case of damage, loss, or theft. Of course, these tapes should reside in a safe place, physically apart from the originals. Considering the time and money involved in recreating large files of data, an organization can easily justify transportation and storage costs incurred to protect the back-up files.

The best time to copy a file depends on the system. For some systems, we need to create back-ups both before and after major processing activities (e.g., printing payroll checks or preparing statements for customers). On the other hand, a real-time system, such as a college student registration system, may require copying databases every day, as well as before and after all weekly or monthly closing runs.

Updating and automatic back-up generation can occur daily, weekly, monthly, or whenever appropriate. Figure 9.3 illustrated an automatic back-up process in which the old transaction file and master file are sufficient to generate a new master file if it is ever lost, damaged, or destroyed.

California's State Medical System routinely uses tape in this manner. The Medical Eligibility History System consists of a master file of over a dozen tapes that contain a 300-plus character record for each of over 12 million medical service recipients. The state updates its tapes weekly, a job that requires about six hours of processing time. The transaction file of new or dropped recipients, plus changes in recipient data takes from two to six reels, depending on the time of year. These tapes reside for 12 months in a cavern near Lake Tahoe on the California–Nevada border, while the new master stays close at hand for the next week's processing cycle.

This type of processing ideally suits tape or disk systems that use an old master to create a new one. If a disk system writes new data over the old, thus making automatic back-up generation impossible, the organization must copy each file individually to a new tape.

USING CASE TO DESIGN FILES

In previous chapters, we examined Valley Blood Center's need for a computerized system to track blood donations by various donors and by hospitals that had units of blood. We drew a data flow diagram of its needs, showing basic processes that convert inputs and outputs.

Complementing the data flow diagram process model is the data model that analyzes the relationships between information stores or files. Data models consist primarily of drawing entity relationship diagrams (ERDs). Entities represent groups of data, such as a file. A line connects entity symbols to show the relationship between them. The relationship is documented with a text description and a specification of cardinality (one to many, one to one, many to many, and so on).

CASE tools support both process and data modeling and allow you to draw either one. For the ERDs, CASE software helps you define the entities, relationships, and attributes, plus cross-check the model for correctness. The tool allows you to add, change, or delete entities, relationships, or attributes in the ERD. It is more than an electronic drawing board because it constantly checks the components for completeness. For example, all CASE tools will check to make sure that each name you assign is unique. If you attempt to name an entity or a relationship with the name of a data store, you will receive an error message.

Each entity or relationship describes the business rules of the system. For example, a donor gives a unit of blood and a hospital transfuses a unit of blood into a patient. Both the Blood Center and the hospital refrigerate units of blood to keep them from deteriorating until the blood can be transfused.

This brief description of Valley Blood Center and its business rules shows the basic entities and relationships. We will need an entity (file) to hold data about donors, donations, patients, and hospitals. Relationships include donates, transfuses, and refrigerates (to make the nouns into verbs). However, a donor in one situation could turn into a patient in another situation. In other words, you could donate a unit of blood (as a donor) and receive a transfusion, perhaps of your own blood (as a patient).

For every entity, relation, and attribute, CASE software systems require you to define each further. For an entity, you must name it, list the alias (if two or more names represent the entity), cite the purpose of the entity, describe the properties, list all the attributes in the entity, and cite the operations performed on the entity. In addition, the CASE software usually allows a comment entry.

In a relationship definition, CASE software requires you to name it, list the alias, define the purpose, specify cardinality (1:1, 1:N, N:M), describe the business rules, and list the attributes.

Additionally, attributes of entities or relationships need complete definition. For each attribute, CASE software makes you name the entity, list the alias, specify purpose, spell out the ranges of acceptable values, show the format, identify the unit or precision, and show any dependencies on other attributes. It will allow a comment entry.

In the Valley Blood Center, we have an entity named donor. Since a donor can turn into a receiver of a unit of blood, we may have an alias, patient. However, the two are really quite different: one is a customer of the Valley Blood Center, while the other is a customer of a hospital. To avoid confusion, we will keep them as two different entities.

The purpose of the donor entity is to keep the facts about donors. The properties are that a donor must relate to guidelines on their age (greater than 18 and less than 65), weight (greater than 105), and condition of their health ("well") to donate.

The attributes of a donor include their Social Security number, name, address, telephone number, date of birth, age, weight, health condition, race, and height. Other facts that we will keep include number of donations made, date of the five most recent donations, and date that the donor made their first donation.

Some of the operations performed on the donor entity include aging the five most recent donations when a new donation is received and adding one to the number of donations.

As you can see, CASE ERDs require you to completely specify all that you know about the data model. These facts are kept in a computer for easy modification and reporting, plus error checking. With all this information maintained by the CASE software, why can't the software generate the file layouts automatically? In fact, many can do this plus much more—as we will see in the next chapter.

SUMMARY

Tapes have existed since the dawn of the computer age. Today, they more frequently provide back-up storage for disk-based systems and personal computers or as a media for software distribution. Few organizations still use tapes as the primary storage medium; new on-line systems continue to erode the popularity of tapes.

Disks provide the best of both worlds: direct and sequential access to data. New on the horizon are compact disks (CDs) and CD-ROMs with high-storage capacity and low costs. However, they suffer from slow access times.

Regardless of the storage medium selected, the analyst must specify a storage method: EBCDIC, ASCII, packed decimal, or true binary. The first two can store alphanumerical data, while the last two permit only numeric data.

Files are classified as master, transaction, back-up, temporary, or program. Each type has its own special purpose.

To describe a file, the analyst assigns it a name, lists the data items for each record, chooses a data type (alphanumeric or numeric) for each field in the file, stipulates the storage method of the data (EBCDIC, ASCII, binary, or packed decimal), estimates the number of records that the file can hold, and classifies the file as either master, transaction, back-up, temporary, or program. Finally, the analyst determines how record keys will logically relate records to their files.

KEY TERMS

Master file	Straight binary
Transaction or detail file	Sequential file
Back-up file	Blocking factor
Temporary or scratch file	Buffer
Suspense file	Cylinders
Table file	Indexed file
Program file	Direct file
Alphanumeric data	Hashing algorithm
Numeric data	Record key
EBCDIC	Fixed-length
ASCII	Variable-length
Collating sequence	Record count
Packed decimal	Back-up

QUESTIONS FOR REVIEW AND DISCUSSION

Review Questions

1. What are the the seven types of files?

2. What does 753 look like in EBCDIC, ASCII, true binary, and packed decimal?

3. What are the advantages and disadvantages of sequential, indexed, and direct file organizations?

4. How would you lay out a record for disk or tape files for a purchase in an accounts payable system?

5. What are two control methods for files?

6. What are the elements of a file design?

7. What is the difference between a compact disk (CD) and a typical hard disk?

Discussion Questions

1. List the three types of data storage techniques.

2. How long is the interblock gap?

3. How long does it take to retrieve data from a hard and floppy disk?

4. How long does it take to retrieve data from internal memory?

5. How much memory does it take to store the word "WORD" using the ASCII coding system?

6. What is the storage capacity of a 2,400-foot reel of tape storing data at 1,600 characters per inch?

7. What are the two parts of a variable-length record?

Application Questions

1. If a record requires 180 characters, is blocked 30 records to a block, and 12,000 records are in the file, how much 1,600-character-per-inch tape is required to store the file?

2. Sketch the bit patterns to store the number 256, using

   a. EBCDIC

   b. ASCII

   c. Packed decimal

   d. True binary

3. How many records must a system process to update a master file of 4,000 records if the transaction file contains 1,000 records and the indexed storage technique is used?

4. How many records must a system process to update a master file of 4,000 records if the transaction file contains 1,000 records and the direct storage technique is used?

5. How many records must a system process to update a master file of 4,000 records if the transaction file contains 1,000 records and the sequential file storage technique is used?

Research Questions

1. Is there any mathematical relationship between uppercase and lowercase letters in the EBCDIC and ASCII system?

2. Is there any mathematical relationship between numeric characters in the EBCDIC and ASCII system?

3. Write the EBCDIC and ASCII binary patterns for

   a. A

   b. a

   c. 6

   d. z

   e. Z

4. Draw the ERD for the Valley Blood Center.

After listing the elements, Frank writes the other specifications for the file. Storage media identifies the type of device that the file will require. For all of Frank's files, he chooses disk. File type states the category of the file; CUSTOMER is a master file. File order tells the logical organization of the file, which is Indexed by Customer-id.

Primary key identifies which element uniquely identifies each record, which is Customer-id. Foreign key is a secondary element in the file that allows a logical relationship to exist to another file; there are no foreign keys in CUSTOMER.

The record length says how much space each record takes (160 bytes). The blocking factor tells the number of logical records per physical record and is not usually pertinent for a disk file. CUSTOMER is not blocked. Estimated length gives a record count for the file, which is 5,000 and growing as new customers are added on a daily basis.

Back-up requirements identify the frequency of making copies. Frank wants a daily back-up of the very important CUSTOMER file.

Security identifies the the information that should be kept private because of sensitive, legal, moral, or ethical concerns. CUSTOMER contains some data that Frank feels are private, such as telephone numbers and financial data.

Frank repeats this process for the TAPE file (Figure 9.10). Every tape will have its own identifier, a tape-id. Frank envisions that the bar codes that View Video will affix to the tape will serve as the Tape-id.

Figure 9.10 File data dictionary for View Video's TAPE file.

System:	Video Cassette Tape Rental
File Name:	TAPE
Analyst:	Frank Pisciotta
Date:	11/05/93

Element/Entity Name	Length	Storage Method	Comments
Tape-id	10	Packed decimal	
Title	30	ASCII	
Year-made	4	Packed decimal	Format YYYY
Rating	4	ASCII	
Category	12	ASCII	
Date-purchased	6	Packed decimal	Format YYMMDD
Date-last-rented	6	Packed decimal	Format YYMMDD
Times-rented	5	Packed decimal	
Cost	7	Packed decimal	Format 9(05)V99
Income-generated	7	Packed decimal	Format 9(05)V99
Tape-status	1	ASCII	

Storage Media:	Disk
File Type:	Master
File Order:	Indexed by Tape-id
Primary Key:	Tape-id
Foreign Key:	None
Record Length:	92
Blocking Factor:	None
Estimated Length:	6,000 and growing
Back-up Requirements:	Daily
Security:	None

The RENTAL file tracks each tape rented by a customer (Figure 9.11). RENTAL is a detail file related back to CUSTOMER and TAPE. Tape-id allows access to the master TAPE record for this rental, in case we need to know the title. Customer-id allows access to the master customer record, in case we need to know the name of the borrower.

PAYMENT is the scratch file that we will pass on to BusinessWorks PC (Figure 9.12). On a daily basis, we create PAYMENT by writing a record to it after checking in a tape. At the end of a day, we will post the payments to customer accounts in BusinessWorks PC. When the posting is completed, the file is erased.

CLERK is another master file holding data about the View Video staff that checks out or in videotapes (Figure 9.13). The purpose of this file is to allow us to translate a clerk's initials into his or her formal name.

STORE is our single table file (Figure 9.14). We keep fairly static data in STORE. Once in a while, we will have to change an element in the file; for example, when we increase prices. The file holds pertinent data for reporting purposes and Frank builds this file in the event that the software is resold to another video store. The new owners would then need to change the STORE file so that all the reports show their store's name and address instead of View Video's.

Mark and Cindy Stensaas wanted to show the names of various movie stars and Frank builds a master file to hold each star's name and a star-id (Figure 9.15). This file allows a star's name to exist a single time and will thus keep spelling errors to a minimum.

Figure 9.11 File data dictionary for View Video's RENTAL file.

System:	Video Cassette Tape Rental
File Name:	RENTAL
Analyst:	Frank Pisciotta
Date:	11/05/93

Element/Entity Name	Length	Storage Method	Comments
Customer-id	7	Packed decimal	
Tape-id	10	Packed decimal	
Date-rented	6	Packed decimal	Format YYMMDD
Date-returned	6	Packed decimal	Format YYMMDD
Rental-fee	7	Packed decimal	Format 9(05)V99
Amount-paid	7	Packed decimal	Format 9(05)V99
Clerk-initials	3	ASCII	

Storage Media :	Disk
File Type:	Detail
File Order:	Random
Primary Key:	Customer-id
Foreign Key:	Tape-id
Foreign Key:	Clerk-initials
Record Length:	46
Blocking Factor:	None
Estimated Length:	2,000 per day
Back-up Requirements:	Daily
Security:	None

Figure 9.12 File data dictionary for View Video's PAYMENT file.

System:	Video Cassette Tape Rental
File Name:	PAYMENT
Analyst :	Frank Pisciotta
Date:	11/05/93

Element/Entity Name	Length	Storage Method	Comments
Customer-id	7	Packed decimal	
Date-returned	6	Packed decimal	Format YYMMDD
Rental-fee	7	Packed decimal	Format 9(05)V99

Storage Media:	Disk
File Type:	Scratch
File Order:	Random
Primary Key:	None
Foreign Key:	None
Record Length:	20
Blocking Factor:	None
Estimated Length:	2,000 per day
Back-up Requirements:	Daily
Security:	None

Figure 9.13 File data dictionary for View Video's CLERK file.

System:	Video Cassette Tape Rental
File Name:	CLERK
Analyst:	Frank Pisciotta
Date:	11/05/93

Element/Entity Name	Length	Storage Method	Comments
Clerk-initials	3	ASCII	
Last-name	20	ASCII	
First-name	20	ASCII	
Middle-initial	1	ASCII	

Storage Media:	Disk
File Type:	Master
File Order:	Indexed by clerk initials
Primary Key:	Clerk-initials
Foreign Key:	None
Record Length:	44
Blocking Factor:	Not applicable
Estimated Length:	50 records
Back-up Requirements:	Daily
Security:	None

The second half of the Stensaases' request is met by PERFORMANCES (Figure 9.16). This detail file allows us to retrieve a star's name (we have the star-id) relative to a specific videotape (we have the tape-id). Thus, for videotape identification number 0000032189, we find all the records in PERFORMANCES with this tape-id and backtrack to the star's name in STAR via the star-id.

Figure 9.14 File data dictionary for View Video's STORE file.

System: Video Cassette Tape Rental
File Name: STORE
Analyst: Frank Pisciotta
Date: 11/05/93

Element/Entity Name	Length	Storage Method	Comments
Store-name	40	ASCII	
Street-address	30	ASCII	
City	14	ASCII	
State-abb	2	ASCII	
ZIP-code	5	True binary	
Phone-number	9	Packed decimal format	AAAPPPDDDD
Daily-rate	7	Packed decimal format	9(05)V99

Storage Media: Disk
File Type: Table
File Order: Not applicable
Primary Key: Not applicable
Foreign Key: Not applicable
Record Length: 107
Blocking Factor: None
Estimated Length: 1 record
Back-up Requirements: When changed
Security: None

Figure 9.15 File data dictionary for View Video's STAR file.

System: Video Cassette Tape Rental
File Name: STAR
Analyst: Frank Pisciotta
Date: 11/05/93

Element/Entity Name	Length	Storage Method	Comments
Star-id	4	Packed decimal	
Last-name	20	ASCII	
First-name	20	ASCII	
Middle-initial	1	ASCII	

Storage Media: Disk
File Type: Master
File Order: Indexed on Star-id
Primary Key: Star-id
Foreign Key: None
Record Length: 45
Blocking Factor: None
Estimated Length: 6000 records
Back-up Requirements: Daily
Security: None

BORROWER finishes the file dictionary definition (Figure 9.17). In a video store, multiple borrowers can rent tapes on a single account. This file keeps the individual names of the borrowers apart from the master owner in the CUSTOMER file. As an example, Ruth Williams may allow her children—Janet, Karen, Kathleen, and James—to borrow on her account, customer-id number 6527201. Ruth is the master owner and her children's names are kept in BORROWER with customer-id 6527201.

Figure 9.16 File data dictionary for View Video's PERFORMANCE file.

System:	Video Cassette Tape Rental		
File Name:	PERFORMANCE		
Analyst:	Frank Pisciotta		
Date:	11/05/93		

Element/Entity Name	Length	Storage Method	Comments
Star-id	4	Packed decimal	
Tape-id	10	Packed decimal	

Storage Media:	Disk
File Type:	Detail
File Order:	Random
Primary Key:	Star-id
Foreign Key:	Tape-id
Record Length:	14
Blocking Factor:	None
Estimated Length:	Approximately 3 times the number of records in TAPE
Back-up Requirements:	Daily

Figure 9.17 File data dictionary for View Video's BORROWER file.

System:	Video Cassette Tape Rental		
File Name:	BORROWER		
Analyst:	Frank Pisciotta		
Date:	11/05/93		

Element/Entity Name	Length	Storage Method	Comments
Customer-id	7	Packed decimal	
Last-name	20	ASCII	
First-name	20	ASCII	
Middle-initial	1	ASCII	

Storage Media:	Disk
File Type:	Detail
File Order:	Random
Primary Key:	Customer-id
Foreign Key :	None
Record Length:	48
Blocking Factor:	Not applicable
Estimated Length:	Approximately 3 times the number of records in CUSTOMER
Back-up Requirements:	Daily
Security:	None

With his file design in hand, Frank seeks out Mark and Cindy and explains each file, the data elements, and file relationships and asks for comments or errors. Cindy asks him how she would associate a specific borrower with a rental. For example, how would she know if Kathleen or Ruth Williams (both with customer-id 6527201) is the renter of a tape? How about multiple copies of a tape with the same title? Will each one have its own tape-id? If so, how can she tell if there are extra copies of that tape that have been not checked out when a customer requests a specific tape?

Frank realizes that he has missed some key parts to his file design and sits down to repair them. He bring back his corrections to Mark and Cindy. They approve them and with his new file designs reviewed and corrected, Frank can now move to the next step of the design phase of the system life cycle.

Working With View Video

Reviewing a design with a user often uncovers errors or omissions in a design. Frank discovered that fact in his review with Mark and Cindy. Users know the application inside out, often working with it for years, and the analyst should ask for and use their expertise.

Case Study Exercises

In our discussion of View Video's files and Frank's two major problems, we did not reveal the solutions that Frank produced. Answer the following questions:

1. What file modifications are necessary to keep track of the specific borrower for a tape?
2. How do you solve the problem of multiple tapes with the same title?
3. Did Frank forget any other files?
4. Some video stores price tapes differently. How do you modify the file designs for varying prices on tapes?

Database Design

GOALS

After reading this chapter, you should be able to do the following:

- State the advantages and disadvantages of a database management system
- Define the three basic database systems
- Explain the schema concept
- Name and classify four commercial database management systems
- Explain the purpose of a query language
- List the four types of utility software that accompany most database management systems
- List three common rules for relational database design

PREVIEW

As an alternative to designing traditional disk and tape files, an analyst may choose a database manager for data storage and retrieval (Figure 10.1). Even so, the analyst must still determine what data the computer must keep and what identifiers users will employ to retrieve the data. With a database manager, you develop a description of the data files (known as a schema), which you define from the report formats, data collection screens, and the data dictionaries developed earlier. If you select a database manager (DBMS), the schema—rather than disk or tape file design—documents your design and becomes part of the system specifications.

A DBMS is a software system that helps manage disk-based data. Instead of concentrating on the physical aspects of the file (blocking, cylinders, tracks, overflow area, and sorting), analysts and programmers can focus on the logical aspects of the data and their processing needs.

Most high-level languages do not have a standard set of commands and phrases that allow them to "talk" to the DBMS. Instead, each vendor implements the tie between the high-level language and the DBMS with their own nonstandard statements. Regardless, this powerful software system is heavily used in application programs and will grow even more in the 1990s.

DATABASE MANAGEMENT SYSTEMS

Traditional file systems, often called **flat files,** require that each application within a system retain responsibility for its own data. In other words, a data item such as an employee number or name, though common to both payroll and personnel systems, appears separately in each. If the employee name ever changes, someone would have to key that change into every file in which it resides. Databases centralize such data to make this duplication of effort unnecessary (Figure 10.2).

Every computerized information system requires not only data, but the structure (physical location) of that data. A **database management system (DBMS)** collects and structures related files so that many users can easily retrieve, manipulate, and store data.

A number of computer manufacturers and software developers market database management systems. These software packages enable users to organize large, complex bodies of data into useful, accessible, and compact forms—thus avoiding the awkward and inefficient file management techniques of the past. Since database systems merge data into one pool (Figure 10.3), any change to one file automatically affects all other relevant files. Programs to retrieve, update, add, or delete data in the database can involve batch, on-line, or mixed processing.

Figure 10.1 Database design is the third stage. During database design, we write the schema for the physical database.

A DBMS precludes the need for the programmer to worry about such time-consuming details as where or how the computer physically stores the data because the DBMS can retrieve the data at will. Now the analyst can focus on higher priorities: the system itself and users' needs.

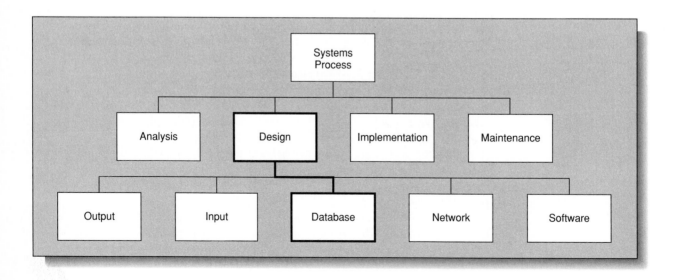

Advantages of a DBMS

DBMS concepts represent an evolutionary step from direct, sequential, and indexed sequential methods. The innovation lies in separating the definition and control of the database from the specific application system, thus making logically connected files accessible to all programs. This results in more efficient data processing, simpler systems development, and lower programming costs.

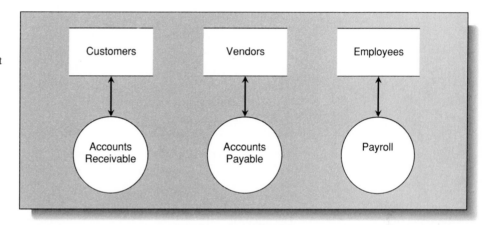

Figure 10.2 Traditional file systems make each system responsible for its own data. Since different systems do not share a database, a change in one system necessitates a separate change in another.

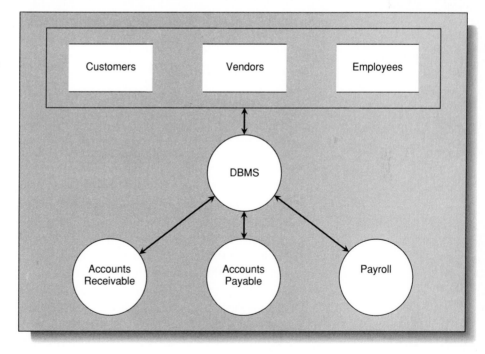

Figure 10.3 Database management systems centralize the data so that different systems can share it.

Specific advantages of databases include:

- *File consolidation:* Pooling data reduces redundancy and inconsistency and promotes cooperation among different users. Since databases link records together logically (even if they are physically separated), a data change in one system will cascade through all the other systems using that data.
- *Program and file independence:* This feature separates the definition of the files from their programs, allowing a programmer to concentrate on the logic of the program instead of on precisely how to store and retrieve data.
- *Access versatility:* Users can retrieve data in many ways. They enjoy the best of both worlds: sequential access for reporting data in a prescribed order and random access for rapid retrieval of a specific record.
- *Data security:* A DBMS usually includes a password that controls access to sensitive data. By limiting access to read only/write only specified records or even fields within records, passwords can prevent certain users from retrieving or altering data.
- *Program development:* Programmers must use standard names for data items rather than inventing their own from program to program. This lets the programmer focus on the desired function.
- *Program maintenance:* Changes and repairs to a system are relatively easy to accomplish.
- *Special information:* Special purpose report generators can produce reports with minimum effort.

Disadvantages of a DBMS

There are definite trade-offs to a DBMS. Some disadvantages include:

- *Additional hardware:* Using a DBMS may require the purchase of additional memory and/or disk drives. Memory is used to hold the DBMS software, while extra disk drives keep the special files that the DBMS requires. However, this is offset by savings due to less redundancy in the data.
- *Staff retraining:* Programmers who are unfamiliar with DBMS concepts and terminology will need special training to orient them to the new DBMS environment.
- *Software cost:* Software vendors charge for their DBMS software and in some cases, the DBMS software cost can exceed $100,000.
- *Special staff:* The organization may need to hire a DBMS expert to supervise and manage the DBMS. Usually called a "Database Administrator," this person establishes rules regarding who is permitted to use the data, monitors the data security against unauthorized intrusion, and advises analysts and programmers about the best ways to use the DBMS.

Despite these disadvantages, many organizations are switching from flat file systems to a DBMS. They see the advantages as outweighing the disadvantages, especially over the long term.

TYPES OF DATABASES

Traditional flat file processing does not automatically relate records, thus forcing a programmer to build those relationships. This becomes unnecessary in a database environment. Databases relate data sets according to one of three models: hierarchical, network, or relational.

In conventional file systems, groups of bytes constitute a field, one or more fields form a record, and two or more records make a file. In a database environment, a group of bytes constitutes a data item or segment, a collection of data items forms a data entry, and a series of data entries makes a data set. The combined data sets form the database itself. Going back to flat file terminology, a data item is a field, a data entry is a record, and a data set is a file.

Hierarchical

Hierarchical models are relatively old, developed in the 1960s, but they are well understood by software developers. The hierarchical model views data in a treelike structure, much like the organization chart of a business. It also resembles a family tree, with different branches representing grandparents, parents, and children. In this model, data sets are categorized as master (owner or parent), while others serve as detail (member or child). The key to understanding the hierarchical model is to understand that a detail data set is subservient to one master data set. A master can have multiple detail data sets; however, detail data sets can have only a single master. The hierarchical model follows the basic family unit: parents can have multiple children; however, a child has only one set of parents.

Besides the master detail idea, hierarchical data sets exhibit the 1:N concept. Read as "one to N," this concept states that for every master data entry (record) it is possible to have N (meaning zero to many) detail data entries. However, for each detail data entry, there is one and only one master data entry. Duplicate master data entries are prohibited.

The diagram in Figure 10.4a illustrates a hierarchical relationship: the "Parts-Master" master data set rules the "Parts-Detail" or detail data set. Rectangles often represent master and detail data sets (Figure 10.4b). Master data sets may govern more than one detail data set, but a detail can serve only one master (Figure 10.4c). For example, Parts-Master governs two detail data sets, Parts-Detail and Sales-Detail, both of which allow for multiple data entries.

Data entries to Parts-Master would include the following data items: product number, description, unit cost, unit of measure, and quantity on hand. Entries to Parts-Detail would include product-number, quantity ordered, date of order, and code number. The product number, which appears in both sets, permits the system to relate data entries logically (Figure 10.4c).

Figure 10.4 Hierarchical data sets employ the master detail relationships. Each record in the master file has a key field (product-number) that is repeated in each detail file. Arrowheads visually show the number of records in the detail file belonging to a record in the master file. A single arrowhead means there can be at most one detail record for each master record (a.); two arrowheads mean there can be more than one detail record for this master record (b. and c.).

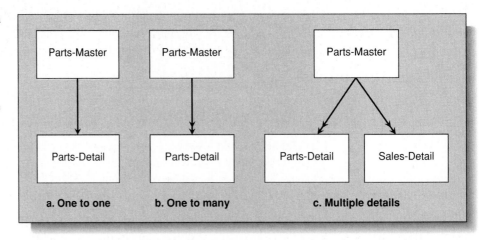

Parts-Master   Parts-Master   Parts-Master

Parts-Detail   Parts-Detail   Parts-Detail   Sales-Detail

a. One to one   **b. One to many**   **c. Multiple details**

The arrows in these diagrams indicate the links that the DBMS has established among the data sets. A single-headed arrow indicates a one-to-one relationship, each master governing one detail. Double-headed arrows indicate a one-to-many relationship: one master governs many detail records.

To access any data entry in the detail data set, the programmer must start at the master data entry and navigate through each detail data entry one at a time. For example, consider the case of Fleet Feet's accounts receivable, where we might have a master data set of customers and a detail data set of purchases that customers have made. We want to list each customer and their purchases. We start by retrieving the first master customer data entry, then sequentially accessing all detail purchase data entries for that customer. Upon reaching the last detail data entry for the customer, we access the second customer master data entry and repeat the process for this customer's purchase.

One of the problems with this model concerns deletions: How do we remove a master record? To solve this problem, the programmer must navigate through all the details for the master, removing each before removing the master itself. While not a logically complex programming problem, it does show one of the main disadvantages of this model. Some examples of hierarchical DBMSs are IBM's DL/1-DOS/VS and Intel Systems Corporation's System 2000.

Network

One major limitation of the hierarchical model is the requirement of a single master data set over a detail data set. In real-life information systems, data do not usually exist in this manner. Some situations call for two, three, or more master data sets linked to a detail or detail data set. Developed in the late 1960s (after hierarchical models), **network** or CODASYL database relationships function much like hierarchical ones, except in this case, detail data sets can serve more than one master data set (Figure 10.5). A master data set does not have to have any detail data sets, but a detail data set must have at least one master data set. The network database is not a communications network, nor is it a networked database; rather, it is a term that describes the model of master to multiple detail data sets.

Figure 10.5 In the hierarchical model, masters could govern several detail data sets, but each detail could have only one master. In networks, however, details can have more than one master. Therefore, the network structure can handle very complex information relationships.

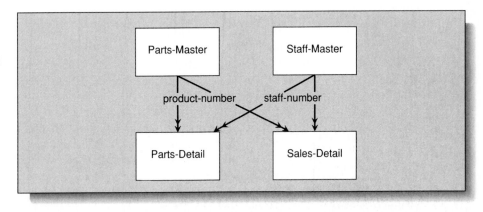

As with the hierarchical model, the network model exhibits the 1:N concept. Each master data entry can have multiple detail data entries. In fact, the hierarchical model is a simple version of the network model; it exhibits the same 1:N relationship, but is restricted by the fact that details are owned by a single master.

Suppose Fleet Feet management wants an inventory system that it can use to track which staff member sold which parts. Such a system would require two master data sets: Parts-Master and Staff-Master. While Parts-Master would be identical to its representation in the hierarchical model, Staff-Master would demand the following data items or segments: staff-number, staff-name, year-to-date sales, and last-year's-sales. Parts-Detail and Sales-Detail contain one new data item, staff-number, which allows the system to link both detail data sets to both master data sets. Now the system can generate several reports: sales records for individual staff members, totals for each part sold, and so on. The system can retrieve the relevant data from Parts-Detail and Sales-Detail by using the product-number or the staff-number as a key.

The problem of deletions in the network model is the same as in the hierarchical model. Each detail is removed individually before the master is deleted. Making the issue even more complex is the fact that when a detail is owned by another master data set, it also may need modification when a deletion is made. Fortunately, most network DBMSs take care of this issue for the programmer and they need not worry about modifying a second, third, or fourth master.

Network DBMSs overcome the single owner limitation of the hierarchical model; however, a substantial amount of navigation is still required between master and detail data entries. Examples of network DBMSs include Cincom System Inc.'s TOTAL, Cullinet Software's IDMS, Hewlett Packard's very popular Turbo-IMAGE, and IBM's IMS.

The chief drawback to both the hierarchical and network models is their inflexibility. All the masters, details, and the links between them are defined at design time. It is difficult (although not impossible) to change the design if the analyst misses a field, makes one too small, wants to rename a field or file, or needs to add or delete a master detail relationship. The analyst thus must have his or her design right the first time when using these two models, which is a rare occurrence.

Relational

Relational data sets order data in a table of rows and columns and differ markedly from their hierarchical or network counterparts. They do not set up master-detail data sets (Figure 10.6), but instead store data in two-dimensional tables, much like the traditional file system with its records, fields, and files. The table's rows, called "tuples" (rhymes with "couples"), contain records and its columns contain fields, called attributes. The full two-dimensional table represents a file or "relation." Note that the Vendor table file in Figure 10.6 contains four tuples and three attribute names. Now, if we wish to know the name of the supplier of black chairs, the relational DBMS searches the type and color columns of the Supplies table, finds supplier number 45, then scans the Vendor table for number 45, which turns out to be Campbell's.

Developed in the late 1970s, a relational DBMS can perform several basic operations:

- Create or delete tables
- Update, insert, or delete tuples
- Add or delete attributes
- Copy data from one table into another
- Retrieve or query a table, tuple, or attribute
- Print, reorganize, or read a table or tuple (utility operations)
- Join or combine tables based on a value in a table

Users of a relational DBMS can create a new relation from two or more existing tables, thus permitting them to manipulate data in creative ways. For instance, we could make a

Figure 10.6 Two inventory relational tables: Supplies and Vendor. The vendor identification appears in both tables for cross-referencing purposes.

SUPPLIES

Product Code	Type	Color	Quantity On Hand	Cost $	Vendor Identification
2162	Easel	Brown	7	89.95	62
2977	Chair	Brown	12	329.62	41
3195	Chair	Black	6	379.47	45
6377	Desk	White	8	816.21	45
7422	Lamp	Red	19	19.27	75
8654	Clock	Walnut	4	205.55	62

VENDOR

Vendor Identification	Supplier Name	Purchases This Year
41	Walkers Office Supply	573.92
45	Campbell's Office Supply	42,764.26
62	Associated Office Supply	5,599.19
75	Rose Office Supply	366.72

new table that lists furniture by type, giving us a table showing chairs, desks, lamps, clocks, and easels.

Because of their versatility and usefulness, relational DBMSs will dominate the 1990s. In the 1970s and early 1980s, all DBMSs required a powerful computer and suffered slow response time when a large number of users simultaneously tried to access the same database. The current emergence of very fast and cheap processors have made relational DBMSs a practical reality. In fact, several relational DBMSs now exist for the IBM PC, IBM PS-2, and Apple Macintosh lines of personal computers, among others. With the advent of even more powerful computers in the future, relational databases should become increasingly accessible, easier to learn, and thus more popular. Examples of relational DBMSs include Relational Technology's INGRES, IBM's SQL/DS and DB2, Oracle Corporation's ORACLE, and Borland's Paradox and dBASE IV.

Hierarchical and network DBMSs require the analyst to specify particular pathways through the database. As we will see in the next section, these pathways must exist to permit the database to locate and extract database records. Relational DBMSs differ in that their pathways are dynamic. The analyst can easily add a new column, change a column's name, change its width, define new tables, or rename existing tables with the relational model.

DEFINING THE PHYSICAL DATABASE

Defining a traditional file processing or flat file involves describing the record formats: deciding how large to make the fields, their order in the record, selecting blocking factors, choosing a mode of data storage, and specifying a storage medium (disk or tape).

Defining a database requires similar decisions. We call such a description a **schema**, and we write it in a special format called the **data definition language (DDL)**.

Each database management system uses a different DDL. Burroughs' DBMS, DMS-II (Data Management System II) uses a definition language called DASDL (Data and Structure Definition Language). Hewlett Packard's data definition language for its DBMS, Turbo IMAGE, is DBDL (Data Base Definition Language). IBM's DB2 is SQL (Structured Query Language).

Before writing a schema for a database system, the analyst must learn the rules of the DDL in question. See Figure 10.7; some of the rules in Oracle's SQL (and other versions as well) include:

- Data sets, entries, and items require names that must start with a letter and can include the digits 0–9 plus some special characters (such as _ , $, and #).
- Each data item must have a data type. Some common data types are char, date, and number. Char and number data types require lengths.
- A comma signifies the end of a data element's definition.
- Parentheses surround all the data elements in a set.

Let's continue with our inventory system example using the Oracle version of SQL, which falls into the relational model. First, we define the data entries and structures for two data

RELATIONAL DBMS—WHAT'S IN A NAME?

In Shakespeare's Romeo and Juliet, the two lovers are thwarted by a family feud, prompting Juliet to wish Romeo had a different last name, so family rivalry wouldn't keep them apart. She asks the very practical question, "What's in a name?"

Today's DP and MIS managers might well ask the same question when they regard advertisements for relational database management systems. Does a new name mean a really new or different product? In the case of relational DBMSs, the name does mean something special.

Dr. Edgar Codd, a member of the IBM project team working on System R (a Relational Store Interface that manages devices, space allocation, transaction consistency and recovery, and more), proposed the "relational" theory in the early 1970s. System R's goal was to seek the optimum access paths for data to travel and to offer users a logical view of data, easy manipulation of data, and a high level of security and integrity.

To improve upon the hierarchical and network DBMSs developed in the early 1960s, Codd began searching for a simpler approach, which he ultimately found, but which required IBM's largest mainframe computer and limited use to only a few users at a time. Yes, Codd's relational theory was viable, but it demanded sophisticated hardware that lay beyond the grasp of most organizations, until IBM introduced SQL, a relational product that evolved from System R, but which could function on smaller computers.

Since Codd borrowed the term *relational* from mathematical theory, a whole new mathematical jargon sprang up around the approach, bestowing new terms upon old DP concepts: a record became a tuple, a field an attribute name, and a file a relation.

A pure, or close to pure, relational system following Codd's theoretical constructs should meet many criteria, among them the rule that information be represented by values in two-dimensional tables with data accessible by value only. Doing so allowed users to create new tables from existing ones by matching column values. The approach also dictated that data item definitions be centrally stored in an integrated dictionary, that extended information retrieval be available through on-line relational query facilities, and that information be accessible via distributed DP facilities. A user organization could then build applications with high-level languages in an interactive environment, with eventual on-line processing executed through a resource-conserving teleprocessing monitor.

Sound complicated? It is, but you should bear in mind the basically simple concept that relational systems allow users to dynamically change associations in files to suit their own purposes within certain authorized limits. This means that users can select fields from files to create new ones, then store data a new way. Most importantly, they can create new associations without getting bogged down in technical details.

So we see that the name *relational* simply refers to a better way to manage information. Users can add data to files without affecting previously existing data, and they can blend data from two different files to form what looks like a third file, all in two-dimensional, flat file or table form.

The relational DBMS model obeys strict mathematical constructs and took intense effort by IBM and many universities to complete, yet it resulted in a beautifully simple result. In terms of the time it takes programming staffs to master a DBMS technique, the traditional hierarchical approach requires about a year or more of study and experimentation, the network type nine months to a year, and the relational system only about three months.

The database management systems of the '90s must accommodate the growing needs of a broad range of users for easy learning and operation. Such systems must be flexible, providing logical access and reporting, and they must allow for continuing growth.

Already many suppliers are marketing products that offer relational capabilities. Some even claim to be almost pure implementations of Codd's theory. Despite continuing controversy about the approach, however, it should steadily gain stature and acceptance in the future.

sets—Parts-Master and Parts-Detail—listing four data items (product-number, description, unit price, and quantity-on-hand). Parts-Master has six items (product number, quantity ordered, date ordered, date delivered, amount, and supplier number) for each entry in Parts-Detail.

Figure 10.7 The Data
Definition Language used by
Oracle's SQL. Analysts indent
and use parentheses to make
the SQL easier to read.

```
CREATE TABLE PARTS_MASTER
    (
    PRODUCT_NUMBER            CHAR(8) NOT NULL PRIMARY KEY,
    DESCRIPTION               CHAR(30),
    UNIT_PRICE                NUMBER(9,2),
    QUANTITY_ON_HAND          NUMBER(6) DEFAULT 0
    )

CREATE TABLE PARTS_DETAIL
    (
    PRODUCT_NUMBER            CHAR(8) NOT NULL,
    QUANTITY_ORDERED          NUMBER(6),
    DATE_ORDERED              DATE DEFAULT SYSDATE,
    DATE_DELIVERED            DATE
        CHECK (DATE_DELIVERED>=DATE_ORDERED),
    AMOUNT                    NUMBER(9,2),
    SUPPLIER_NUMBER           CHAR(8) NOT NULL
    )
```

To the right of each data item, we write CHAR or NUMBER and a number in parentheses to indicate the type of data we want stored and the number of characters or digits involved. For example, product number contains eight (8) digits, and unit price contains eleven (11) digits total—nine to the left of the decimal point and two to the right. Date data type says the field will hold a value that contains a date and time relative to the century, year, month, day number, hour, minute, and second.

Other entries in the Create Table are primary key, default value, null (or not null), and check. Primary key identifies this field as the one that will uniquely identify this row in the relational table. Default value is the value the field will take if no other value is assigned to it. Not null mandates that this field have a value all the time. Check verifies that a logical relationship is true before altering a column's value. In this case, we want to make sure that the dates are not out of order.

Having written the schema to define the database, the analyst must now compile it (Figure 10.8). The compiling process checks the schema for syntax errors (such as missing commas or parentheses and misspellings), allocates space on the disk, and builds all the files necessary for managing the database.

DATA MANIPULATION LANGUAGES

To access the database and add, change, or drop rows, the database management system uses special interface commands provided by the **data manipulation language (DML)**, a built-in feature of database management systems languages. A user wanting to retrieve the row for Product Number 3412 from Parts-Master, for example, would write the following SQL statement:

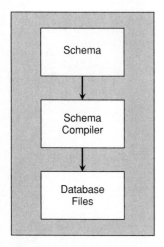

Figure 10.8 During the database definition process, the analyst writes the schema, which is compiled by a schema processor that builds the actual disk files.

```
SELECT *
    FROM PARTS_MASTER
    WHERE PRODUCT_NUMBER = '3412'
```

The word SELECT tells SQL to examine PARTS_MASTER to locate and retrieve product number 3412. The asterisk specifies to SQL that all columns are wanted from the row found. If no row exists for product number 3412, SQL will return an error message and the user can determine that this is not a legal product number.

We call the language enhancements that permit manipulation of the database **host languages**. If a DBMS supports its own language (other than COBOL, C, Pascal, or FORTRAN), we call the DBMS **self-contained**. Oracle's version of SQL is both host and self-contained. The SELECT statement above is written in SQL. In COBOL, we would write the same statements as:

```
EXEC-SQL
    WHENEVER NOT FOUND GO TO 1125-NOT-FOUND
END-EXEC.

EXEC-SQL
    SELECT * INTO :PRODUCT-NUMBER, :DESCRIP, :UNIT-PRICE,
        :QTY-ON-HAND
    FROM PARTS_MASTER
    WHERE PRODUCT_NUMBER = '3412'
END-EXEC.
```

The EXEC-SQL tells the COBOL compiler that it will see an SQL statement next. The WHENEVER NOT FOUND identifies the logical events the programmer desires if there is no row that the subsequent SELECT can locate. END-EXEC tells the COBOL compiler that the SQL statement started earlier is finished.

The COBOL SELECT has a list of variables preceded by colons. These variables, called host variables, will hold the data values retrieved by the SELECT. Within the COBOL program, the programmer must use these data names to access the values. A similar structure for a SELECT is possible in C and other languages.

COMMERCIAL VERSIONS OF SQL

Of all the relational DBMSs on the market, three especially stand out: IBM's SQL/DS and DB2 for the minicomputer and mainframe marketplace, Borland's dBASE for personal computers, and Oracle Corporation's ORACLE SQL for mainframes, minicomputers, and personal computers. All three DBMSs have gained wide acceptance, with dBASE having sold over a million copies.

IBM's SQL/DS, DB2 (Structured Query Language/Data System) and QMF (Query Management Facility) trace their roots to the experimental mid-1970s relational system called System R. SQL (pronounced "sequel") appears to have set the standard for relational database managers. After IBM published the specifications for SQL/DS in the mid-1970s, many competing vendors developed SQL/DS look-alike systems. Today, Digital Equipment Corporation, Hewlett Packard, and many personal computer manufacturers offer their relational DBMSs for their hardware and operating systems.

A product called dBASE has become the premiere database manager for personal computers. Originally written in 1979 by Wayne Ratliff (and named Vulcan after the home planet of Star Trek's Mr. Spock), the product did not attain commercial success until George Tate acquired rights to it, renamed it dBASE II (there never was a dBASE I), and marketed it under the Ashton-Tate logo. dBASE III+ eclipsed the earlier dBASE II by being faster, more powerful, and easier to use. Late in 1988, hierarchical dBASE III+ was superseded by relational dBASE IV, which offered more functions than its predecessor, including SQL commands. In 1991, Borland International, Inc. bought Ashton-Tate and dBASE.

In sharp contrast to IBM DB2 and dBASE IV, Oracle Corporation has taken a different approach. Whereas dBASE is PC-based and DB2 rooted in IBM, Oracle's SQL strives for vendor independence. The software operates on such varied platforms as IBM's 6000 series of minicomputers, Apple's Macintoshes, IBM's PC/PS-2 and clones, Hewlett Packard's UNIX computers, Sun's Workstations, and Digital Equipment's VAX series of machines. Oracle's claim is that an Oracle application can run on any platform with no changes. This platform independence makes it an ideal pick for software developers that want to sell their product on a wide variety of systems and minimize or eliminate costly conversions.

SQL RULES AND SYNTAX

Regardless of which version an organization uses, SQL follows the relational rules of tables, columns, and rows. Remember that rows correspond to records, columns to fields, and tables to a file. Also, a relational DBMS does not use the master and detail data sets that characterize network or hierarchical DBMSs. SQL uses its own data definition language (DDL) and data manipulation language (DML).

SQL makes it easy to add or delete columns to or from a relation, and also simplifies the building of indexes. For example, if a user desires to add a preferred vendor column to PARTS_MASTER, he or she merely enters a single command in the form of the clause:

```
ALTER TABLE PARTS_MASTER ADD PREFERRED_VENDOR CHAR(10)
```

Immediately upon receiving this clause, SQL adds a new column to all rows in PARTS_MASTER naming it PREFERRED_VENDOR.

Deletions of a column(s) are just as easy by using the following clause:

```
ALTER TABLE PARTS_MASTER DELETE QUANTITY_ON_HAND
```

From now on, the column is deleted from the table, any data in the deleted column are lost, and all references to it will cause an error to take place.

There are other uses for the SQL's ALTER statement. We can increase the width of a column by using:

```
ALTER TABLE PARTS_MASTER
     MODIFY (DESCRIPTION CHAR(40))
```

This ALTER changes the width of a column (DESCRIPTION), from the original value in a CREATE—which was 30—to a new value, 40. Other versions of ALTER allow shrinkage of a column or changing the data type from NUMBER to CHAR.

ALTER can remove constraints that we built with our CREATE statement:

```
ALTER TABLE PARTS_DETAIL DROP CONSTRAINT DATE_ORDERED
```

This ALTER removes our original constraint on DATE_ORDERED from SYSDATE to having no constraint at all.

If a table needs renaming, SQL has this statement:

```
RENAME PARTS_DETAIL TO PARTS
```

and from now on, the table is known as PARTS.

Removing a table is a very dangerous operation and one that not everyone should have the capability to do. If authorized to do so, you would write:

```
DROP TABLE PARTS
```

and it is gone, along with all the data it held, all the column names, and all the indexes.

An index in SQL can function according to sequential access or unique values. A sequential index provides an ordered pathway through a table in either ascending or descending order. First, the user declares an index with a CREATE command:

```
CREATE INDEX BY_PRODUCT_NUMBER ON
     PARTS_MASTER (PRODUCT_NUMBER)
```

When using BY_PRODUCT_NUMBER, SQL retrieves rows (records) in ascending order by PRODUCT_NUMBER. A user can drop the index by using:

```
DROP INDEX BY_PRODUCT_NUMBER
```

Only authorized users should be allowed to do this, however.

The second type of index ensures unique values for selected columns. In an inventory system, for example, no two parts should ever have the same PRODUCT_NUMBER. We create this type of index by using:

```
CREATE UNIQUE INDEX PROD_NUM ON
      PARTS_MASTER (PRODUCT_NUMBER)
```

The word UNIQUE specifies individual part numbers for all rows in PARTS_MASTER.

To manipulate the database, the user employs SQL's data manipulation language (Figure 10.9). FROM clauses specify the table targeted for the command and WHERE clauses dictate which logical condition SQL will use to locate the row to which the command relates. Other SQL commands create a new temporary relation consisting of entries from other tables, make a user view (which is a subschema of a table), and remove duplicate rows. Built-in functions permit calculations of average, maximum, minimum, sum, and count values. Other functions allow date comparisons in order to find out which of two dates is the earliest or how many days have elapsed between two dates.

The GRANT command provides SQL's security. It can revoke a user's access to certain commands or afford them "read only" access. It can also limit their access to views or to specific columns and rows of any table.

SQL features a unique operation that can take records from multiple tables or relations and bring them together to form a new data set. This operation is called JOIN and is one of the most powerful statements in SQL's vocabulary. While not a separate statement, a typical SQL JOIN operation on our supplies and vendor tables might read:

```
SELECT PRODUCT_CODE, TYPE, COLOR, QUANTITY_ON_HAND,
       SUPPLIER_NAME, VENDOR_IDENTIFICATION, PURCHASES_THIS_YEAR
    FROM SUPPLIES, VENDOR
    WHERE SUPPLIES.VENDOR_IDENTIFICATION =
          VENDOR.VENDOR_IDENTIFICATION
```

Figure 10.9 IBM'S SQL/DS data manipulation commands allow the user to add, change, delete, select, and display data from tables.

Command	Purpose	Example
SELECT	Obtain a row from a table	SELECT DESCRIPTION, UNIT_PRICE FROM PARTS_MASTER WHERE PART_NUMBER = '1060'
UPDATE	Modifies a row or rows in a table	UPDATE PARTS_MASTER SET UNIT_PRICE = 34.95 WHERE PART_NUMBER = 123
INSERT	Adds a new row to the table	INSERT INTO PARTS_MASTER VALUES ('1440', 'PEN', .39, 100)
DELETE	Removes a row from a table	DELETE PARTS_MASTER WHERE PART_NUMBER = '4096'
JOIN	Matches rows in two tables based on contents of related columns	SELECT * FROM VENDORS, OPEN_ITEMS WHERE VENDORS.VENDOR_NUMBER = OPEN_ITEMS.VENDOR_NUMBER

The JOIN statement gives us a table we might think of as:

Product Code	Type	Color	Quantity On Hand	Vendor Identification	Supplier Name	Purchases This Year
2162	Easel	Brown	7	62	Associated	5,599.19
2977	Chair	Brown	12	41	Walkers	573.92
3195	Chair	Black	6	45	Campbell's	42,764.26
6377	Desk	White	8	62	Associated	5,599.19
7422	Lamp	Red	19	75	Rose	366.72
8654	Clock	Walnut	4	45	Campbell's	42,764.26

The JOIN operation is controlled by what are called "JOIN fields"; not every record may have a common JOIN field. In these cases, those records cannot take part in the JOIN operation. In our example, all records had a JOIN field and became a part of new logical table.

If we wanted to make this table that was built with SQL's JOIN concept into a real table, we would write:

```
CREATE TABLE SUPPLIES_VENDOR AS
    SELECT PRODUCT_CODE, TYPE, COLOR, QUANTITY_ON_HAND,
        SUPPLIER_NAME, VENDOR_IDENTIFICATION, PURCHASES_THIS_YEAR
        FROM SUPPLIES, VENDOR
        WHERE SUPPLIES, VENDOR_IDENTIFICATION =
        VENDOR.VENDOR_IDENTIFICATION
```

When SQL executes this command, it will create a new table named SUPPLIES_VENDOR with the columns named in the SELECT. All the data selected will be brought into the new table. This is an ideal operation when an application needs to export data to another system, such as to a spreadsheet operating on a personal computer.

The last SQL command is VIEW. A VIEW is a logical representation of another table or combination of tables. It allows the analyst to reduce the apparent complexity of the table. VIEWs can restrict access to specific columns (a vertical VIEW) or to specific rows meeting certain predefined criteria (a horizontal VIEW).

VIEWs let the analyst tailor the appearance of the tables so that users see it from a different perspective. They allow the analyst to restrict access to data so that users see what they need to see. As an example, suppose we want a user to know only the PRODUCT_CODE, COLOR, and TYPE from the SUPPLIES table. In SQL, we write:

```
CREATE VIEW PROD_FACTS AS
    SELECT PRODUCT_CODE, COLOR, TYPE
        FROM SUPPLIES
        WHERE PRODUCT_CODE > 0
```

VIEWs allow the analyst to rename columns, make a table look smaller, change the name of a table, logically join tables for user convenience or understanding, and add security to the system so that specific users see only the data that are relevant to them.

A VIEW is easily deleted with the DROP command. To remove our VIEW, the command is

```
DROP VIEW PROD_FACTS
```

Most operating systems will respond with a message confirming the removal of the VIEW.

VIEWs are not all positive. The two major disadvantages are performance and update problems. By adding a VIEW, the SQL DBMS must translate all inquires from their logical tables into their physical tables, which adds time to the execution of the inquiry. When a user tries to update a row(s) of a VIEW, the DBMS must convert the logical update into physical SQL UPDATE commands on row(s).

SQL is a powerful software database manager with all types of capabilities built into a few statements. This short look at some of SQL's capabilities does not reveal all that is possible. Like viewing an iceberg, you have seen only the 10% that is immediately visible. The 90% that is harder to see contains even more power.

One of SQL's chief advantages is that users can operate it from a terminal. Programmers can write SQL commands embedded within a high-level language such as COBOL, Pascal, or C. It simultaneously functions as both host and self-contained language. Regardless of how the user wants to tailor SQL, its consistent language rules make it equally accessible to terminal—as well as high-level language—users.

Performance poses the greatest challenge to SQL. Relational DBMSs often require a significant amount of internal computer memory and central processor attention. However, as the price of memory declines, computer speed increases, and costs of all hardware products drop, SQL—as well as all other relational databases—will become even more attractive.

Standardization is the second challenge. COBOL achieved dominance partly due to the fact that it is a truly standardized language. As software developers add features to SQL, it is mandatory that they do so in a way that makes SQL standardized across all product lines.

QUERY LANGUAGES

Data manipulation languages require a lot of technical training. To help untrained users locate and retrieve data, many DBMSs offer special-purpose languages, called **query languages** (such as QBE, INQUIRE, ADASCRIPT, or NATURAL). Query languages allow casual inquiry of the database in language that more closely resembles English and thus simplifies data retrieval, updating, addition, deletion, or reporting. An example of a query program to print the Product Number, Description, and Quantity-on-Hand for Parts Master appears in Figure 10.10.

Query and their host languages give the user a set of tools for referencing, reporting, and manipulating information from a database without becoming programmers. They allow users to enter new data, change existing data, and delete data from the database.

Most query languages allow users to generate reports on the printer or CRT screen. When using query languages in this capacity, the user must specify:

Figure 10.10 Query provides a way for noncomputer personnel to access the database. It is well suited for interactive use on a remote terminal for uses such as rapid prototyping and production reporting.

```
REPORT
LINES=60
H1, "List of Parts on Hand by Part Number", 60
H1, "Page:", 74
H1, PAGENO, 80, SPACE A2
H2, "Product Code", 12
H2, "Description", 25
H2, "Cost $", 50
H2, "Quantity on Hand", 68, SPACE A1
S, PART-NUMBER
D1, PRODUCT-CODE, 12
D1, DESCRIPTION, 40
D1, COST $, 50, E1
D1, QUANTITY-ON-HAND, 68
E1, "$$$,$$$,$$$.99"
END
```

a. Query routine named **"LISTING"** to print a report.

```
>FIND PARTS-MASTER.PRODUCT-CODE > 0
>REPORT LISTING
```

b. Query commands to locate and print all product-numbers greater than zero in the Parts-Master data set.

```
List of Parts on Hand by Product Code                    Page:      1

Product Code      Description              Cost $      Quantity on Hand
2162              Easel                   $89.95                     62
2977              Chair-Brown            $329.62                     41
3195              Chair-Black            $379.47                     45
6377              Desk-White             $816.21                     45
```

c. Report printed by the query routine for the Parts-Master data set.

- Header information
- Sorting of data
- Totals and subtotals
- Edit masks (e.g., American dates in mm/dd/yy format)
- Line spacing and page skipping
- Averaging, counts, standard deviations, and other statistics
- Printing of data on one or many lines

The query language converts this description into a report "automatically," without the need for a program written in a high-level language such as COBOL, Pascal, or C.

UTILITIES

Many DBMSs also incorporate sophisticated utility routines to help users perform common tasks. Some utility tools provide a means for copying the database to disk or tape, thus creating a back-up in case of hardware failure. Others log database changes to tape or disk, so users can rebuild the database if it suffers accidental damage. Still others purge (erase) the database, move it to another disk drive, increase the database's capacity, or reload it.

Most DBMS software also permits users to change a data entry's name, add or delete a data entry, or change the name of a data set without redoing every program accessing the database. Other utilities allow users to upload ASCII files data from popular personal computer-based database managers, word processing programs, or spreadsheets into relational tables. It is also possible to do the opposite: download selected rows to a personal computer.

NORMALIZATION

Once the analyst finishes deciding on the various tables, indexes, and attributes, yet another step remains: normalization. A set of design criteria and five rules, **normalization** helps database designers prove that their logical design is correct and optimal. Normalization is the idea that as each of the five rules is applied, undesirable features are removed from the design.

Each successive rule implies that the prior ones are met. For example, if the design meets the third rule or third normal form (often abbreviated 3NF), it should have already met 1NF and 2NF. Normalization implies splitting tables (relations) into two or more tables with fewer columns. Most designs try to reach 3NF and a few 4NF, but not many reach 5NF (or need to).

The five normalization rules are

1NF.   Each row or column must have a single value with no repeating values.

2NF.   Every nonkey column must depend on the primary key.

3NF.   No nonkey column can depend on another nonkey column.

4NF.   An entity cannot have a 1:N relationship between primary key columns and nonkey columns.

5NF.   Break all tables into the smallest possible pieces to eliminate all redundancy within a table.

As an example of normalization, let's return to our continuing example of the Valley Blood Center and their donors, donations, hospitals, and patients. Let's look at each rule and what it is really saying.

The 1NF rule implies that you cannot subdivide an entity (column) and that you cannot have values that repeat in a row. 1NF really says that a column in a table with a donor's name (first, middle, and last) is probably a violation of 1NF. To meet the 1NF rule, split

Cheryl Ruiz felt ecstatic when she learned that her company, AAA Real Estate Brokers, had just decided to purchase IBM's SQL/DS relational database management software. Although Cheryl extensively experimented with sequential file systems, and had never used a DBMS before, she was eager to try one out. The IBM sales representative had recommended that the company retrain one member of its computer services staff to become the in-house database expert, and Cheryl hoped her boss, Salvatore Gianna, would pick her.

Though Cheryl regularly read articles in *Computerworld* about DBMSs, she only understood them superficially. She knew some of the buzzwords—hierarchical, network, relational, host language, query language—but how they really worked was a mystery to her.

At last, Sal Gianna called Cheryl into his office to discuss her career at AAA Brokers. Cheryl had earned her bachelor's degree in business 15 years earlier, married, then gone back to her local community college to study computers. Since joining AAA Brokers as a programmer four years ago, she had won praise from Sal and her peers for establishing rapport with users and by frequently updating her skills. In fact, it had been Cheryl who finally convinced Sal to switch from punched cards to terminals in an effort to increase the productivity of AAA Brokers' programmers and analysts. She had been right. Programmer productivity and job satisfaction had benefited and staff turn-over had declined.

"I admire your enthusiasm for the new DBMS," Sal said when Cheryl sat down in his office.

Cheryl nodded her head. "It's the future. I only wish I knew more about it."

"Well, you're going to get your chance. The IBM training center in Boston is offering 3 two-week classes on SQL/DS beginning next month and I want to enroll you."

To prepare for the class, Cheryl asked IBM's training manager in Boston to recommend books on database; she bought and read several before the class began.

The six weeks of classes flashed by. In addition to reviewing her own books, Cheryl went through four IBM SQL/DS manuals. When she returned to AAA Brokers, her former admirers gave her a cool reception. What had she done? One day, she overheard a programmer tell another analyst, "Cheryl thinks she's a wizard now. We're just a bunch of outdated hacks." Cheryl felt crushed. She hadn't expected her friends to become jealous and insecure. But she decided to do something about it.

The next day, she invited AAA Brokers' staff of ten to an informal breakfast meeting to discuss SQL/DS. "You'll probably catch on much faster than I did," she said.

All ten showed up early and surrounded Cheryl at a terminal to which she had attached three monitors so everyone could see. When her colleagues saw SQL/DS running, they were impressed. "This works better than I thought," said one.

At the end of the week, the group treated Cheryl to lunch. "To our new database manager!" toasted the formerly embittered programmers. Cheryl laughed when everyone lifted their glasses. Learning new technology was fun, but sharing it with others was even more rewarding.

the donor's name column into three columns: one for the donor's first name, the second for the donor's middle name, and a third for the donor's last name. Having a donor's name in three columns does not mean that an application program cannot bring them back together to appear as a single name. Strings such as "Vice President of Marketing" or "12 Mill Street" are normally single-column data items.

The second half of 1NF implies that a row cannot store multiple values that have a common characteristic. As an example, suppose we want to keep the dates of donations by a donor. This rule says that we can't keep these dates in the same row with the rest of the donor's data. Instead, we should place the data in another table (along with the donor's number, which will allow us to relate the donations back to the donor). This rule makes a lot of sense when you think that a donor could have from zero to hundreds of donations. If we did not have this second half of 1NF, we would have to reserve space in each row for possibly a hundred or more donations. Since most donors would not use this much space, we would waste it.

The 2NF rule says that the other columns in a table must relate to the key column; if they don't, a change is needed. Each table has a primary key field (e.g., donor number or their Social Security number) and all the other columns have meaning relative to that column. A donor's home telephone number is of little value unless we associate it with their donor number and it is therefore dependent on the primary key.

The 3NF is an expansion of 2NF in that it says we do not want to have two or more non-key columns that depend on each other. A very simple example of 3NF is carrying a donor's city, state abbreviation, and ZIP code as individual columns in the donors table. We can deduce the donor's city and state if we know the donor's ZIP code. The solution to making the table meet 3NF is to carry the ZIP code in the table and have another table with columns of ZIP code, city, and state abbreviation data. 3NF also maintains consistency when data change. For example, in 1991, the town of Manchester, Mass (ZIP code 01944) changed its name to Manchester-by-the-Sea. With this ZIP code town name table, we make a single change. Without 3NF, every address we store for this ZIP code would require a change. This design reduces the amount of data we have to carry about each customer, but we may pay a penalty (for example, it becomes clumsier to look up the city and state abbreviation when we need to print or display a customer's address). This is a trade-off that the analyst must monitor.

4NF requires us to review the tables and what we are using them to represent. Suppose our design called for combining the donors table and the donations table into a single table. We have now formed a variable-length record with the root holding the donor data and the segment occurrences holding the donations data. Further, we do not know how many donations there are; anywhere from zero to many (0:N). 4NF says that this is a poor design. Instead, we should have our original two tables: donors and donations. To relate donor to donations, we should have a common column key: the donor-id. The advantages of placing repeating data items in their own table include reducing disk requirements, eliminating unused reserved space, and using less complex, single-purpose tables.

The 5NF rule is quite simple. The rule results in many small tables specific for a single purpose, rather than a few large tables that carry data we do not often require.

Our definition of the five rules was written in everyday English. Their formal definition is quite mathematical. For example, C.J. Date's book, *An Introduction to Database Systems, Volume I* (4th Edition published by Addison-Wesley, 1990), writes the 3NF as:

"A relation R is in third normal form (3NF) if and only if it is in 2NF and every non-key attribute is nontransitively dependent on the primary key."

The other definitions are written in this format, but we chose to explain them by using working definitions.

New rules for normalization appear in technical journals now and then. The Boyce/Codd rule reads (again from Date's book):

"A relation R is in Boyce/Codd normal form (BCNF) if and only if every determinate is a candidate key."

As specialists in database systems conduct further research, new rules similar to Boyce/Codd will assist analysts in making better designs.

Coupled with normalization is denormalization, the reverse procedure to normalization. In some situations, a database in 3NF may digress to 2NF or 1NF purely for reasons of improving hardware and software performance on production systems. As relational systems evolve, denormalization may no longer have to occur.

CASE AND DATABASE DESIGN

In our look at CASE software, we drew data flow and entity relationship diagrams (ERDs); wrote definitions of attributes, entities, and relationships; listed cardinalities; and specified report formats and contents. We have also seen that the CASE software performs all types of checks to make sure that there are no errors or omissions and that everything is fully defined. Of course, the CASE software cannot detect a logic error on the part of the analyst or the fact that the analyst may have forgotten something important. That still remains in the hands of analysts themselves.

All the entries we have made in the CASE software are stored in a database (usually relational) and are ready to help the analyst. If the CASE software really has all this information, it should seem clear that it can write the SQL statements (or equivalent ones for other databases) to define each table and attributes in the table. It can also write the SELECT statements to access the various tables.

If we had a hypothetical CASE system with our Valley Blood Bank's donors and donations entity relationship diagrams, all the attributes defined within the CASE software would create the SQL CREATE statements in Figure 10.11.

The common key column between DONOR and DONATION is DONOR_ID. Between DONOR and ZIP_CODE, the common key column is ZIP. The SQL statement to retrieve all the rows in DONATION belonging to the person whose identification is 565342312 and then to present them in order from most recent donation to oldest donation is

```
SELECT * FROM DONATION
    WHERE DONOR_ID = 565342312
    ORDER BY DATE_DONATED
```

To also get this person's city and state abbreviation, we write:

```
SELECT * FROM DONOR, ZIP_CODE
    WHERE DONOR_ID = 565342312 AND
        ZIP_CODE.ZIP = (SELECT ZIP FROM DONOR
            WHERE DONOR_ID = 565342312)
```

This rather formidable SELECT is quite easy if you read it from the inner parentheses out. We start reading the SELECT beginning from the one inside the parentheses. It returns the ZIP code of our desired person to use in the outer SELECT that chooses the proper row in DONOR and the associated row in ZIP_CODE.

The DONOR and DONATION tables are normalized at 2NF. Notice that the design keeps the last donation date as a row in DONOR. In fact, that same data are kept in DONATION. Similarly, both DONOR and DONATION have the blood type. A person has only one blood type. Why keep the blood type and donation date in both places? The answer may lie in performance and common usage. When a donor wants to make a donation, the business rules specify that the person may not have made a previous donation in the last three

Figure 10.11 SQL CREATE statements generated by CASE software.

```
CREATE TABLE DONOR
        (
        DONOR_ID                NUMBER(9) NOT NULL UNIQUE PRIMARY KEY,
        LAST_NAME               CHAR(20),
        FIRST_NAME              CHAR(20),
        MIDDLE_INITIAL          CHAR(02),
        STREET_ADDRESS          CHAR(40),
        ZIP                     CHAR(09) NOT NULL,
        PHONE_NUMBER            NUMBER(10),
        BLOOD_TYPE              CHAR(04) NOT NULL,
        DATE_OF_BIRTH           DATE,
        WEIGHT                  NUMBER(03),
        MEDICAL_STATUS          NUMBER(04),
        UNITS_DONATED           NUMBER(03),
        DATE_LAST_DONATED       DATE
        )

CREATE TABLE DONATION
        (
        DONOR_ID                NUMBER(09) NOT NULL,
        DATE_DONATED            DATE NOT NULL,
        WHERE_DONATED           CHAR(06),
        BLOOD_TYPE              CHAR(04) NOT NULL,
        STATUS                  CHAR(04)
        )

CREATE TABLE ZIP_CODE
        (
        ZIP                     CHAR(09) NOT NULL UNIQUE PRIMARY KEY,
        CITY                    CHAR(20) NOT NULL,
        STATE                   CHAR(02) NOT NULL
        )
```

months. To look up this fact requires a SELECT to DONATION and a calculation on how long ago this donation was made: a costly set of events. When a unit of blood is taken, we also need to keep track of the blood type. Why keep looking back to the DONOR to find this fact? The analyst may choose to back off from 3NF to 2NF for performance and safety issues. Again, the analyst may have to make a trade-off decision, balancing a theoretically clean database design with a more functional database that users really need.

The CASE software may also identify the key fields as primary, composite, or foreign. DONOR_ID in DONOR, and ZIP in ZIP_CODE are **primary keys** (they have the UNIQUE clause).

DONATION contains no primary keys since a donor may have multiple donation rows for the many times that person has given blood. However, DONOR_ID and DATE_DONATED taken together do uniquely identify a row in DONATION. When it takes a combination of columns to make a row unique, those two columns are called the **composite key**.

ZIP in DONOR is a **foreign key**. It is a column in DONOR whose value matches the primary key in some other table, ZIP_CODE. A table can have more than one foreign key if it is related to more than one other table, but this is not the situation in our example. For instance, BLOOD_TYPE in DONATION could represent a foreign key to another table where BLOOD_TYPE was the primary key. In this hypothetical table, we may store statistics about various blood types.

The CASE software has again helped the analyst perform his or her duties. It took the theoretical descriptions and merged them into useful results. Once more, the CASE system allowed the analyst to focus on the logical parts of the database design and on the wishes of the user.

DATABASE CONTROLS

Databases require all types of controls. For traditional file systems, we saw two control methods: record counts and back-up. Though both apply to database systems, a database environment can include three additional ones: transaction logging, access security, and mirror databases.

Back-Up

Despite their sensitivity to the need for back-ups, many professionals overlook copying a system's programs and documentation. Computer installations that take great pains to provide back-up or transaction logging facilities for the data itself may neglect to do so for the processing programs. What purpose would back-up data serve without programs to process them?

The procedure for copying programs and system documentation parallels that for files: the organization builds a duplicate disk or tape that contains copies of its programs and documentation. As organizations more frequently develop programs and documentation with computer text editors or word processors, they must copy them to tape for storage in a safe vault or at some remote location.

Transaction Logging

Accidental destruction of any on-line system's files by a hardware, software, or user failure is a reality. To solve this problem, newer and more sophisticated operating systems and database management systems perform **transaction logging**. With it, special software (usually "built-into" the DBMS or operating system) automatically copies the old and new records, plus the transaction record to tape or disk every time a user adds, alters, or deletes a record. An organization can use such log files to recreate lost or destroyed database files. Log files also provide clues to the frequency and types of file changes made within a system. On a terminal-based transaction logging system, the software can keep track of the time, the terminal, and the user making the changes.

The analyst needs to establish a log maintenance cycle for the database. For example, suppose the database is a daily one. This means that we would keep one day's worth of transactions in the log file. At the end of the day, the log file is written to tape and a new log file starts up for the next day. The length of the cycle depends on two factors: the amount of time required to recover the database from the back-up, plus the amount of time to rebuild the database from the log file. The more often the entire database is backed-up, the smaller the log maintenance cycle. In some situations, a legal requirement for audit purposes, for example, may dictate the back-up cycle.

Access Security

Another type of processing control restricts system access to approved users. Access security not only determines which users can enter a database, but how they may use it. When analysts build their schema, they may assign each user a restricted read-only or a more liberal read-and-write access status.

Some database managers employ individual passwords to restrict file access. For example, a bookstore clerk in Silva's bookstore might have the capability to ask the computer if the store has a specific book in stock and to alter the inventory record of books after making a sale, but the same clerk might not have the authority to change the book's retail price because the store owner wants to maintain strict personal pricing control.

Access security can occur at the database, data set, and data-entry level. Database security keeps unauthorized users out of the entire database unless they know the right "password." Data set security prohibits some users from reading or writing certain data sets. Data entry security limits access to the field level, not letting a user read or write specific data entries in specific data sets in a database.

SQL implements access security with GRANT and REVOKE. The GRANT statement allows access of various types to some or all tables. For example,

```
GRANT SELECT UPDATE TO FURNITURE_SUPPLIERS TO CLERK
```

allows a user named CLERK to use SQL's SELECT and UPDATE statements. However, the user is barred from INSERT and DELETE. Taking away a user's privileges is the purpose of REVOKE. For example,

```
REVOKE UPDATE FROM CLERK
```

removes the CLERK user from altering the database. However, he or she can still issue the SELECT command.

These two SQL statements—coupled with the CREATE VIEW—are extra security provisions built into SQL. Other security provisions can include restriction to certain columns or rows and passwords to gain initial access to the system and software.

Mirror Databases

Transaction logging and back-ups are good maintenance procedures. However, if a database is accessible all the time and cannot suffer a failure, even for back-up or recovery from some other type of failure, then a new maintenance procedure is required. A database with constant access or high availability can still have protection.

The **mirror database** consists of two identical databases on perhaps two separated computer systems. One database is housed on the primary system and is constantly accessible to users and application programs. The second, or "mirror," database may reside on the other computer system. Tandem Computer Corporation was one of the first commercial firms to offer mirror databases and places the databases on separate drives on one computer system. Each drive/database has its own path to eliminate any hardware component as a single cause of system failure.

Establishing mirror databases implies two identical copies of the database. All transactions on the primary system are written in parallel to the "mirror" system. The primary system is never brought down for maintenance.

The mirror database is a fundamental element in those environments in which the database requires 100% availability. Some excellent examples of mirrored databases are on-line real-time applications such as airline reservation systems and bank automated teller machines (ATMs).

DESIGNING FLEET FEET'S DATABASE

With her reports and data collection screens designed, Fleet Feet analyst Peggy Adams-Russell is ready to design her database. Earlier, she drew the logical system as a data flow diagram (Figure 10.12), showing six data stores:

1. *Apadjst:* Stores data on vouchers authorized for payment and allows for adjustments to an invoice

2. *Aptran:* Keeps track of each invoice and voucher issued by Fleet Feet

3. *Vendor:* Holds all the data about each Fleet Feet vendor

4. *Aponitm:* Stores data on all vouchers authorized for payment

5. *Apcheck:* Tracks each check Fleet Feet writes, to whom it was written, and what it was written for

6. *Apchkrg:* Holds data on checks that were cashed and have cleared the bank

Along with a verbal description for each data store, Peggy decides on primary key fields. She chooses to assign each vendor a vendor number, a transaction receives an invoice number, and checks get a check number.

Peggy writes the SQL CREATE TABLE statements for all six relational tables, naming each column and assigning data types and lengths (Figure 10.13). She places primary and foreign key fields as the first few fields in each table and orders the others so that they are grouped logically together.

While not a mandatory part of database design, Peggy estimates the amount of disk storage space that the new accounts payable system will need. She lists each table, along how many rows she thinks the table will have at the end of the first year, and the length of each row:

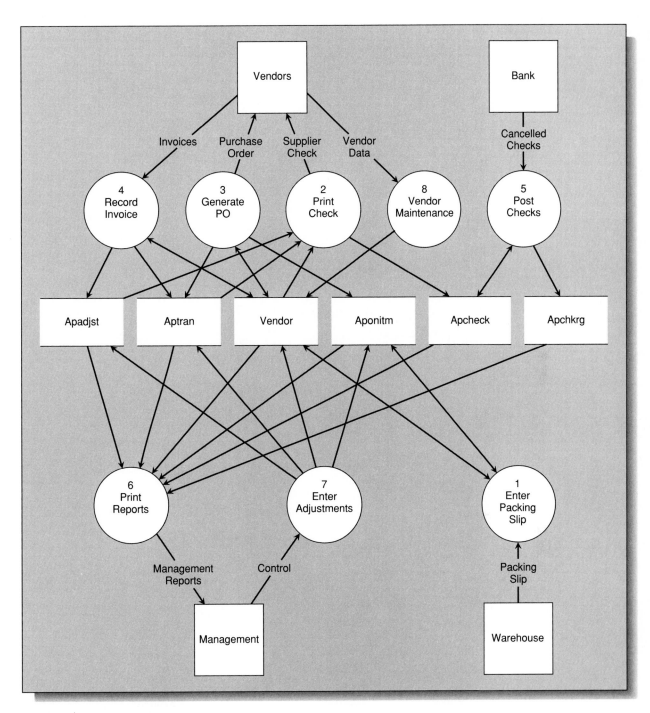

Figure 10.12 Data flow diagram for Fleet Feet's accounts payable system, showing the six data stores that the system needs.

Figure 10.13 The CREATE TABLE statements that define Fleet Feet's relational database.

```
CREATE TABLE VENDOR (
        VENDOR_NO                  CHAR(06) NOT NULL,
        VENDOR_NAME                CHAR(30) NOT NULL,
        VENDOR_ADDRESS_1           CHAR(30) NOT NULL,
        VENDOR_ADDRESS_2           CHAR(30),
        VENDOR_ADDRESS_3           CHAR(30),
        TYPE                       CHAR(04),
        TERMS_DESCRIPTION          CHAR(16),
        DUE_DAYS                   NUMBER(04),
        DISCOUNT_DAYS              NUMBER(02),
        DISCOUNT_PCT               NUMBER(02),
        STATUS_CODE                CHAR(02),
        PURCHASES_YTD              NUMBER(8,2),
        PURCHASES_LAST             NUMBER(8,2),
        DISCOUNT_YTD               NUMBER(8,2),
        DISCOUNT_LAST              NUMBER(8,2)
        )
CREATE TABLE APTRAN (
        INVOICE_NO                 CHAR(08) NOT NULL,
        VOUCHER_NO                 CHAR(08),
        PO_NO                      CHAR(08),
        INVOICE_DATE               DATE,
        INVOICE_AMOUNT             NUMBER(8,2),
        DUE_DAYS                   NUMBER(06),
        DISCOUNT_DATE              DATE,
        DISCOUNT_PCT               NUMBER(02),
        DISCOUNT_AMOUNT            NUMBER(8,2),
        ACCOUNT_NUMBER             CHAR(10)
        )
CREATE TABLE APCHECK (
        CHECK_NO                   NUMBER(08) NOT NULL,
        VENDOR_NO                  CHAR(06),
        CHECK_TYPE                 CHAR(02),
        CHECK_DATE                 DATE,
        CHECK_AMOUNT               NUMBER(8,2),
        INVOICE_NO                 CHAR(08),
        INVOICE_DATE               DATE,
        INVOICE_AMOUNT             NUMBER(8,2),
        DISCOUNT_AMOUNT            NUMBER(8,2)
        )
```
(continued)

Table Name	Number of Rows	Row Length in Bytes	Space Required
Vendor	250	200	50,000
Aptran	7,000	82	574,000
Apcheck	6,000	75	450,000
Aponitm	5,000	75	375,000
Apadjst	7,000	125	875,000
Archkrg	6,000	50	300,000

Figure 10.13 *(continued)*

```
CREATE TABLE APONITM (
      VENDOR_NO                 CHAR(08),
      CHECK_NO                  NUMBER(08),
      INVOICE_NO                CHAR(08),
      VOUCHER_NO                CHAR(08),
      ORIG_INVOICE_AMT          NUMBER(8,2),
      ORIG_DISCOUNT_AMT         NUMBER(8,2),
      PARTIAL_PAYMENT           NUMBER(8,2),
      CURRENT_DISCOUNT          NUMBER(8,2)
      )
CREATE TABLE APADJST (
      VENDOR_NO                 CHAR(08),
      CHECK_NO                  NUMBER(08),
      INVOICE_NO                CHAR(08),
      VOUCHER_NO                CHAR(08),
      INVOICE_BALANCE           NUMBER(8,2),
      INVOICE_DATE              DATE,
      INVOICE_DISCOUNT          NUMBER(8,2),
      OLD_DUE_DATE              DATE,
      DUE_DATE                  DATE,
      CHECK_DATE                DATE,
      CHECK_AMT                 NUMBER(8,2),
      DISCOUNT_TAKEN            NUMBER(8,2),
      CASH_ACCOUNT              CHAR(10)
      )
CREATE TABLE APCHKRG (
      VENDOR_NO                 CHAR(08),
      CHECK_NO                  NUMBER(08),
      INVOICE_NO                CHAR(08),
      CHECK_DATE                DATE,
      DATE_RECONCILED           DATE,
      ACCOUNT_NO                CHAR(10)
      )
```

Peggy multiplies her crude row lengths by the number of rows and adds together the space requirements for each table. From these calculations, Peggy finds the size of the accounts payable database well under 3 million bytes (3MB), within the available space on Fleet Feet's computer.

With her tables defined, Peggy reviews each relative to the five rules for normal forms. Every table meets 1NF; there are no repeating groups and all columns are single valued. Similarly, the design meets 2NF, with every nonkey column relating to the key column. But it fails 3NF! She has a vendor's ZIP code buried within her address columns. This is an easy repair job and she writes a CREATE TABLE statement for a ZIP code table, inserting a ZIP code row in the Vendor table:

```
CREATE TABLE ZIP_CODES (
      ZIP                    CHAR(09),
      CITY                   CHAR(40),
      STATE_ABBREVIATION     CHAR(02)
      )
```

Reviewing 4NF and 5NF, Peggy decides that she needs a table for TYPE:

```
CREATE TABLE VENDOR_TYPES (
    TYPE              CHAR(04),
    TYPE_DESCRIPTION  CHAR(24)
    )
```

The entire discount storage seems odd to Peggy. Isn't a discount related to a purchase—or is it related to a vendor? She talks to Fleet Feet staff and they agree that each vendor has its own "standard" discount, but on occasion, vendors receive special discounts. Peggy decides to make a TERMS table that explains each TERMS_CODE:

```
CREATE TABLE TERMS (
    TERMS_CODE         CHAR(04) NOT NULL,
    TERMS_DESCRIPTION  CHAR(24) NOT NULL,
    DISCOUNT_DAYS      NUMBER(02),
    DISCOUNT_PERCENT   NUMBER(02)
    )
```

She removes all discount references from Vendor and Aptran, placing a TERMS_CODE column in each instead. This allows each vendor to have a standard discount, plus one associated with individual purchases. With these corrections, Peggy revises her database CREATE TABLE statements.

The last phase of her database design concerns indexes. What indexes does each table require? Peggy talks to Fleet Feet staff to see how they use the data and discovers that the best plan is to index Vendor by VENDOR_NO, Aptran by INVOICE_NO, and Apcheck by CHECK_NO, the primary keys. For each of the other three tables, she builds indexes on the same three columns: VENDOR_NO, INVOICE_NO, and CHECK_NO, the foreign keys. She also indexes ZIP_Codes by ZIP, Vendor_Types by TYPE and Terms by TERMS_CODE. Her overall design has 9 tables and 15 indexes.

Still not finished, Peggy reviews her design with Fleet Feet staff. While quite technical, she draws pictures of the tables and the connecting columns, explaining what is contained in each table and the names of the various columns. Some minor suggestions are made in table and column names, as well as in the sizes of a few columns. After making the requested changes, Peggy is done with this stage. Next up is network design.

SUMMARY

Database managers enhance the power of disk-based systems, freeing analysts from the physical details of data storage so that they can concentrate on the logical aspects of the system and user needs.

There are three types of DBMSs: hierarchical, network, and relational. Hierarchical systems have master-detail records in which each detail belongs to a single master. In network DBMSs, details obey more than one master. Relational DBMSs allow users to view data as two-dimensional tables that are logically connected with a common key.

To describe the database, the analyst writes a schema, which defines the data items for each record and shows how the records will be linked.

Three widely used database management systems are IBM's SQL/DS and DB2, Borland's dBASE, and Oracle Corporation's ORACLE SQL. Users of minicomputers and mainframes are gravitating toward IBM's SQL and its lookalikes, personal computer users have made dBASE (and its clones) a top seller, and Oracle successfully runs on many platforms and operating systems.

Query languages allow novice users to manipulate the database. Most of these languages permit users to enter data, change them, delete them, and report data in a wide variety of formats.

Security is built into most DBMSs, rather than added on as an afterthought or appendix as in flat file systems. Passwords are used to allow users access to specific data sets, entries, or items.

DBMSs offer many advantages. They consolidate files independent of the system or program. They are versatile, fast, secure, and efficient; lessen program development time; reduce program maintenance; and fulfill special needs more easily than older file storage techniques. Drawbacks include the need for additional hardware and staff training—relatively high costs.

Some futurists forecast the imminent death of traditional flat file processing. They say relational DBMSs are the wave of the 1990s.

KEY TERMS

Flat file	Self-contained
Database management system (DBMS)	Query languages
Hierarchical model	Normalization
Network model	Primary key
Relational model	Composite key
Schema	Foreign key
Data definition language (DDL)	Transaction logging
Data manipulation language (DML)	Mirror database
Host language	

QUESTIONS FOR REVIEW AND DISCUSSION

Review Questions

1. What are the three basic database models?

2. What is a schema?

3. What are four types of utility software that accompany most database management systems?

4. What are three advantages and disadvantages of a database management system?

5. What are the names and classifications of four commercial database management systems?

6. What are three common rules that govern relational database design?

7. Why do we have a query language?

Discussion Questions

1. What do the letters in each of the following abbreviations stand for?

   a. DBMS

   b. DML

   c. DDL

2. List three reasons why a firm would select a database manager over the traditional methods of data storage.

3. What two operations can occur on relational DBMSs?

Application Questions

1. How many records must a system process to update a master file of 4,000 records if the transaction file contains 1,000 records and a DBMS is used to store records?

2. How many master data sets can a detail data set belong to in a

   a. Relational DBMS?

   b. Network DBMS?

   c. Hierarchical DBMS?

Research Questions

1. Find the names of five other database management systems now in use and classify them as hierarchical, network, or relational. If any of them are advertised as "relational," check its features to make sure they fit the definition of a relational DBMS.

2. Which type of database manager seems the most versatile?

3. Some database managers are classified as inverted or linked. What do these two terms mean?

After reviewing View Video's desired report formats and talking with Mark Havener at Manzanita Software (developers of BusinessWorks PC), analyst Frank Pisciotta is faced with a dilemma. Use a database manager or a flat file system? BusinessWorks PC is built around a flat file system with an indexed sequential access software product. The videotape check-out/check-in is new and he has made no commitment to any product. Frank's database choices are many, varying from dBASE, Paradox, R:Base, FoxPro, and other similar products to Oracle SQL. All have similar prices, are relational, operate on PCs, allow networking, and provide database controls, especially access security. dBASE products are self-contained, while Oracle is both self-contained and host language.

Frank considers all the trade-offs and picks Oracle SQL. It was a hard choice, but his research found that the platform independence, supplementary software to generate reports and data collection screens, and the ease of linking to BusinessWorks PC were the deciding criteria.

Frank reviews his flat file design. He needs 9 files:

1. CUSTOMER: Stores data about each View Video customer
2. TAPE: Hold data about each tape that View Video rents
3. RENTAL: Holds data about individual tape rentals
4. PAYMENT: Holds data about every payment for transfer to BusinessWorks PC for account posting
5. CLERK: Keeps all the data about each clerk that checks out or in a videotape
6. STORE: Keeps all the data about View Video itself
7. STAR: Holds the names of movie stars that appear in various tapes
8. PERFORMANCE: Holds the individual performances made by a movie star in a videotape
9. BORROWER: Holds additional legal borrowers for a customer

With relational concepts and normalization in mind, he adds a tenth table for ZIP codes. This table stores the customer's ZIP code in the customer table, and the city and state abbreviation in the ZIP_Code table.

The RENTAL and PERFORMANCE tables are a sticky issue. Why are they needed? Frank reviews the tables in his mind. His entity relationship diagram (ERD) drawn during the design stage of the systems process showed a many-to-many (M:N) relationship between CUSTOMER and TAPE. A customer will rent many tapes and a tape can be rented by multiple customers. Similarly, a many-to-many relationship exists between TAPE and STAR. A tape has many stars appearing in it and a star can perform in many different tapes.

RENTAL breaks the many-to-many relationship, establishing a one-to-many (1:N) between CUSTOMER and RENTAL and a second one-to-many between TAPE and RENTAL. Likewise, PERFORMANCE breaks the many-to-many relationship into two one-to-many relationships. Good relational database design wants these complex M:N relationships dissolved into 1:N relationships. RENTAL and PERFORMANCE do just that.

Frank bumps into difficulty with two issues: clerks and multiple copies of a title. How will he track the clerk that checks out a tape, along with the clerk that checks in the tape when the customer returns it? After some thought, he decides to place two sets of clerk initials in the rental table, each with a different attribute name.

Accounting for multiple copies of a tape with the same title is harder. His analysis statistics showed that Mark and Cindy Stensaas wanted to buy 5,000 tapes representing 3,000 different titles. His flat file design called for each tape, regardless of title, to have its own record (row in a relational design), taking 92 bytes of space. This means that the file uses 460,000 bytes of disk space ($92 \times 5,000 = 460,000$). Suppose he splits the file into two tables, one called Title and another called Tape. Title would hold the tape's name, rating, category, and date released. Tape would keep the specifics about this particular tape. He would thus have 3,000 rows in Title and 5,000 rows in Tape. The primary key field is the name of the tape; not a very good choice, but Frank can't think of a better alternative. He calculates the length of each row in Title as 50 bytes and in Tape as 74 bytes. He then calculates the length of Title as 150,000 ($50 \times 3,000 = 150,000$) and Tape as 370,000 ($74 \times 5,000 = 370,000$). This design uses more disk space, but allows for easy access to multiple copies of a tape.

With his new tables in mind, Frank sketches the 11 tables he now needs (Figure 10.14). He lists the common key columns between the tables to show how they logically link together.

After completing his drawing, determining their relationships, and establishing keys, Frank writes an Oracle SQL CREATE statement. This statement defines each table, attributes, attribute data types, lengths, and constraints (Figure 10.15).

The Oracle SQL, which follows the standard implementation of SQL, shows some new SQL options. For example, the definition of Last_name:

```
Last_name   char(20)
      check (Last_name = initcap(lower(Last_name))),
```

specifies that the attribute name is Last_name of data type character or string with a length of 20 characters. The check statement is a constraint that restricts the range of valid values for

Figure 10.14 The relational tables of View Video's videotape database. The table keys (in lowercase letters) logically link the tables.

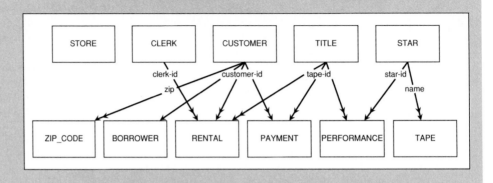

a column. In this case, the constraint says that SQL is to convert the entry to all lowercase characters and then uppercase the first letter in each word. Thus, when a user enters the name STENSAAS, Oracle converts it to stensaas, then converts this to Stensaas. If the user enters Stensaas to start with, the constraint ends up with the same value, Stensaas. The same constraint clause is applied to First_name.

The Middle_initial definition shows another constraint clause:

```
Middle_initial      char(01)
      check (Middle_initial = upper(Middle_initial)),
```

This clause uppercases the user's entry.

Income_generated shows seven digits to the left of a decimal point and two to the right:

```
Income_generated      number(07,02) default 0,
```

A comma shows the decimal point and the default value is numeric zero.

Frank reviews the design relative to the five rules of normal forms. The design passes 1NF: all columns have a single value and no repeating values. It passes 2NF since all non-key columns depend on the primary key column. It fails 3NF because the Borrower and Customer tables both contain customer names. But these are the realities of the video world and Frank is not concerned. His design is very close to 3NF.

With his database design in hand, Frank can now move to the next stage: network design.

Working With View Video

Relational databases and SQL work together to help users, designers, and programmers store, retrieve, process, and report a system's data. The CREATE statement is quite powerful because it constructs the columns in the database, giving them names, data types, lengths, and placing necessary constraints. The SELECT statement simplifies retrieving data from the database by using a simple format that's based on English.

Case Study Exercises

Write CREATE and SELECT statements for the following:

1. Modify the CREATE TABLE title to include a new column named Price_category, whose values are limited to "A," "B," "C," "D," and "E."

2. Create a new table called Prices with two columns: Price_category and Amount.

3. What is the purpose of the new Price table?

4. Write a SELECT statement to retrieve all the tapes rented by a specific customer.

5. Write a constraint phrase to limit the category column in the Title table to Western, Science Fiction, Drama, Action, Comedy, Children, Mystery, and Musical.

6. Create an index for the Star, Performance, Rental, Customer, and Borrower tables.

Figure 10.15 The Oracle CREATE statements for View Video's relational system.

```
CREATE TABLE customer
    (
    Customer_id number(07) not null primary key,
    Last_name                 char(20)
        check (Last_name = initcap(lower(Last_name))),
    First_name                char(20)
        check (First_name = initcap(lower(First_name))),
    Middle_initial            char(01)
        check (Middle_initial = upper(Middle_initial)),
    Street_address            char(30),
    ZIP                       number(05) not null,
    Date_joined date default sysdate,
    Income_generated          number(07,02) default 0,
    Date_last_rented          date,
    Tapes_rented              number(05) default 0,
    Credit_card char(16),
    Credit_card_type          char(06),
    Customer_balance          number(07,02) default 0
    )

CREATE TABLE zip_code
    (
    ZIP        number(09) not null primary key,
    city       char(14) not null
        check (city = initcap(lower(city))),
    state_abb  char(02) not null
        check (state_abb = upper(state_abb))
    )

CREATE TABLE title
    (
    Title                     char(30) not null primary key
        check (Title = initcap(lower(Title))),
    Year_made                 number(04) not null
        check (Year_made between 1900 and 1999),
    Rating                    char(04) not null
        check (rating in ('G    ', 'PG ', 'PG13', 'R   ',
            'X   ', 'None')),
    Category                  char(12) not null
    )

CREATE TABLE tape
    (
    Tape_id                   number(10) not null primary key,
    Title                     char(30) not null
        check (Title = initcap(lower(Title))),
    Date_purchased            date,
    Date_last_rented          date,
    Times_rented              number(05) default 0,
    Cost                      number(07) default 0,
    Income_generated          number(07,02) default 0,
    Tape_status char(01) default ' '
    )
```
(continued)

Figure 10.15 *(continued)*

```
CREATE TABLE rental
      (
      Customer_id number(07) not null,
      Tape_id                    char(10) not null,
      Date_rented date not null
             check (Date_rented <= sysdate),
      Date_returned              date not null
             check (Date-returned >= Date-rented),
      Rental_fee                 number(07),
      Amount_paid number(07),
      Clerk_initials_out         char(03),
      Clerk_initials_in          char(03)
      )

CREATE TABLE payment
      (
      Customer_id number(07) not null,
      Date_returned              date
             check (Date_returned <= sysdate),
      Rental_fee                 number(07)
      )

CREATE TABLE clerk
      (
      Clerk_initials             char(03) not null primary key,
      Last_name                  char(20)
             check (Last_name = initcap(lower(Last_name))),
      First_name                 char(20)
             check (First_name = initcap(lower(First_name))),
      Middle_initial             char(01)
             check (Middle_initial = upper(Middle_initial))
      )

CREATE TABLE store
      (
      Store_name                 char(40) not null,
      Street_Address             char(30) not null,
      City                       char(14) not null,
      State_abb                  char(02) not null
             check (State_abb = upper(State_abb)),
      ZIP                        number(09),
      Phone_number               number(09) not null,
      Daily_Rate                 number(07)
      )
```

(continued)

Figure 10.15 *(continued)*

```
CREATE TABLE star
     (
     Star_id                    number(04) not null primary key,
     Last_name                  char(20)
          check (Last_name = initcap(lower(Last_name))),
     First_name                 char(20)
          check (First_name = initcap(lower(First_name))),
     Middle_initial             char(01)
          check (Middle_initial = upper(Middle_initial))
     )

CREATE TABLE performance
     (
     Star_id                    number(04) not null,
     Tape_id                    number(10) not null
     )

CREATE TABLE borrower
     (
     Customer_id number(07) not null primary key,
     Last_name                  char(20)
          check (Last_name = initcap(lower(Last_name))),
     First_name                 char(20)
          check (First_name = initcap(lower(First_name))),
     Middle_initial             char(01)
          check (Middle_initial = upper(Middle_initial))
     )
```

Network Design

GOALS

After reading this chapter, you should be able to do the following:

- Describe a batch processing system
- Define an on-line system
- Describe an on-line transaction processing system
- List two types of network configurations
- Explain the difference between an RFP and an RFQ
- Design an evaluation form for hardware or software acquisition

PREVIEW

As we have seen, systems revolve around people: their organizations, equipment, and data—and their procedures for collecting, processing, and distributing information based on that data. This chapter will focus on the many factors that an analyst should take into account when making decisions about hardware and software (Figure 11.1).

While we depict the selection of the network and the modes of processing as taking place after many other design activities, it really is a parallel event. The analyst must take into account system operation during output, plus make input and database decisions. If you think back to these phases of design, we really were also making decisions about the network. Users enter customer or vendor numbers and the system responds with the customer or vendor name. Fleet Feet decided to pick a relational database, mandating an interactive mode of operation, which affects the network.

During the last decade, hardware prices have fallen dramatically. Capabilities once restricted to the largest and most sophisticated million-dollar machines come now on a single-circuit board. For less than $10,000, organizations can now acquire previously unaffordable machines that are capable of servicing many terminals simultaneously. The recent proliferation of powerful personal computers further expands an analyst's options, especially when the analyst considers linking them to larger machines.

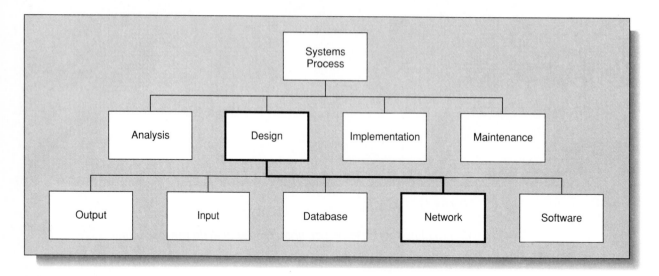

Figure 11.1 Network design is the fourth stage. During network design, we decide the systems mode of operation and order hardware and software.

How do we process the data in order to achieve the desired output? Should we collect data and hold them for future input into the computer? Should we store data in a computer in raw form for later processing? Should we process data as the organization collects them? In this chapter, we will consider the answers to these questions as we examine the modes in which computer hardware processes data, including the cost and performance merits of each of these modes.

Having settled on a mode of operation, we will then explore how the analyst specifies and acquires equipment for the new system. In some instances, a new computer may be necessary. In others, acquiring a terminal, additional disk storage, personal computer, or even laser printer may make sense.

MODES OF PROCESSING DATA

Depending on users' needs, budgets, and the equipment on hand or on order, the analyst selects the most appropriate mode of data processing. A system can function in any of four basic modes: batch, on-line, on-line transaction processing (OLTP), distributed. The fourth mode, distributed processing, has more to do with the coordination of the subsystems than with the actual mode of processing itself. For example, a distributed system may employ both batch and OLTP, or it may involve only one of the other three basic modes.

Various applications lend themselves best to one method; others demand two or more. In a bank, for example, mail deposits are processed overnight, while deposits made in person are processed immediately so that the customer can receive an instant and up-to-date account balance. Similarly, a company may require several systems, each of which needs its own mode of processing. For example, the reservation system of an airline must process data from several locations immediately, while statistical information concerning the promptness of take-offs and landings can wait until later in the month.

Batch Processing

In a **batch processing** environment, the organization collects data over a period of time, then processes them in groups at specified intervals. Most payroll systems operate in batch mode, collecting data on hours worked over a specific time period (such as a day, a week, or a month) before issuing paychecks at the end of that period. Similarly, an airline route assignment system gathers crew members' requests (called bids) each month for certain flights, then determines which people will work which flights the next month. Such a once-a-month system lends itself nicely to a batch mode of operation.

As a result of recent technological advances, the operation of batch systems has changed a lot. Formerly, an operator would enter data onto cards, which a machine would read before the computer processed the data. Today, a data-entry person may collect the data through a personal computer, then store it on a floppy or hard disk for eventual uploading to a mini- or mainframe computer (Figure 11.2). The computer may reside far from the data-entry people, who transmit data to the distant computer by means of a data communications (or telecommunications) link, telephone wire, microwave radio, or satellite.

The main advantages of batch processing are its simplicity and relatively low cost. Data entry, processing, and output can each proceed separately. Furthermore, some applications do not demand immediate feedback. For many years, batch processing was the only option available to the analyst.

Figure 11.2 An organization using batch processing may do it all in one physical location or it may send data from a remote source to a central computer for processing.

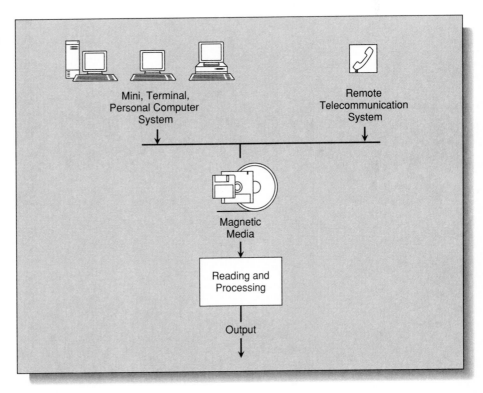

Mini, Terminal, Personal Computer System

Remote Telecommunication System

Magnetic Media

Reading and Processing

Output

However, the lack of timely data argues against batch processing. Organizations that need to base decisions on the most up-to-date data cannot afford to wait a week or even a day for current reports. Therefore, many organizations have abandoned batch processing for one of the more modern modes.

On-Line Processing

In an **on-line processing** environment, users with their PCs or terminals can inquire about data at any time other than during the updating of their data. Users can update data by changing existing—as well as adding new and deleting old—information. On-line systems are more sophisticated and complex than their batch counterparts because they include terminals and update features.

The design of an on-line system involves two steps. First, the analyst designs a system for data collection and processing. Then, he or she designs the inclusion of personal computers or terminals for inquiries (Figure 11.3).

Figure 11.3 On-line systems can use batch data input, and have ongoing user inquiry capabilities.

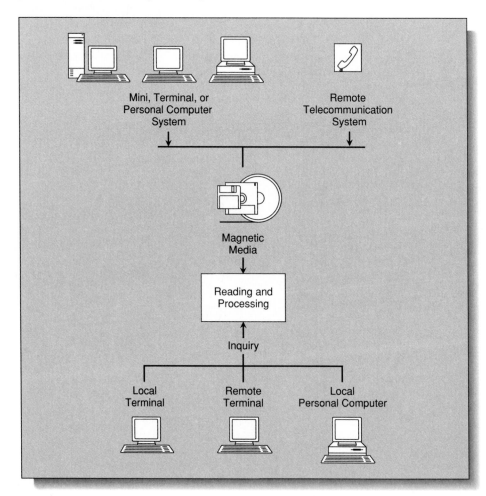

On-line output can also occur in a batch mode, perhaps with reports generated in the evening when the demand on the computer declines. The Federal Bureau of Investigation's (FBI's) National Crime Information Center implemented one of the first major on-line systems. Still in use, it allows local law enforcement agencies to send data concerning arrests and trial results to FBI headquarters in Washington, DC, where the FBI enters it into the system. In return, agencies can access the system at any time to inquire about and retrieve useful information about a suspect (such as criminal history, identifying marks, and physical description).

Many organizations that need timely data have opted for on-line systems, although they typically demand additional hardware and software. Hardware for on-line systems must support nearby or remote personal computers or terminals; software must permit the computer to communicate with local or remote devices. On-line systems continue to grow in popularity because they provide users with quick access to data.

On-Line Transaction Processing (OLTP)

On-line transaction processing (OLTP) refers to an interactive system on which users enter data for immediate processing and access updated data at will (Figure 11.4). The system updates data quickly enough to control an ongoing activity.

For example, an airline reservation system immediately updates information about available seats the instant that a customer purchases a ticket. During check-in, when a passenger reserves a seat, the system again operates in an OLTP mode by setting aside the requested seat, thus making it unavailable to another passenger. Many rapid transit districts also rely on OLTP systems for train control, as does the National Aeronautics and Space Administration for satellite launching.

Hardware for OLTP systems can include terminals, with a variety of computers tied to a central computer. The computer may do all of the processing and storing of data locally, or it may only store and process selected data locally before sending them on to a central location, where they will join other data for coordinated processing.

Figure 11.4 On-line transaction processing (OLTP) systems involve data input at will or in batches and allow for constant updating and inquiry.

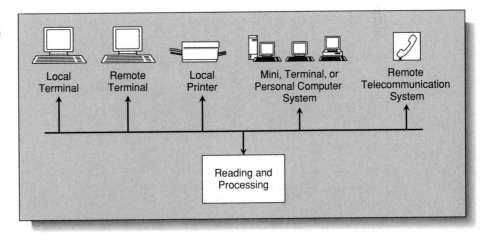

Data security poses a special problem to analysts who are designing on-line or OLTP systems. Since users can retrieve and alter data at will, the system must guard against unauthorized manipulation. To accomplish this, analysts may establish passwords or physically restrict personal computers and terminals from unauthorized use. They may also assign some users limited read capabilities, while allowing others both read and write access. Use of certain portions of the database can also be confined to certain users. Data regarding employees' pay rates, for example, may be accessible only to the payroll or personnel departments.

Another problem involves setting priorities for use of the data. Since two or more users might request identical data simultaneously, or one user might try to process a record already being processed by someone else, analysts must design a way to establish priorities among authorized users. For example, two airline ticket agents in different cities may try to assign the same seat to two different passengers at the same time. To resolve this conflict, an analyst might assign a higher priority to the terminal in the city where the flight originates and have the system tell the other terminal that the seat in question is not available.

An OLTP provides up-to-the-minute data, but installing one usually represents a big investment. Terminals and personal computers geographically separated from the central computer add to the system's cost, as does the additional hardware required to drive the system when in interactive mode.

DATA COMMUNICATIONS

Batch, on-line, and OLTP systems all require linking together devices that can "talk" to each other. The movement of data over short or long distances implies **data communications**. While appearing quite simple, data communications require having both a sender and a receiver of the data. Between the sender and receiver are converters and channels (Figure 11.5). Converters translate the data from digital (binary 0 and 1) form into analog form and back again. Channels deliver the data between converters. Software controls all the facets of the communication.

Sender or receiver devices include PCs, terminals, bar-code scanners, point-of-sale devices, automated teller machines, and any other device capable of collecting or receiving data. All of these devices must have a communications outlet or port to allow it to connect to the converter.

Converters often take the form of a **modem** (Figure 11.6a), short for modulator and demodulator. Since a telephone line requires an analog signal, modems convert outgoing digital signals into analog signals and incoming analog signals into digital signals (Figure 11.6b).

Figure 11.5 Data communications require having a sender as well as a receiver. Between them are converters and channels.

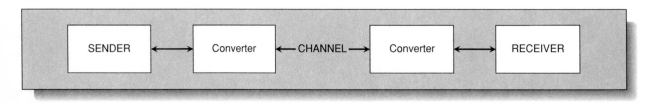

TOO MANY CONTROLS

Dick Couzens, an analyst at Solar Systems, Inc. (a supplier of electronic components for missile systems) prided himself on his data-entry designs. He loved this part of his job and relished every opportunity to display his skills. He got a lot of chances because Solar, as a brand-new operation, required many sophisticated systems. For every new one, Dick designed sight verification and check digits for all key fields and validations for all other data items. When the manager of data-entry operations, Mary Odom, scheduled an afternoon meeting to discuss data-entry routines for a new payroll system, Dick reviewed his designs. Dick's new data-collection system had undergone all kinds of tests to make sure that all data entered were correct. Expecting to win high praise for his many data-entry and error-checking routines, Dick eagerly anticipated the meeting.

When he strolled into Mary's office, however, Dick received a nasty shock. Mary barely exchanged pleasantries before she exploded with a list of complaints: "This data-entry system is a time-consuming pain in the neck! It's too complex! Training wastes the data-entry staff's time, the prompts confuse everyone, and the system's responses make people feel stupid. Whenever someone makes a mistake, 'You goofed!' flashes on the screen! And why does the screen go blank for ten seconds or more for no apparent reason?"

Dick had tried to inject a little humor into the system, but no one seemed to be laughing. What had he done wrong? Weren't his controls the best possible?

Overwhelmed by Mary's outburst, Dick suggested that they watch one of Mary's staff enter data, but Mary insisted that Dick enter a few transactions himself. "You play data-entry operator for 15 minutes—then come back and tell me how you like it!" she said. Reluctantly, Dick agreed. After 20 minutes on a terminal, he began to understand Mary's complaints.

Despite the fact that he had designed the system himself, it took Dick over ten minutes to become comfortable with it. Then, after Dick finally entered a transaction, the computer took almost a minute to return the cursor for the next entry because it was so busy performing a multitude of class, range, check digit, and presence tests. Such a delay could waste over 50 percent of an operator's time, which was understandably unacceptable to Mary. Next, Dick tried to complete a more complicated transaction. When "You goofed!" lit up his screen for the third time, he knew why Mary's operators felt patronized.

Sheepishly, Dick walked back to Mary's office to apologize. "I violated one of the laws of design, keeping it simple. I guess I got carried away with controls."

Mary smiled. "Like hot peppers, too much of anything can be worse than none at all."

After agreeing to overhaul the data-collection system, Dick retreated to his office down the corridor. He paused at the door of fellow analyst Susan Caldwell's office and peered at a little poster she had tacked there: KEEP IT SIMPLE TO MAKE IT FASTER. Well, he told himself, I hope I never make *that* mistake again.

Modems come in a variety of shapes, sizes, speeds, and transmission methods. Some will automatically dial, answer, or redial.

Once the signal is converted by the modem, it is placed on a delivery channel. Channels are like highways of data; like highways, they may have single or multiple lanes. Telephone twisted-pair wire, coaxial cable, microwave, satellite, and fiber optic cables are common types of communication channels.

Each channel type has specific advantages, as well as disadvantages. Twisted-pair wire is the oldest and slowest type; it carries the least amount of data per unit of time. Coaxial cable can manage as many as 100,000 simultaneous telephone calls. Microwave and satellite are ideal for long-range data communications. Fiber optics use pulses of light or no light in a hair-thin glass fiber. In theory, a fiber optic cable can carry 100 trillion (100,000,000,000,000) bits of data a second.

Figure 11.6 Converter modems are translators that change the data from one format to another.

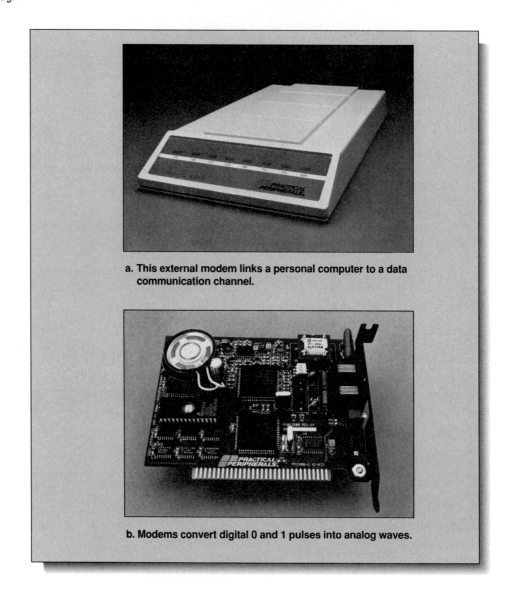

a. This external modem links a personal computer to a data communication channel.

b. Modems convert digital 0 and 1 pulses into analog waves.

Knowing the basic components of data communications is only the beginning. Besides senders, receivers, converters, and channels, two major codes are in use, at least two different methods exist for transmitting data along the channel, and the channel must follow specific rules during transmission.

The two codes in wide use are ASCII and EBCDIC, which should be familiar to you by now. Both codes specify specific bit patterns that stand for predefined characters. In ASCII, a unique pattern of 1 and 0 bits represents each character that we wish to send. Of particular interest are the first 32 codes that perform special functions for controlling data-transmission facilities. For example, the ASCII code 10, which we name LF, stands for

Line Feed. Other codes such as STX, ETX, ENQ, ACK, and EOT stand for Start of Text, End of Text, Enquire, Acknowledge, and End of Transmission. These 32 special codes keep data communication devices talking with each other in order to understand what is sent, when to start sending, and when to stop sending.

The two different methods of transmitting data along the channel are asynchronous and synchronous. **Asynchronous** sends data character by character, with special characters telling the sender and receiver when to pay attention and when to relax. Start and stop bits surround the seven or eight bits that represent a character. The start bit tells the receiving device that a character follows. The stop bit or bits mark the end of a character.

Synchronous transmission sends data in a group or block, with special characters at the beginning and ending of the block. Synchronous is faster and more efficient than asynchronous transmission, but requires that sender and receiver have perfect timing, since there is no time to relax. Synchronous does not need or use start and stop bits, thus reducing the amount of data that need transmitting. The trade-off is that both the sender and receiver must be timed identically, or "synchronized."

Sometimes signals include an extra parity bit for error detection. If characters are sent as seven bits, an eighth bit is also sent to set the patterns to match any one of five schemes: none, mark, space, even, or odd. None means that no parity checking is used. Mark always sets the parity bit to 1. Space sets the parity bit to zero. Even means that the parity bit is set to 1 only when it is needed to make the total number of 1 bits an even number. Odd parity is the opposite of even, setting the parity bit only to establish an odd number of 1 bits. Parity is thus an automatic check the hardware can use to better guarantee the accuracy of the data.

With the method established, the data communication devices must now consider the transmission mode: simplex, half-duplex, or full-duplex. Simplex mode provides for one-way transmission. A radio is a simplex device; data are sent from the station to the radio. Half-duplex mode allows for two-way transmission, although only one way at a time. CB radios are half-duplex; you can either send or receive, but cannot do both at the same time. Full-duplex allows for simultaneous two-way transmission. Separate transmission signals prevent simultaneous data communications from interfering with each other.

The rules the channel follows for maintaining communications are called **protocols**. Protocols define how to start the transmission, which control codes are sent and their order, and which error-detection techniques to use. Protocols are sometimes built into the hardware by the manufacturer. If two devices do not use the same protocol, they cannot communicate.

Among the most popular protocols are one of IBM's Binary Synchronous Communications (BSC or bisync) known as BSC 3270, 2780, or 3780. Other popular protocols are IBM's SDLC (Synchronous Data Link Control), the International X.25, and the High-Level Data Link Control (HDLC). For personal computers, popular protocols include XON-XOFF, XMODEM, MODEM7, and X.PC.

Communications software (such as Crosstalk) is the final requirement for successful data communications. This software manages an operating environment, gives an operator the opportunity to select transmission speeds and set protocols, and allows for capturing, displaying, manipulating, and storing data. Most of today's communications software is Window-based with detailed selection lists so that operators can mark their choices.

NETWORKS

Most business systems of the late 1950s and early 1960s ran in batch mode because the other possibilities did not exist. However, as hardware and software advanced technologically from the mid-1960s to mid-1970s, on-line systems finally became feasible. OLTP systems did not find their way into consumer and business applications until the mid-1970s, when less expensive minicomputers made real-time systems affordable.

By the late 1970s and early 1980s, advances in data communications enabled organizations to distribute data electronically to many different locations. Some of the movement to distributed systems resulted from the advent of inexpensive personal computer systems, which can function as independent computers and/or act as terminals, collecting data and storing them on floppy or hard disks. PCs can also perform a dedicated task such as monitoring all the devices linked together. In the 1990s, the move to distributed decentralized personal computer-based systems will continue.

Many businesses vitally depend on regular communication among employees for timely decisions and planning. But, given the decentralized nature of some organizations, those who must exchange information may reside in another room, building, city, state, or country. Thus, after we have installed the data communications hardware/software and picked compatible transmission speeds/protocols, we still need to join the devices together. A group of interconnected communications devices is called a **network**.

At the heart of distributed processing lies the notion of a network linking together machines from different manufacturers. If a user at one location needs data from another location, the different machines must communicate that data, and they must have the facility to update duplicate data that are generated at two or more sites within the system.

A distributed system, which stores data in various locations and allows users to access all or part of a database, may process data in a batch, on-line, or OLTP mode. Regardless of the mode, each local computer still provides collective resources for local users.

Topologies

Four popular methods for configuring the various computers, terminals, and personal computers in a distributed system are the completely connected, star, ring, and bus networks. Called topologies, each presents trade-offs in cabling, message switching, and hardware requirements.

A **completely connected network** provides a direct path or link between any two stations (Figure 11.7). For example, if a system includes 8 stations, it will need 28 links. Consequently, this type of network becomes unwieldy in situations that have a large number of stations.

Star networks tie one or more personal computers, terminals, or satellite computers to a central computer (Figure 11.8). All data transmission or reception passes through the central computer or "server," which routes it to the appropriate station. Star topology requires a lot of cable, but no complex sending protocols.

Figure 11.7 A completely connected network links all subsidiary terminals or computers to each other.

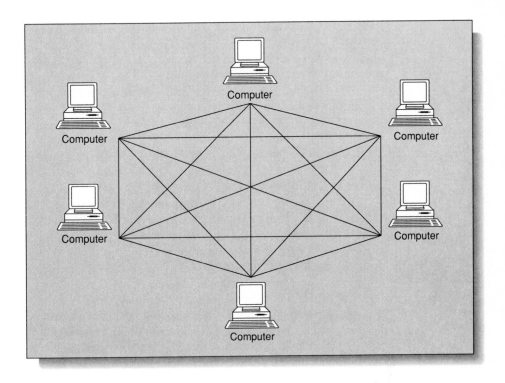

Figure 11.8 A star network links all subsidiary terminals or computers to a central or host computer whose only job may be to tie the computers together and to route data from one to another.

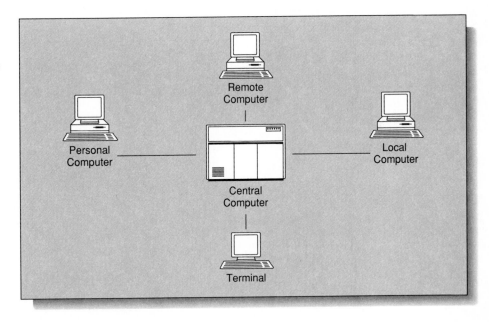

In a **ring network**, data flow in one direction from station to station (Figure 11.9). When a station originates a message, it places an address (the name of the destination station) with the message. Then each station, in turn, can check incoming messages and addresses, accepting those that match and sending those that do not to the next station. If a message progresses all the way around the ring without being accepted, it stops at the originating station. Every station listens to and repeats every other station's message, a fact that results in low efficiency of the links. A failure by any station in the ring causes failure of the entire ring.

Bus networks work similar to a water pipe that projects through an apartment building (Figure 11.10). Apartment dwellers can tap on the pipe when they want to send a message. All dwellers can listen, but only one may tap at a time. If two dwellers tap at the same time, the messages become garbled. In a computer system using this approach, a single length of cable runs past every device. Short cables or stubs attach workstations to the main cable. Bus topology is the most popular and easiest to install. However, when many devices are on the network, the potential for simultaneous access is greater. When two or more devices access the network at the same time, a collision occurs and access is attempted again by each device at random intervals.

In 1980, Xerox, Digital Equipment Corporation, and Intel announced a commercial bus network, Ethernet. To detect garbled data streams, the devices in the Ethernet system listen at the same time as they send messages and stop sending when they encounter interference.

IBM's Token Ring local area network (LAN) uses the ring approach with token passing (Figure 11.11). A **token** is a stream of characters of fixed length and format that constantly flows around the ring. Each token contains a header, trailer, priority, source, destination, data, and type. Headers and trailers identify the beginning and end of the stream of characters, source and destination reveal the origin and destination of the

Figure 11.9 A ring network ties computers together. Data may travel from one machine to another through more than one machine before they reach their destination. A host computer is not needed. Each computer in the ring may have terminals.

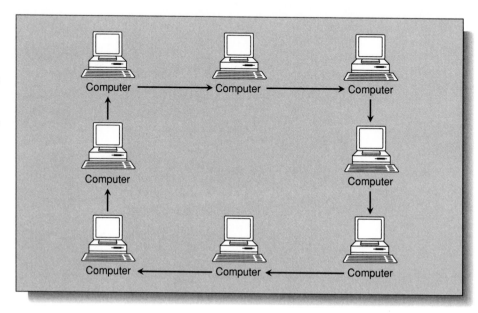

Figure 11.10 Bus networks have terminals or computers tapping into a central line. Each piece of equipment communicating on the line has a unique identification code.

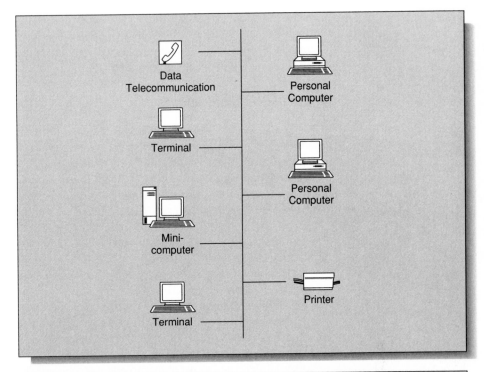

Figure 11.11 IBM's Token Ring local area network (LAN) uses the ring structure. Multiple rings can be joined at bridges. Gateways permit access to outside computers or to data communication links.

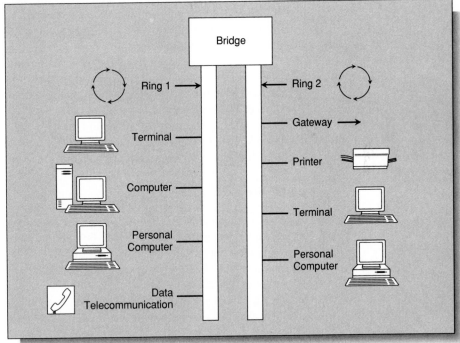

character stream, the data portion holds the message itself, and the type identifies whether the token is free to accept a message or not. Access to the ring is possible only when a station is in possession of an all-clear message that passes around the ring like a token. This access protocol is called token passing.

Network Considerations

The analyst should weigh the following six factors when considering networks: capacity, performance, reliability, flexibility, security, and resource utilization.

Capacity refers to the number of users that the system can support at each site simultaneously. A star network may have an upper limit as to the number of devices it can support based upon the constraints of the central host computer. Ring networks are easily expandable when a new device is inserted between two others, but eventually performance degrades and ultimately becomes unacceptable to users.

Performance is a measure of how quickly the system can respond to a user's request. In ring structure, a message may have to travel all the way around before the user receives his or her response. Completely connected networks respond almost immediately because all devices are linked to each other.

Reliability implies the quality of the equipment in terms of how often it is likely to crash due to a hardware or software malfunction. Star networks fail when the central computer fails, but can remain functioning when any other device fails. Bus and completely connected networks do *not* fail when any device fails.

Flexibility relates to the ability to add new components (such as PCs, terminals, and printers) without undue problems. Completely connected networks require that a new device must link to every other device. On the other hand, a star network requires only a single connection—to the host computer.

Security has to do with protection against unauthorized access to data or from interference among users. In all network configurations, security can occur at the device level with "passwords" permitting access to those who know the proper entry code.

Resource utilization refers to the ability to keep track of time consumed by specific users and system failures for accountability reasons. For a star network, resource utilization is tracked by the central or host computer. All other types of networks require tracking by each device or by one device dedicated to tracking utilization.

Local Area Networks (LANs)

The millions of personal computers sold have provided businesses with powerful tools for manipulating and managing their data. Originally, these PCs operated independently. However, their networking ability quickly started owners thinking about tying them together. One of the easiest types of networks is the **local area network (LAN)**. Most LANs are a take-off on the bus topology. LANs link PCs together within a limited but geographically shared area, typically within 2,000 meters.

LANs allow PCs to share peripherals such as modems, high speed printers, and large disk drives. LANs can also enable PCs to leave messages when a user is unavailable. The heart of the LAN is a **server**, usually a dedicated device that allows PCs to share another device. A printer server permits the sharing of printers, a file server shares disk or tape storage facilities, and a communications or gateway server allows a LAN to communicate with other networks.

Many LANs link their devices together with coaxial or fiber optic cables; others use twisted-pair wire. Coaxial or fiber optic LANs have greater transmission speeds and allow the greatest amount of simultaneous communication traffic with voice, data, and even video on a single cable. Twisted-wire uses inexpensive, in-place telephone-type connections that have slower speeds. This method does, however, permit voice—as well as data—on the network, but not at the same time.

Advantages and Disadvantages

Networked systems offer several advantages. Most importantly, they give local users access to an organization's centralized data. In most cases, networked systems also send and receive data quickly and efficiently. Users at local sites enjoy immediate access to their own data, and they can obtain system-wide data within a reasonable amount of time.

On the other hand, networked systems do pose several disadvantages. First, they require extra hardware for data communications. Second, machines from different manufacturers must be made compatible. For example, IBM and Unisys mainframe computers store data in EBCDIC; HP, DEC, and Data General minicomputers use ASCII. However, PCs made by all these companies use ASCII. For these mainframe, minicomputer, and PC machines to talk to each other, software must translate ASCII to EBCDIC and vice versa.

Third, if the same data exist at two or more separate locations, costs will rise. For example, if the names and addresses of registered voters are stored in county government computers, as well as in state computers, the data are duplicated. When a voter moves, both computers must update their files to hold the new address. Not only is this duplicate data inefficient, but the address needs changing in both places, costing extra computer time.

Ever since the advent of the minicomputer and the PC, analysts have debated the pros and cons of centralized versus distributed information systems: costs, efficiency of operation, responsibility, coordination, communication, security, local control, and decision-making. Regardless of an analyst's own biases, however, comparisons of costs and benefits indicate that the 1990s will see increasing reliance on distributed or network systems, especially as we solve their software and hardware problems: minimizing data communication costs, providing adequate security for data communication links, and developing better methods for tying the computers to each other.

HARDWARE AND SOFTWARE ACQUISITION: RFQS AND RFPS

By this point in the systems process, the analyst has chosen the system mode and, in the case of a distributed system, a link joining personal computers, terminals, or other devices. If the system is brand-new or requires some new hardware or software, the analyst must now set about the task of obtaining it. Usually, new applications are expected to run on the existing hardware or software unless a clear impracticality is shown. If this is the

LANS AND DISTRIBUTING THE WORKLOAD

Since the dawn of the microcomputer revolution, unprecedented computing power has become accessible to vast numbers of office, factory, management, and data processing workers. These machines are proliferating throughout every level of virtually every organization and creating a pressing need for users to be able to communicate with one another. Sending, receiving, and sharing information stored in computers and computerized devices has become an essential goal of analysts in all information-handling environments.

Fortunately, significant technological advances have increased our ability to link various computing resources, whether they lie across the office or across the nation. While research has long been conducted in the area of trans-continental communications, large organizations have increasingly demanded improved communications between offices in a building, between the floors of a building, or within a complex of nearby buildings; hence the current focus on what we call local area networks.

A local area network (or LAN) links hardware and software elements located within close geographic proximity. It may include the transmission not only of data, but of voice, video, and facsimile copies.

Unlike transcontinental hook-ups, which must rely on satellites or phone lines, a local area network operates without the need to rely on external influences not under an organization's control. Furthermore, LANs cost less and are less complex. This independence has spawned a new technology, the Common Bus Local Area Network, where processing devices link to a common medium, such as a coaxial cable or standard multi-strand cable. While the components, or "nodes" in such a network are fully connected with one another, their transmissions do not pass from node to node. Rather, messages are broadcast to the entire "net" simultaneously without the intervention of the nodes.

The Common Bus approach eliminates the need for a central controller, thereby enabling communications to rely only on the nodes directly involved with a given message. Such an approach offers significant advantages over any other scheme for the configuration of a local area network because workers wishing to communicate with one another can do so independently.

An organization obtains the following benefits from the Bus Network Architecture:

- Dynamic reconfiguration of nodes and functions (reconfiguration without interruption of service)
- Reliable operation resistant to malfunction (no exposure to net controller failure)
- Relatively low cost of installation (a fraction of the cost of other methods)
- Very fast transmission of messages (200 KB to 10 megabit per second)

The Common Bus approach will probably dominate the local area communications field through the end of the 1990s.

Forthcoming microprocessor products, whose prices will rapidly drop, will routinely offer LAN capability, thereby making the new technology available and affordable to an increasingly wide range of organizations.

The ideal LAN provides any user access to any computer resource in the LAN, regardless of location. To achieve this goal, analysts must ensure that communication among different devices that demand different protocols remains completely transparent nor effect the user. In addition, analysts must make sure the organization can install new resources with no more effort than simple device definition and physical attachment to the network.

As organizations scramble from the Information Age to the Communications Age, they will look for ever greater speed, reliability, and flexibility in the local area networks. For the time being, the Common Bus will provide those results.

case, the analyst may look at an upgrade (more memory, disk space, and/or faster CPU) rather than a new system.

Analysts may approach acquisition in one of two ways: by compiling lists or by specifying functional requirements. A list contains exact specifications in the following areas:

1. *Memory*—measured in bytes
2. *Disk storage*—measured in millions of bytes
3. *Printer speeds*—measured in lines per minute
4. *Terminals*—numbers, types, and capabilities
5. *Programming languages*—COBOL, C, Pascal, FORTRAN, and so on.
6. *Operating system*—disk-based with a certain number of concurrent operations, database managers, utility sorts or merges, file copy routines, and editors

Vendors such as IBM, Unisys, Apple, Hewlett Packard, or Digital Equipment Corporation can recommend what equipment in their product lines would fulfill the requirements specified on a hardware list.

When following a functional requirements specification approach, an analyst lists necessary functions and asks vendors to recommend appropriate equipment. Functional descriptions look like hardware lists, except that they omit exact measurements. For example, a list for a new full-scale accounting system might look like this:

1. *Memory*—sufficient to support 3 active database systems, 4 concurrent COBOL compiles, 10 other program terminals, and 12 data-entry terminals.
2. *Disk storage*—sufficient to store all operating system programs, 1,000 COBOL source programs, the data for a large accounting system, and room to expand 25 percent in each of the next 5 years.
3. *Printer*—fast enough to print all the outputs of a large accounting system within a reasonable number of hours.
4. *Terminals*—memory able to store at least 5 pages of program listing. Must have a 10-key pad, screen that tilts, and separate keyboard.
5. *Programming languages*—COBOL, C, Pascal, and FORTRAN.
6. *Operating system*—disk-based with at least 65 active terminals, a relational database manager, full screen text editor for program development, data dictionary, query language, 4th generation language, CASE software, program generator, and spooler.
7. *Application software*—fully integrated multi-user accounting system, including accounts payable, accounts receivable, payroll, inventory, and general ledger.

This second approach, while it may appear vague, nevertheless provides vendors with enough detail to make informed recommendations. If using the first approach, the analyst calculates system requirements based on accurate and up-to-date knowledge of the state-of-the-art in hardware or software. However, vendors may have new and exciting products waiting in the wings, products that even the most astute analyst may not have encountered yet. Vendors know their hardware and software best and can often quickly select the right amount of memory or disk space to fulfill the purchaser's need. For example, one vendor's database manager may require 100MB of disk space for a given application, while another may need 250MB to support it. If the analyst specifies 100MB of disk space, the second vendor's hardware will not suit the application. Conversely, if he or she specifies a 250MB disk, the first vendor is placed at an unfair cost disadvantage.

When developing certain potentially costly applications, the analyst may expand either the hardware list or the functional requirements to include more detail. Regardless, the lists form the basis for either a **request for quotation (RFQ)** or a **request for proposal (RFP)**. RFQs usually accompany hardware lists, while RFPs generally go along with functional requirements specifications.

To evaluate vendors' responses to an RFQ or RFP, most analysts have a mandatory set of criteria and a point system, such as the following:

1. The analyst lists the major categories (such as software capabilities, languages, and database requirements) with each category weighted according to its importance to the system (Figure 11.12). For instance, a newly formed organization with no established system will not need to convert data from one form to another, so it will not assign many points to conversion. By the same token, an organization that demands only new software need not mention hardware requirements.

2. Next, the analyst breaks each category down into its components, with each component carrying a percentage of the entire category's weight. For example, preinstallation provisions might include five components: computer time necessary, training of operations staff, training of programming staff, training manuals, and CASE training. The total weight of this category would be 4.3, with 2.3 allotted to computer time and 2.0 allotted to training.

3. The analyst completes a bid evaluation form by category for each vendor, including all components and their weights. In Figure 11.12, there are eight categories, so there would be eight forms for each vendor.

4. The analyst rates each vendor's ability to provide the needed component on a scale from 1 to 5. These component ratings are multiplied by the respective weights, then their points are totaled (Figures 11.13 and 11.14).

Figure 11.12 A sample vendor evaluation point system, evaluating responses on eight different criteria. The allocation of points occurs prior to sending the RFQ. Each major category is broken down into subcategories on the bid evaluation form.

Criteria	Total Weighting
1. Management Summary	6.0
2. Hardware Configuration	25.0
3. Software Capabilities	21.5
4. Demonstration	12.5
5. Space, HVAC	9.0
6. Conversion	15.9
7. Support Services	7.8
8. Pre-Installation Provisions	4.3

Figure 11.13 Bid evaluation form for a software vendor's response (each vendor is numbered). Each component is rated by an evaluator (who is assigned an evaluator number) on a scale from 1 to 5, then multiplied by its weighting to receive the points. Points are totaled and a bonus system may be included to add extra value if the analyst feels that it is appropriate.

Bid Evaluation Form
Software Capabilities

VENDOR NUMBER: 45
EVALUATOR: 18

Evaluation Component	Score	x	Weight	=	Points
1. 1985 ANS COBOL	5		3.5		17.5
2. FORTRAN-77	0		1.5		0.0
3. C	5		1.5		7.5
4. Pascal	3		1.5		4.5
5. Relational database manager	4		2.5		10.0
6. Text editor	5		2.0		10.0
7. UNIX operating system	4		3.0		12.0
8. Data communication software	4		2.5		10.0
9. System software described in detail	4		1.0		4.0
10. Software operational for at least 1 year	5		1.0		5.0
11. Utility systems software	4		1.0		4.0

Other Considerations

12. Program generator	2		0.5		1.0
13. CASE software package	3		2.0		6.0

Note: Two bonus points awarded for full-screen editor (instead of a line editor).

Total Weighting	23.5	
Subtotal of Points		91.5
Bonus Award		2.0
Total Points		93.5

Notes: Points awarded as follows: 0 = unacceptable, 1–3 = acceptable, 4–5 = outstanding. Bonus points not to exceed 15% of subtotal points.

Figure 11.14 Bid evaluation form for pre-installation provisions.

Bid Evaluation Form
Pre-Installation Provisions

VENDOR NUMBER: 45
EVALUATOR: 18

Evaluation Component	Score	x	Weight	=	Points
1. Computer time provided	3		2.0		6.0
2. Training of operations staff	3		1.0		3.0
3. Training of programming staff	2		1.0		2.0
4. Training manuals provided	3		.3		.9
5. CASE training	3		1.0		3.0

Note: 1 bonus point awarded for on-site training.

Total Weighting	5.3	
Subtotal of Points		14.9
Bonus Award		1.0
Total Points		15.9

Notes: Points awarded as follows: 0 = unacceptable, 1–3 = acceptable, 4–5 = outstanding. Bonus points not to exceed 15% of subtotal points.

5. To strengthen the objectivity of the process, several people may evaluate the vendors' responses. Individual scores are added together to create a composite score, which in turn is divided by the cost of the vendor's equipment or software to obtain a point-cost ratio (Figure 11.15). In this case, if we were to select the vendor solely on the basis of cost, we'd pick vendor number 27; if we based our decision on points, we'd pick vendor 63. However, since the point-cost ratio evaluates both capabilities and costs, vendor 27 clearly wins the order (the lower the point-cost ratio, the better the value).

Having chosen a vendor, the organization begins negotiating a contract. Although most vendors offer standard contracts, any organization purchasing new equipment should ask its purchasing and legal departments, controller, and the manager of computer services to scrutinize any written documents before anyone signs them. In most organizations, only the highest level executives can enter into legally binding agreements.

REAL-TIME CASE TOOLS

Batch, on-line, and OLTP processing are all supported by CASE tools. While most computer usage from the 1960s to the mid-1980s was predominately batch or on-line, the 1990s bring increasing reliance on real-time systems. As a consequence, CASE suppliers have had to incorporate real-time modeling in their products.

Batch and on-line systems are time-driven, whereas real-time processing is event-driven. Time-driven systems are those in which the software needs to perform a specific task on some predetermined day, hour, or minute. For example, payroll systems print paychecks at the end of the pay period, weekly, every other week, or at the end of a month. Similarly, accounts receivable systems print bills at the end of a month that recount the purchases and payments for the prior month. Schools issue grades at the end of a semester or quarter.

Event-driven systems are those in which the occurrence of a transaction triggers the software to perform a task. Most events are sensed from the outside environment and the system must respond to this outside stimuli.

Figure 11.15 Completed evaluation of each vendor's response, showing costs, points earned, and a point-cost ratio. Vendor names are confidential so that name brands won't affect the evaluation process.

Vendor Number	Total Five-Year Cost	÷	Points	=	Point-Cost Ratio
9	$630,531		313.85		2009
27	416,867		313.36		1330
45	565,999		302.47		1871
51	646,095		324.27		1992
63	539,769		338.24		1596

As an example, consider mouse-based personal computers such as the Macintosh or MS-DOS Windows-based software systems. The operating system needs to monitor the movement of the mouse, repositioning the cursor as it moves horizontally or vertically. At the same time, the operating system must check on the mouse button(s) and test whether or not a button was pressed. While it is monitoring the mouse and buttons, the operating system must also monitor the keyboard to see if the user depressed a key. Depending on the event, the computer then performs a specific and unique task.

Other examples of event-driven systems include traffic signals that sense when vehicles are waiting for a traffic light to change and a missile-tracking system that alters the course of the missile in flight, depending on sensors or directions from a radio controller.

CASE software implements these real-time events with a modification to traditional data flow diagrams. These alterations introduced three new symbols: control processes, control data flows, and control data stores. The DFD represents these three new symbols with dashed lines (Figure 11.16). This system models the cruise-control system found on many automobiles. The system turns on cruise-control and resets the automobile's speed by adjusting the throttle position.

The two control processes are named "Control Speed" and "Monitor CC Status," with accompanying control data flows named Cruise, Resume, A (for "activate"), and T (for "trigger"). Multiple activate and trigger data flows with the same names are allowed for control activities in these DFDs. A control data flow with a label A specifies that the process pointed to is activated or turned "on." A trigger is a signal sent that indicates process completion. Control data flows do not carry data, but simply signal that this process should "start operation" or that it is "done."

Figure 11.16 Cruise-control data flow diagram with dashed line process and data flows.

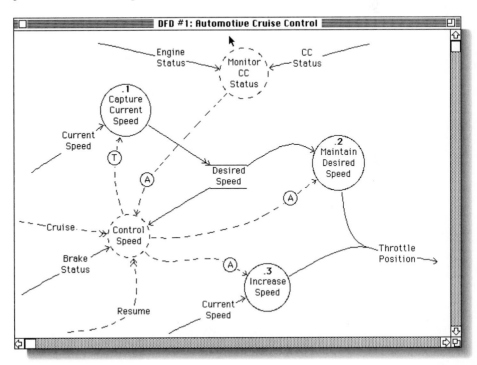

A control process is a monitor that supervises all the other processes and calls them up at the right time. Control processes may read or write to data stores and can reference control data stores, which hold data specific for control purposes.

Most CASE systems allow the DFD author to create these three new symbols, along with traditional DFD symbols. In creating a process, the DFD author is asked to select from a normal process or from a control process. Likewise, when selecting a data flow, the author can pick a solid line value bearing data flow or a dashed line bearing event flow. The software provides the same types of checks on the diagram to ensure accuracy and allows the DFD author to delete control processes or create child control processes.

Control processes are defined and described with a **state transition diagram (STD)**. These pictures list the rules that tell how the process moves from one state to another. In our cruise-control example, we would define three states: idle, cruise, and brake. Let's suppose we want to maintain a speed of 30 miles per hour (mph). The car starts at the idle state, with brake status on and speed at zero. We then set the brake control to off, apply the accelerator until the speed reaches 30 mph, and set the cruise-control to on. The rules say that the system should back off on the accelerator if the cruise-control is set and the current speed is greater than 30 mph. If the speed falls below 30 mph, the system will apply the accelerator until the speed reaches 30 mph again. We will maintain cruising when the cruise-control is set and the speed stays at exactly 30 mph. Similarly, most CASE systems have interactive windows that allow the STD author to compose and edit rules, specifying states, activations, and triggers.

Each CASE tool supports OLTP process control systems somewhat differently. However, as we move onward in the 1990s, these types of systems and CASE tools will play an ever-increasing role in systems analysis and design.

DESIGNING FLEET FEET'S NETWORK

Fleet Feet's current network topology is the star. At the center of the hub is their Hewlett Packard computer. Terminals and personal computers (DOS and Macintosh) are wired directly to the HP computer. Three 2400 baud modems are available for outsiders to "dial-up." The modems allow remote access to the computer for other stores or for Fleet Feet staff who are traveling.

The central HP machine serves as a printer and a file server. Users can store and retrieve data on the HP's 670MB hard-disk drive. They can also access laser or dot-matrix line printers that are attached to the computer.

All of Fleet Feet's software works in real-time mode. When a sale is made, inventory is adjusted, general ledger information updated, and commissions noted. Their other accounting software also is interactive, updating files as transactions are entered.

Installing the new accounts payable system should pose only minor problems for analyst Peggy Adams-Russell. Her research determined that there is adequate free disk space, the relational database manager is "owned" by Fleet Feet, and printing time is minimal for various reports and purchase orders.

Peggy's main problem centers around the two new terminals or personal computers that the AP system will need. She reviews the situation with Fleet Feet owner Sally Edwards. They come up with three alternatives: terminals, DOS personal computers, or Macintosh PCs. Terminals are inexpensive (less than $650), but offer no other capabilities. DOS PCs are more expensive (about $1,000), but they allow users to operate word processing and spreadsheet application software. Macintosh PCs are the most expensive (about $2,000), offer word processing and spreadsheet software, and are more visual and easier to learn. All three types of devices support Hewlett Packard's terminal emulation, will run the software that Peggy designed, and will operate on the star network topology Fleet Feet now has in place.

Sally and Peggy decide to rule out a terminal. All staff members regularly use application software packages. They also decide against the Macintosh because store owners have DOS PCs and are very comfortable with them. Their Macintosh machines are now used for store floor plan layouts, their monthly newsletter (*Footnotes*), advertising layouts, and race announcements.

Peggy is left with DOS-based PCs and arranges to purchase two from a local retail store. She then calls Jack Richter to install cables to the workstations where the machines will reside.

Her network design finished, Peggy is ready to start software design—the last phase of all the design activities.

SUMMARY

Computerized information systems can operate in a variety of modes. This chapter examined the modes of systems operation (batch, on-line, on-line transaction processing, and distributed) and methods for linking terminals to computers (completely connected ring, star, and bus networks). It also showed how each mode affects data input, processing, and output. In many cases, personal computers offer an excellent alternative to terminals and centralized computers.

Batch processing systems dominated the early age of computerization and are still used today. In a batch environment, an organization collects and processes data in groups. After all data are prepared, the computer can read and process them at a specified time. After processing, information becomes available to users, usually in the form of screen displays or printed reports. Batch processing can satisfy users' needs in situations in which the delay between data collection and processing will not hamper an organization's decisions.

Though on-line systems may also collect data in batches, users enjoy terminal access as soon as processing ends. Users can inquire about, but not change, data. On-line systems became popular in the late 1960s and satisfy the needs of users who must have immediate access to data, but can wait for data entry and processing at night or during off-hours.

On-line transaction processing systems process data immediately after entry. They fulfill the needs of users who demand the most up-to-date information. Often, OLTP systems require a computer that can support many terminals simultaneously.

The move from batch to on-line and OLTP systems demands communication among remote devices. Organizations can link two or more computers in a network. Distributed processing and local area networks (LANs) will undoubtedly gain in popularity.

With control systems established, input and output designs completed, and the schema for the database written, the analyst next evaluates what equipment is needed. Descriptions of needs appear in a request for quotation (RFQ) or a request for proposal (RFP), which goes to potential vendors. The analyst weighs the vendors' responses by balancing features and costs, usually with a prepared evaluation form.

Having established the mode of system operation, the placement of terminals and computers, and necessary data links, the analyst can now move to the next phase of the systems process: program definition and module design.

KEY TERMS

Batch processing
On-line processing
On-line transaction processing (OLTP)
Data communications
Modem
Asynchronous
Synchronous
Protocol
Network
Completely connected network

Star network
Ring network
Bus network
Token
Local area network (LAN)
Server
Request for quotation (RFQ)
Request for proposal (RFP)
State transition diagram (STD)

QUESTIONS FOR REVIEW AND DISCUSSION

Review Questions

1. What is meant by the term *batch processing*?
2. What is meant by the term *on-line*?
3. What is meant by the term *on-line transaction processing*?
4. What are two types of network configurations?
5. What is the difference between an RFP and an RFQ?
6. What would an evaluation form for hardware or software acquisition look like?

Discussion Questions

1. List and briefly describe the four modes of system operation.
2. List two types of networks.
3. What do the following abbreviations stand for?
   a. RFP
   b. RFQ
   c. LAN
4. What are three advantages of on-line systems?

5. What are the six measures used to evaluate the capabilities of an OLTP system?

6. What are the advantages of a batch processing system?

7. What are three disadvantages of an OLTP system?

Application Questions

1. What modes have systems typically followed in their development life cycles?

2. What is the difference between a centralized and a decentralized data processing system?

3. Diagram three types of networks.

Research Questions

1. Design a bid evaluation form useful for acquiring a line printer.

2. Design a bid evaluation form useful for acquiring a disk drive.

View Video's RFQ for Equipment

F rank Pisciotta has always looked forward to the network design stage because it marks a major milestone; he is almost done with design activities. View Video's LAN of personal computers running the video check-out/check-in software and BusinessWorks PC will operate in both batch and OLTP modes.

As clerks check out or check in a tape, their data are validated and verified before becoming instant updates to the relational Oracle database. Management and the accounting office can inquire into the database whenever the need arises.

At the end of every day, the financial data are passed to BusinessWorks PC for batch updating. It is not important to Mark and Cindy Stensaas to have the BusinessWorks PC accounting information updated in real-time since the financial reports it can generate are needed only weekly and monthly.

The new video check-out/check-in system design provides for bar code readers at each station, including readers for stations in Mark and Cindy's office area. Given the relatively small size of this investment, probably less than $3,000, Frank decides to contact bar code reader vendors by telephone, asking each to send sales brochures and price quotations within two weeks. Frank has worked with vendors before and feels confident that their brochures and telephone responses will permit him to complete his evaluation form without any difficulties (Figure 11.17).

In the next two weeks, Frank receives extensive descriptive literature, which he reads carefully in order to complete the evaluation forms that he has initiated. These help him to objectify the many factors that will influence his ultimate decision. Since vendors offer a variety of special enhancements, Frank includes an optional features section, where he can award additional points for capabilities exceeding his minimum standards. Completed forms in hand, Frank studies each bar code reader and rejects those not meeting minimum standards (Figure 11.18).

Frank eventually chooses a unit from Telzon. It is the lowest price with the most points, giving it the lowest point-cost ratio. According to Frank's criteria, it represents the best long-term buy.

Working With View Video

Use of bar code readers, such as the ones that View Video has considered, is becoming more common in many aspects of the business world. Bar code readers are cheap, very accurate, small in size, easy to interface, and increase productivity. They come in a variety of types, ranging from those that resemble a pen, pistol, or the familiar table-top models found in supermarkets.

Case Study Exercises

1. List three other applications we see as consumers that make use of bar code readers.
2. Why would one organization select "pistol" style bar code readers while another picks "pen" style readers?
3. Many types and styles of bar codes exist. Make a list of three different types of bar codes.
4. Write a short paragraph explaining how bar code readers "read."

Figure 11.17 Frank Pisciotta's bid evaluation form for bar code readers. A vendor must meet all minimum requirements; if not, Frank will eliminate that vendor from further consideration.

View Video Bar Code Reader Evaluation Form

Vendor:	Model Number:
Price:	Monthly Maintenance:

Minimum Requirements   **Status (Y or N)**

1. Pencil (not pistol) shaped
2. Uses PC serial/parallel port
3. Upload software routines provided
4. Delivery within 60 days
5. Reads standard codes (UPC-A, UPC-E, EAN-12, EAN-8, ZIP+4, Standard 3 of 9)
6. Over 1,000 installed
7. Weight less than 2.0 pounds
8. 1 year warranty

Optional Features   **Score x Weight = Points**

9. Weight less than 1.0 pounds   5.0
10. Programmable   5.0
11. Infra-red (not cable) connection   4.0
12. Ergonomic design   4.5
13. Repair policy includes loaner   5.0
14. Software for PC included   5.0

Total Weight for Optional Features   28.5

Subtotal for Optional Features

Minimum Requirements Met   100.0

Total Points

Point-Cost Ratio

Notes: Points awarded as follows: 0 = unacceptable, 1–3 = acceptable, 4–5 = outstanding.

Figure 11.18 View Video's responses on bar code readers are evaluated by Frank.

View Video Bar Code Reader Evaluation Form

Reader Name/Model	Unit Price	Minimums Met	Optional Features	Total Points	Point-Cost Ratio
Scanteam 6100	349.00	Yes	Yes	115	3.03
VTI 102	300.00	Yes	Yes	110	2.73
Symbol LRT 3800	698.00	No	NA	NA	NA
DNT 540	398.00	Yes	Yes	120	3.32
Denso BHS-6000	598.00	Yes	Yes	120	4.98
Telzon 210	298.00	Yes	Yes	122	2.44

Software Design

GOALS

After reading this chapter, you should be able to do the following:

- Define a module
- State the rules for writing a module
- List the three module control structures
- Explain four criteria for good module design
- Cite three widely used programming languages and appropriate usage of each
- Describe the components of a program specification
- Explain the purpose of a design review

PREVIEW

In the earlier phases of system design, the analyst designs the outputs, database, inputs, and network—then decides on new hardware. Now the analyst must define necessary software (Figure 12.1). This final phase of design is divided into four tasks: program definition, module design, program specification, and the design review.

Program definition demands careful consideration of the priority, purpose, and function of each program. After identifying the programs, the analyst carefully divides each into specific tasks or modules whose functions are represented with pseudocode, Warnier-Orr diagrams, or one of the other program logic tools.

After identifying modules, the analyst concentrates on the details of each module, paying particular attention to the ways in which they relate to one another. The full collection of program definitions and modules completes the program specifications.

With modules specified, the analyst can consider which language the system might use. For many organizations, there is no choice; a single language is used for all programs, even though it might not ideally suit a new application. In other organizations, the analyst may choose among many languages, allowing the analyst some latitude when selecting one for a particular application.

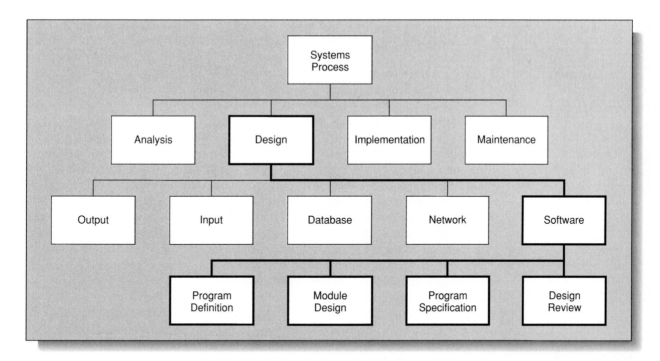

Figure 12.1 Leveling of design reveals five major tasks. The final four tasks require the analyst to specify necessary programs that management can review before development.

Finally, the entire system undergoes a walkthrough and a design review. Similar in scope and purpose to the earlier analysis review, a design review gives management and users a final chance to scrutinize the proposed system, making sure that it fulfills all of their needs and involves acceptable costs and schedules. Even at this late date, management can still reject the system or request design modifications before authorizing commencement of development.

The entire design phase ends with the writing of the program specifications. The specifications include the output, database, input, network, and software designs.

PROGRAM DEFINITION

Until a few years ago, many analysts often ignored **program definition**, which is a detailed description of each program in the system. Often, this lack of interest was because their training centered more on the details of code than on the more general aspects of program design and the management of programmers. To further complicate the situation, the lack of CASE and structured tools tended to obscure the general flow and logic of programs. Fortunately, CASE and the structured methodology have made it possible for today's analysts and programmers to speak a common program design language.

Creating a list of programs requires the analyst to review the data flow diagram (DFD) of the proposed system, as well as the data dictionary and report formats (from output design), the schema (from database design), and data collection screen formats (from input design). During this review, the analyst isolates the point at which the system produces each report and examines each process activity (represented by a circle in the DFD) as a potential program. (If you've forgotten some of the details of DFDs, you may want to review the full discussion of them in Chapter 2.)

The analyst looks for primitive process symbols, those that are not further subdivided. These processes are ideal candidates for programs.

Any transformation of data, portrayed by the circle or bubble in the DFD, most likely requires a program. Just how many programs will a system need? A study made in the late 1980s showed that the average system produced 26 user reports with 55 programs involving almost 40,000 program source statements. Given the fact that today's users are demanding even more from their systems, it is not surprising to find that the contemporary system requires more and longer programs.

Each report most likely requires a program. Reporting programs typically take data from a file(s), sort the data into order, and display the data to the user (accumulating totals or counting the number of occurrences of a certain event). Figure 12.2 represents the requirements for a system that prints customer statements. Reporting programs in a DFD appear as a circle, with a data flow arrow pointing to a named sink of data (a box).

Often, a reporting program needs to sort or merge data files before presenting a display to the user. Sorting and merging activities may require a set of utility routines based on the computer's operating system sort or merge functions. Some programming languages, such as COBOL, contain SORT and MERGE verbs that a programmer can embed in a reporting program. The need to sort or merge should become apparent as the analyst scrutinizes each process activity. For example, if a user wants customers' statements in order by ZIP code (for cheaper bulk mailing postal rates), the system must sort the file by ZIP code before printing statements.

Process symbols with inward pointing arrows may call for a data collection program, complete with verification, validation, and other data-checking methods. A box (source or sink of data) and an arrow pointing to the bubble clearly identify these types of programs. A customer statement system needs a program to collect customer payment data, check it for errors, and update the customer master file (Figure 12.3).

Still other process symbols in the DFD may call for programs that take data from a variety of origins and pass them on in some converted form. The DFD in Figure 12.4 shows Commission Note created by Record Payments and used by Pay Commission to produce the Commission Report. A named data flow between two processes is typically not a potential program.

Regardless of the contents of a given DFD, the analyst inspects each component closely to determine whether or not it requires program(s) to accommodate activities. A thorough inspection should yield a list of programs that the system will need. Since each process symbol in a DFD contains an identifying number (such as 3 in Figure 12.2) that number can also identify each program (Figure 12.5). The collection of program definitions leads to module design.

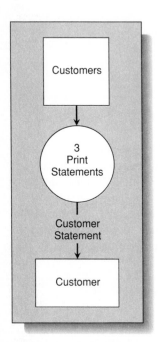

Figure 12.2 Data flow diagrams show the analyst where the system's programs are likely to be found. Reporting programs take data from a data store to a process symbol leading to a sink box.

Figure 12.3 Data collection
programs in a data flow
diagram show data from a
source to a process symbol.

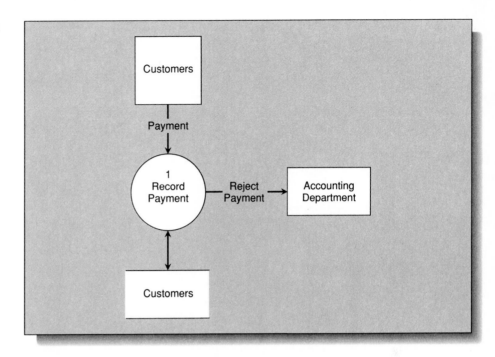

Figure 12.3 Data collection programs in a data flow diagram show data from a source to a process symbol.

MODULE DESIGN

After defining the general purpose of each program, the analyst divides each program into modules, each of which performs a single function and has an identifiable beginning and ending point. Each of these functions forms one distinct module that relates to the other modules in a certain way.

Not every organization uses an analyst to design modules. Some businesses assign module design to programmers and use analysts to review the designs. In a small organization, a programmer/analyst may do it all.

Modules

Just as the collection of rooms result in a house, groups of program modules create a program that turns data into information. Program **modules** are discrete or identifiable single function units. Figure 12.6 shows three modules. Notice how module names reflect their actions: Opening, Print-Data, and Closing.

Figure 12.4 Programs internal to a system exhibit themselves as circles between data flows (arrows), such as "Record Payment" and "Pay Commission."

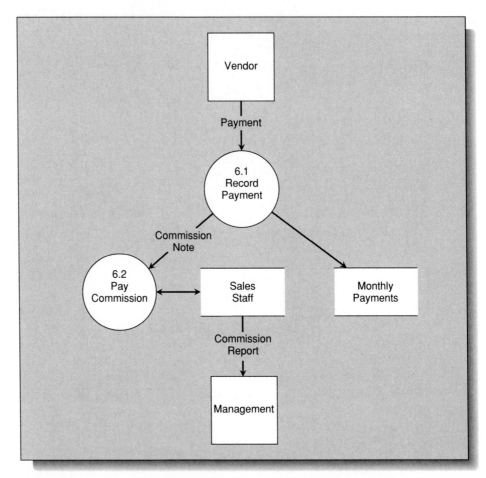

Figure 12.5 For each program in the system, the analyst defines the program by referencing its location back to the data flow diagram.

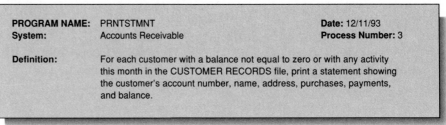

PROGRAM NAME:	PRNTSTMNT	**Date:** 12/11/93
System:	Accounts Receivable	**Process Number:** 3

Definition: For each customer with a balance not equal to zero or with any activity this month in the CUSTOMER RECORDS file, print a statement showing the customer's account number, name, address, purchases, payments, and balance.

To conform to the standards of the structured methodology, each module should contain only one entry and one exit point. Figure 12.7a illustrates a correct module; Figure 12.7b, an incorrect one. Single entry and exit points organize logic to flow smoothly from beginning to end without detours or interruptions.

This "single" concept results in the "single rule." The best modules are single entry, single exit, and single purpose. The goal is for the programmer to write modules that are as independent as possible, perform the required actions on the data, and produce timely, accurate, quality software. The "single rule" and all the tools of the analyst strive to reach this goal.

The analyst must also pay attention to the lengths of modules. Since most terminal screens can display only 24 lines at a time, many CRT applications should limit module lengths to that size. In nonterminal environments, one printed page (containing perhaps 60 lines) could serve as a sensible limit. Module sizes of more than a terminal screen or a page will make it difficult for the programmer to remember what took place at an earlier point in the module.

Dividing a program into modules simplifies the job, making it easier for the analyst and programmer to understand. Furthermore, by examining each module separately, the analyst can more readily spot potential errors in logic. The earlier errors are located, the easier they are to fix.

Control Structures

In 1964, Corrado Bohm and Guiseppe Jacopini demonstrated that all programs can spring from three basic control structures: sequence, decision, and repetition or iteration (Figure 12.8). **Control structures** are patterns for building the logic of a computer program. Regardless of a system's complexity or the advanced techniques required to program it, all of its programs involve combinations of these three structures.

Figure 12.6 Modules are groups of actions with identifiable beginning and ending points.

```
OPENING.
        Open Customer file, Transaction file.
        Read a record from Customer file.


PRINT-DATA.
        REPEAT until no more records in customer file.
                Select all the records in transaction file for this customer.
                IF records are found then
                        Sort transaction records selected into order by transaction date.
                        Print the customer's account number, name and address.
                        Print customer's balance from last month.
                        Fetch transaction record.
                        REPEAT until no more transaction records
                                IF purchase record
                                        Add amount to balance
                                ELSE
                                        Subtract amount from balance.
                                Print the transaction data.
                                Fetch next transaction record.
                        END-REPEAT.
                        Print grand total line.
                END-IF
                Read a record from Customer file.
        END-REPEAT.


CLOSING.
        Close Customer file and Transaction file.
        Stop program.
```

Figure 12.7 Modules should contain single entry and exit points.

```
            Read employee data.
            REPEAT until (there is no more data)
                  Calculate Gross-pay.
                  Subtract Deductions from Gross-pay.
                  Calculate Year-to-date totals.
                  Calculate-Quarter-to-date totals.
                  Print Employee paycheck.
                  Read Employee data.
            END {REPEAT}.

     a.  A correct module.
```

```
            Read-Again.
                  Read Employee data.
                  IF (there is no more data)
                        THEN
                              GO TO Print-Final-Total-Routine.
                  Calculate Gross-pay.
                  Subtract Deductions from Gross-pay.
                  Calculate Year-to-date totals.
                  Calculate-Quarter-to-date totals.
                  Print Employee paycheck.
                  GO TO Read-Again.

     b.  An incorrect module in which the GO TO statement permits two exits.
```

Sequence control structures describe a series of actions that follow one another linearly (Figure 12.9). The payroll calculations leading to the printing of a paycheck exemplify this structure.

Decision control structures describe a situation in which an action depends on which of two conditions has been met (refer to Figure 12.8b). For example, suppose you wish to determine whether or not a customer deserves a discount on a purchase (Figure 12.10a). Beforehand, you would determine two conditions—perhaps "prompt" or "delinquent" payer—that govern your decision about awarding the discount. If the customer falls into the "prompt" category, they receive a 2 percent discount. If they fall into the "delinquent" category, they receive no discount. We also call this decision structure IF-THEN-ELSE because many computer programming languages use these words to implement such decision-making.

Decision control structures can involve more than one action based on a given condition. For instance, water utility customers pay one rate until they use enough water to qualify for another (usually lower) rate. The ultimate bill could include two separate totals: one for the amount consumed prior to meeting the condition and one for the amount consumed thereafter (Figure 12.10b).

Writing a long series of IF-THEN-ELSE commands can lead to errors, with IF, THEN, and ELSE not in the proper groupings. An alternative decision structure, **case**, can depict complex logic. Consider newspaper classified advertising; rates depend on the number of times an ad will run (Figure 12.10c). Though case offers an alternative to writing long or involved IF-THEN-ELSE, it still falls in the decision control structure category.

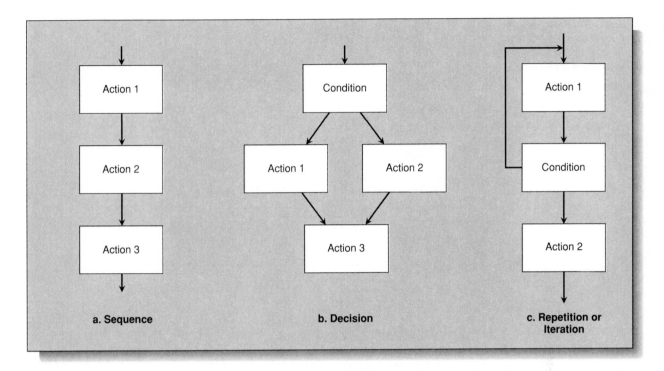

a. Sequence b. Decision c. Repetition or Iteration

Figure 12.8 Bohm and Jacopini postulated three control structures at an international computer science conference in 1964.

Repetition (**iteration** or looping) provides the third control structure (refer to Figure 12.8c). Looping or iteration allow repetition of a group of other control structures as long as a condition remains true. In certain programming languages, WHILE-DO or REPEAT-UNTIL statements allow for an action to repeat a specific number of times or until certain conditions arise. In Figure 12.11, six actions relating to each employee's data repeat until the last employee's data have been processed, at which point the repetition ceases.

Decomposition and Refinement

Most systems decompose into modules, each of which decompose into more modules—which, in turn, may undergo further refinement until all actions meet the single purpose, entry, and exit criteria. As you have seen in earlier chapters, we call this continual breaking down of a system into its elementary components decomposition or refinement.

Figure 12.9 A sequence control structure requires that each action occurs in a set order.

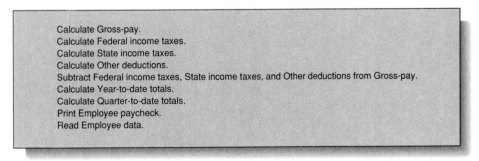

Calculate Gross-pay.
Calculate Federal income taxes.
Calculate State income taxes.
Calculate Other deductions.
Subtract Federal income taxes, State income taxes, and Other deductions from Gross-pay.
Calculate Year-to-date totals.
Calculate Quarter-to-date totals.
Print Employee paycheck.
Read Employee data.

Figure 12.10 A decision control structure tests whether a predetermined condition exists, then acts accordingly.

```
IF (the customer type is 'prompt')
    THEN
            Set the discount percent to 2%
    ELSE
            Set the discount percent to 0%.
```

a. In the decision control structure above, action depends on one condition.

```
IF (Amount-of-water used is less than 1500 cubic feet)
    THEN
            Amount-owed equals $2.00 plus .015 times the
                Amount-of-water
            Add Amount-owed to Total-owed-by-regular-customers
    ELSE
            Amount-owed equals $1.00 plus .01 times the
                Amount-of-water.
    Add Amount-owed to Total-owed-by-large-users.
```

b. This decision control structure allows two actions based on one condition.

```
IF (Number-of-times the ad is run is)

    1 or 2:     Set Rate to     3.75
         3:     Set Rate to     2.96
    4, 5, 6:    Set Rate to     2.33
    7, 8, 9:    Set Rate to     1.68
10 thru 13:     Set Rate to     1.54
14 thru 29:     Set Rate to     1.49
     Else:      Set Rate to     1.43
```

c. Case control structure that allows one of many actions, depending on numerous conditions..

Let's look at an example of decomposition. Fleet Feet's analyst, Peggy Adams-Russell, decides that the shoe store chain needs to keep track of sales contacts (customers, clients, and others who have visited a store) for their monthly newsletter, *Footnotes*. She knows that the system needs separate modules for each function it will need to do. She visualizes modules that will initialize the contacts file, add a new contact, change a contact, delete a contact, print the contacts file, post an inquiry of a Fleet Feet product, and terminate the program. First, she draws a context data flow diagram (Figure 12.12). Next, she develops a series of detail data flow diagrams for each module, starting with the one in Figure 12.13.

Peggy goes on to develop a main selector menu screen (Figure 12.14), displaying the date of the last file update and prompting the operator to select a desired operation. Peggy's main menu screen design reflects the same functions she visualized earlier: initialize, add, change, delete, print, allow inquiry, and terminate. Since Peggy anticipates that Fleet Feet will want to print gummed labels for selective advertising campaigns and issue an analysis report showing which contacts have requested further information some time in the future, she provides for these two further functions.

Figure 12.11 Repetition control structure repeats actions a specific number of times or until a certain condition arises.

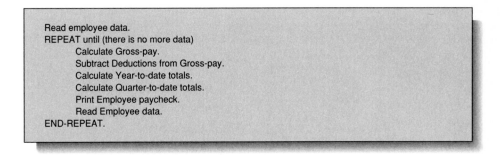

```
Read employee data.
REPEAT until (there is no more data)
        Calculate Gross-pay.
        Subtract Deductions from Gross-pay.
        Calculate Year-to-date totals.
        Calculate Quarter-to-date totals.
        Print Employee paycheck.
        Read Employee data.
END-REPEAT.
```

Peggy would now examine each function and turn it into a module. Peggy reviews the report (in this case, "print contact file") it must produce (Figure 12.15), and uses pseudocode to refine it into its most basic components (Figure 12.16). In some modules ("add a new contact," for example), Peggy will refer to a data collection screen design rather than a report design. Regardless of the module under consideration, she examines the data flow diagram, output and input requirements, lists modules, and describes details of the modules.

Figure 12.12 Context data flow diagram for Fleet Feet's contact tracking system.

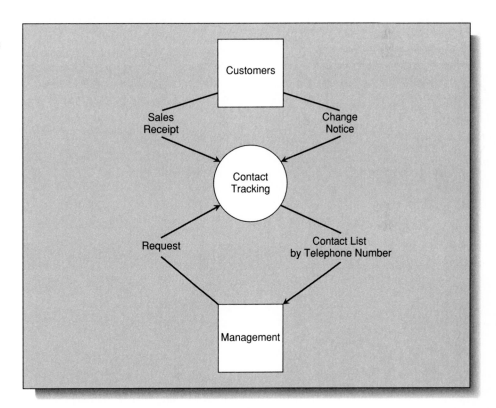

Figure 12.13 Data flow diagram to list contacts by telephone number, only one of six diagrams that Peggy Adams-Russell draws.

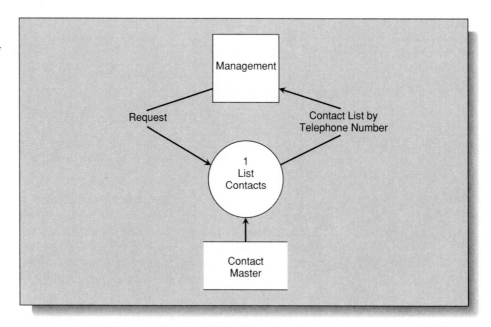

Figure 12.14 Main selector menu for the Fleet Feet Contact System that gives the terminal operator a choice of 7 separate operations. (Functions 7: and 8: remain open for use in future operations.)

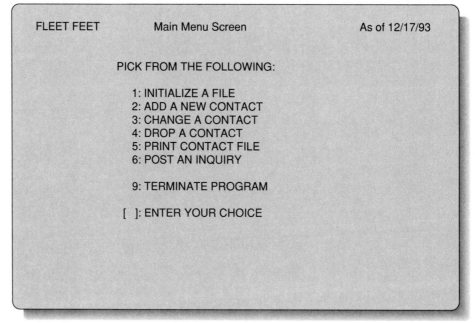

Coupling

Programs are a collection of modules linked together in a hierarchical or treelike manner. Some modules serve as parents, controlling only those child modules directly under their responsibility. We use the term **coupling** to refer to the measurement of control and

Figure 12.15 Output report formats developed during preliminary design paid close attention to user needs.

```
FLEET FEET                       List of Contacts                MM/DD/YY
2408 J Street
Sacramento, CA                                                   Page: 99
Telephone           Contact's Name        Date of     Date of     Number
Number                                    First       Last        of
                                          Contact     Contact     Contacts

(916)782-3412       Siemmers, Rebecca     12/22/91    12/23/92       5
(916)782-5612       Crowley, Marty        12/20/91    12/20/91       1
(916)888-4563       Lee, Virginia         11/12/91    5/05/92        3
```

Figure 12.16 Pseudocode for modules to print list of contacts by telephone number.

Opening.
  1. Open Files.
  2. Access today's date and place it into the heading line.
  3. Initialize page number to 0.
  4. Perform Page Headings Module.
  5. Find or Read first record from CONTACT-MASTER using TELEPHONE-NUMBER-KEY set.

Print-Line-Read-Record.
  1. Build up the detail line by moving the data.
  2. Print the detail line if end of page perform Page-Headings.
  3. Read next CONTACT-MASTER using TELEPHONE-NUMBER-KEY set.

Closing.
  1. Eject paper to a blank page.
  2. Close files.
  3. Terminate program.

Page-Headings.
  1. Print first heading line.
  2. Print second heading line.
  3. Add one to page counter.
  4. Move page counter to third heading line.
  5. Print third heading line.
  6. Print line of dashes.
  7. Print fourth heading line.
  8. Print fifth heading line.
  9. Print sixth heading line.
  10. Print a line of dashes.

interdependence among modules. Coupling allows us to organize modules in a way that reduces the relationships among the modules to the minimum.

After defining and refining all modules, the analyst begins the important job of determining coupling relationships (Figure 12.17). As a general rule, we want to keep the relationships among modules to the minimum. Modules exhibit five types of couplings: data, stamp, control, common, and content.

We refer to independent modules as loosely coupled, dependent ones as tightly coupled. For Fleet Feet's application, second-level modules (Initialize, Add, Change, Drop,

Print, List, and Terminate) do not interrelate and are, therefore, loosely coupled. These modules only share the data that the system has stored about the application. In contrast, the page heading functions within the opening and processing submodules of the PRINT modules are dependent on one another and, therefore, tightly coupled.

Since changes within loosely coupled modules do not always require changes in other modules, they are not as difficult to program. On the other hand, tightly coupled modules can present complicated programming problems because any change within one module may dramatically affect others. Although it would seem desirable to make modules as loosely coupled as possible, in actual practice, very few modules turn out as loosely coupled.

Modules that process data and send it to other modules exhibit data coupling, and they are the most desirable types of modules. Data coupling appears when a module passes a data value to another module for processing; the second module then manipulates the data value. A good example of data coupling exists for page heading modules that print a page number. We pass the page number to page heading module; it adds one to the page number and sends it back. Another example is date validation. We send a date to the module

Figure 12.17 Coupling links the modules together in a logical manner.

MAIN DRIVER MODULE
1. Perform Opening.
2. Perform Print-Line-Read-Record until end of CONTACT-MASTER.
3. Perform Closing.

Opening.
1. Open Files.
2. Access today's date and place it into the heading line.
3. Initialize page number to 0.
4. Perform Page Headings Module.
5. Find or Read first record from CONTACT-MASTER using TELEPHONE-NUMBER-KEY index.

Print-Line-Read-Record.
1. Build up the detail line by moving the data.
2. Print the detail line if end of page perform Page-Headings.
3. Read next CONTACT-MASTER using TELEPHONE-NUMBER-KEY index.

Closing.
1. Eject paper to a blank page.
2. Close files.
3. Terminate program.

Page-Headings.
1. Print first heading line.
2. Print second heading line.
3. Add one to page counter.
4. Move page counter to third heading line.
5. Print third heading line.
6. Print solid line.
7. Print fourth heading line.
8. Print fifth heading line.
9. Print sixth heading line.
10. Print solid line.

and it determines if the date is legal, returning an error message if the date is in error, but returning nothing if the date is correct.

Stamp coupling exists when a module receives more data than it needs. In Fleet Feet's accounts payable example, we will need a module to calculate the discount on an invoice. The only data the module needs are dates, amounts, and the percentage of the discount. To send this module vendor number, name, or address violates the stamp coupling concept. The rule of thumb here is to send only the needed data and nothing else.

Control coupling takes place when a module passes another module a flag or indicator that affects the logic of the called module. The end-of-file switch is a good example of control coupling. When we reach end-of-file, the program does a different activity than when we are not at the end-of-file.

Situations involving common coupling take place when two or more modules manipulate the same data. We usually see common coupling in modules in which the program tests the same value(s) in different modules.

Content coupling takes place when one module transfers to the middle of another module or when one module references the middle of another module. This is the least desirable type of coupling because it violates the single entry, single exit rule of the structured methodology. Content coupling often arises when programmers make incorrect use of the GO TO statement. Fortunately, most modern programming languages make content-coupled modules difficult to program.

Cohesion

Coupling is only one measure of the strength of a module design. The second is **cohesion**, which measures how well the module follows the "single rule." Similar to coupling, we categorize a module's cohesion as functional, sequential, communicational, procedural, temporal, logical, or coincidental.

A functional module is the ideal. These are modules that truly obey the single rule. Some examples are modules that print page headings, sort data into order, and open or close the files that the software needs.

Sequentially cohesive modules are those that take the data from the previous module and perform the next task on the data. Data collection screens exhibit sequentially cohesive modules. They prompt the user for the data. The user then collects them from the input device, validates or verifies the data, and accepts the data or rejects them. Each of these modules takes the data one step further in the collection process.

We find communication cohesion in modules that take the same data and use them for two or more purposes. In our data collection example, once we accept the data, we may update a file and write the data to another file. If the data we are collecting represents a payment from a customer, we will reduce the customer's balance and store the payment data for printing on the customer's statement when their bill is prepared at the end of the month.

Procedural cohesion is one that has many different tasks and the module decides which task to perform. When the payment is received from a customer, we need to reduce their balance, but which balance? Certainly we apply it to their overall balance, but how do we apply the payment to finance charges, an over 90-day balance, and an over 60-day balance? These modules use IF-THEN control structure to decide what to do with the data.

Temporaral cohesion has to do with time. Modules in this classification are those that are executed at the same time and may have no relationship to each other. In our continuing payment example, our opening module opens the file, acquires the date, and initializes the page number to zero. All three can execute at the same time and do not affect each other.

Modules that are logically cohesive usually have multiple functions, not truly following the single rule. In our example of accounts receivable, we need to keep data about our customers. The system will need to have the capability to add new customers, delete them, and change the data we currently have about them (such as an address if they move). If we try to put all these functions into one module, called customer maintenance, performing many different logical tasks becomes too complex, so multiple functions may be needed.

The final and worst category of cohesion is coincidental. Modules are coincidentally cohesive when the tasks they perform are unrelated to each other. Often these modules perform what programmers call "garbage collection," a plethora of tasks that the system needs, but that are different from each other.

So what are the best modules? Those that are data coupled and functionally cohesive. Unfortunately, these are often the hardest to write—but they usually produce the fewest problems to the software developer.

Span of Control

In the case of some modules, regardless of the type of coupling or cohesion they exhibit, a parent may govern a number of child modules. **Span of control** refers to the number of subservient modules controlled by a given parent. Ideal spans of control include 5 to 9 child modules; while spans of 1 or 2 are usually too few, spans of 12 to 15 are usually too many.

Now let's suppose the user wants to show the date in only two of the three possible places (Opening and Print-Line-Read-Record, but not in Closing). If so, a programmer must write a new page heading module or reconstruct the existing one to print the date for only two, rather than three, modules. In this case, the module now contains a decision point and must include some indicator or flag that identifies when and where a heading needs printing. As the new module grows more complex, now requiring two data items (the report date and the flag), it goes from a loosely coupled to a control coupled module.

Each programming language links modules together with its own method. COBOL uses the verb PERFORM (Figure 12.18), while Pascal employs the name of a procedure or function such as OPENING, PRINT_LINE_READ_RECORD, and CLOSING (Figure 12.19). Both COBOL and Pascal enable users to call up other modules while still working within any other given module. For example, a user working within OPENING can call up HEADINGS without abandoning OPENING.

The structured methodology allows programs to call up a module, execute it, and then return to the statement that accessed it. Notice that accessing another module from within a module does not violate the single rule because the accessed module will return the user to the original one.

Unstructured programs resort to the GO TO statement in order to access and return from specific statements from various points within a program (Figure 12.20). However, the GO TO does not obey the single rule module logic rules; as a result, programs constructed with them are said to look like a plate of cooked spaghetti, meaning a messy tangle of statements.

Figure 12.18 A portion of a COBOL program that provides access to modules for Fleet Feet's listing of contacts. COBOL allows access to modules with its PERFORM verb.

```
005000 PROCEDURE DIVISION.
005100 000-MAIN-DRIVER.
005200     PERFORM 100-OPENING.
005300     PERFORM 200-PRINT-LINE-READ-RECORD UNTIL
005350         FINISHED = "DONE".
005400     PERFORM 300-CLOSING.
005450     STOP RUN.
005500
005600 100-OPENING.
005700     OPEN OUTPUT PRINTER-FILE.
005800     OPEN INPUT CONTACT-DATA-BASE.
005900     MOVE CURRENT-DATE TO LINE-3-DATE.
006000     PERFORM 400-PAGE-HEADINGS.
006100     READ CONTACT-RECORD AT END MOVE "DONE" TO FINISHED.
006200
006300 200-PRINT-LINE-READ-RECORD.
006310     MOVE CONTACT-NAME TO DETAIL-LINE-NAME.
006320     MOVE CONTACT-PHONE TO DETAIL-LINE-PHONE.
006330     MOVE FIRST-CONTACT TO DETAIL-LINE-FIRST.
006340     MOVE LAST-CONTACT TO DETAIL-LINE-LAST-CONTACT.
006350     MOVE NUMBER-OF-CONTACTS TO DETAIL-LINE-NUMBR-CONTACTS.
006360     WRITE PRINT FROM DETAIL LINE BEFORE ADVANCING 1 LINE
006370         AT END-OF-PAGE PERFORM 400-PAGE-HEADINGS.
006380     READ CONTACT-RECORD AT END MOVE "DONE" TO FINISHED.
006400
007000 300-CLOSING.
007100     CLOSE PRINTER-FILE.
007200     CLOSE CONTACT-DATA-BASE.
007300
008900 400-PAGE-HEADINGS.
009000     ADD 1 TO PAGE-COUNTER.
009100     MOVE PAGE-COUNTER TO PAGE-NUMBER.
009200     WRITE PRINTER FROM HEADING-LINE AFTER ADVANCING PAGE.
```

LANGUAGE CONSIDERATIONS

After management approves the system design, the analyst selects an appropriate programming language. In many situations, an organization relies on a particular language. In others, the organization's programming staff simply feels more familiar with a certain language. In still others, the organization may have access to only one language compiler.

If none of these types of restrictions exists, the analyst can select the language best suited for the application. For example, whereas COBOL might work best for a system that processes long lists of data containing nonnumeric data (addresses, descriptions, and so on), C, Pascal, or FORTRAN may offer more benefits in a scientific setting in which the system must handle complex mathematical formulas.

Languages vary in terms of capabilities, syntax structure, speed on the hardware, ease of learning, maintenance efficiency, ability to process data efficiently, and cost. While most can get the job done, those that employ structured concepts work best with a system

Figure 12.19 A portion of a
Pascal program providing
access to modules for Fleet
Feet's listing of contacts.
Unlike COBOL's PERFORM
verb, Pascal accesses a
module by using the module's
name.

```
PROGRAM LISTCONTACTS;
VAR
    PAGECOUNTER        : INTEGER;
    DONE               : BOOLEAN;
    INFILE             : TEXT;
    .
    .
    .
{---------------------------------------------------------------------}
PROCEDURE HEADINGS;
BEGIN
    WRITELN (CHR(16)); {ADVANCE PAPER TO TOP OF PAGE}
    PAGECOUNTER:=PAGECOUNTER + 1;
    WRITELN ('PAGE NUMBER: ':10, PAGECOUNTER:4)
END; { HEADINGS }
{---------------------------------------------------------------------}
PROCEDURE OPENING;
BEGIN
    REWRITE (INFILE, 'PRINTER:');
    RESET (INFILE, 'C:CONTACT.DAT');
    WRITE ('ENTER DATE: ');
    READLN (TODAYSDATE);
    DONE:=FALSE;
    PAGECOUNTER:=0;
    HEADINGS;
    READLN (INFILE, CONTACTRECORD)
END; { OPENING }
{---------------------------------------------------------------------}
PROCEDURE PRINT_LINE_READ_RECORD;
BEGIN
    .
    .
    .
END; { PRINT_LINE_READ_RECORD }
{---------------------------------------------------------------------}
PROCEDURE CLOSING;
BEGIN
    CLOSE (INFILE, LOCK)
END; { CLOSING }
{======================= Main Program =======================}
BEGIN  { MAIN PROGRAM }
    OPENING;
    REPEAT
        PRINT_LINE_READ_RECORD
    UNTIL DONE
    CLOSING;
END. { MAIN PROGRAM }
```

Figure 12.20 A portion of an unstructured COBOL program. Note the heavy use of the GO TO statement to control program logic flow.

```
005000 PROCEDURE DIVISION.
005100
005200 FIRST.
005300     OPEN OUTPUT PRINTER-FILE.
005400     OPEN INPUT CONTACT-DATA-BASE.
005500     MOVE CURRENT-DATE TO LINE-3-DATE.
005600     MOVE 1 TO PG-COUNTER.
005700     MOVE PG-COUNTER TO PAGE-NUMBER.
005800     WRITE PRINTER FORM HEADING-LINE
005850         AFTER ADVANCING PAGE.
005900     GO TO READ-AND-PRINT.
006000
006100 ENDING.
006200     CLOSE PRINTER-FILE CONTACT-DATA-BASE.
006300     STOP RUN.
006400
006500 READ-AND-PRINT.
006600     READ CONTACT-RECORD AT END GO TO ENDING.
006700     WRITE PRINTER FROM DETAIL-LINE AFTER ADVANCING 1 LINE
006800         AT END-OF-PAGE ADD 1 TO PG-COUNTER
006900             MOVE PG-COUNTER TO PAGE-NUMBER
007000                 WRITE PRINTER FROM HEADING-LINE AFTER
007100                     ADVANCING PAGE.
007200
009000     GO TO READ-AND-PRINT.
```

designed using a structured approach. C, Pascal, Ada, and COBOL allow structured programming; BASIC and FORTRAN now also have structured versions. The former support modules, the three control structures, and top-down program composition.

IBM licenses the right to use their language compilers. The COBOL compiler license costs about $800 per month (or $9,600 a year) and the assembler about $500 per month (or $6,000 a year) on a medium-sized mainframe. If an organization uses two languages priced in these ranges, its yearly expenses come to over $15,000! As you can see, changing languages requires the analyst to research costs.

The premise of structured programming is that this technique facilitates quicker program development, reduces errors, and permits easier program maintenance than nonstructured programming. One county government reports that its programmers spend over 80 percent of their time maintaining nonstructured systems, while a two-year community college using structured concepts estimates that its programmers spend only 20 percent of their time on maintenance. While the two organizations perform vastly different functions, there is a difference in maintenance expenditures between structured and nonstructured applications.

No hard-and-fast rules can guide an analyst to the perfect choice among these languages and personal preferences can outweigh other factors. Though most business application systems still rely on COBOL, in the future, C should increase in popularity as more people learn it. In stark contrast to third-generation languages such as COBOL and C, an analyst may have the option of using a fourth-generation language (4GL) such as Oracle Corporation's SQL*Forms and SQL* Reportwriter.

GUIDELINES FOR PLANNING MODULES

Strictly speaking, a module is "a bounded contiguous group of statements having a single name and that can be treated as a unit." Loosely speaking, however, a module is like a single block within a pile of blocks. Though at this stage we must identify its inputs and outputs when designing a module, we need not worry about all of its internal intricacies until later in the system process, when it must be programmed. Good module design implements the structured methodology, and the following guidelines can help analysts design good ones:

1. Remember the purpose of any module: to receive data, perform one or more operations on the data, and return output data.

2. Make sure modules obey the single task, single entry, and single exit rules of structured methodology. They should read from the top down, with clear, understandable, and appropriate names.

3. A module should seldom contain more than 40, or less than 5, lines of code. To accommodate fewer lines that may only be invoked once, a programmer can write the instructions into the main line of the calling module.

4. Try to minimize the amount of data that any module can reference. Instead of expecting the system to pass an entire record, restrict it to the individual fields needed for a given operation.

5. Write an initialization module to establish key values for a system.

6. Isolate a system's input/output operations into a small number of modules. Such isolation enhances the portability of the system. If an organization adopts a different data structure, it can easily convert the system by replacing existing I/O modules with new ones.

7. Identify system dependent functions, such as acquiring today's date, and isolate them to a small number of modules so that changes can be made easily to accommodate another computer.

8. Avoid content coupling that leads a module back to itself or to the middle of another module. This violates the single entry and exit concept fundamental to the structured methodology.

9. Try to write loosely coupled modules. Such modules accomplish a single goal (e.g., load a table, look up a table element, build a master file record, or calculate FICA or hours worked).

10. To test the single nature of a module, write a sentence describing what the module does. Does the sentence:

    a. Contain multiple verbs? If so, the module is probably performing more than one function.

    b. Contain words such as Initialize and Clean-up? Such jargon also implies poor definition.

    c. Have a predicate without a single specific object following a strong verb? If so, the module is probably logically complex and needs leveling.

PROGRAM SPECIFICATIONS

Having defined programs, planned modules, selected control structures, decomposed and refined modules, established coupling, and chosen a programming language, the analyst gathers together the outputs from all the other phases of the design activities to form the **program specifications**. Program specifications take the form of a report with the following elements:

- System overview
- System data flow diagrams
- Output report format or design
- Database schema or file design

- Screen layout or input formats
- Program definitions
- Module descriptions

This part of the design process pulls together and summarizes all the materials developed during design. Although analysts understand data flow diagrams, report formats, database specifications, and input/output layouts, programmers may not. If not, they will look to the program modules for keys to coding the desired programs.

The system overview provides background information, as well as system goals and objectives (Figure 12.21). Overviews often include descriptions of data collection methods and systems operations, explanations of each program's purpose within the system, and a schedule for programming—plus the names of the analyst, programmer, and all individual programs.

System data flow diagrams show the system in pictorial view from a logical perspective. Developed during analysis and perhaps refined during design, the DFDs orient the reader to where they are in the system.

Developed at the beginning of the design phase, output report formats also form part of the program specifications. Analysts often describe their report designs on printer layout sheets, although many now use computerized prototyping software or word processors to create report designs.

The fourth component of program specifications is the database schema (Figure 12.22). In this case, the database design was written using Oracle SQL.

Figure 12.21 Program specifications include goals and objectives for the system, as well as relevant background information.

PROGRAM SPECIFICATION

System: CON - Newsletter Contacts
Program Name: CON304
Description :   Print list of contacts by telephone number

Analyst :   Sue Weins
Programmer: Craig Smalley
Date Assigned: 10/22/93
Date Completed: 11/09/93

Background: Fleet Feet sells shoes and associated supplies to walkers and runners. The owner wants to track customers and send them a copy of Fleet Feet's monthly newsletter, *Footnotes.* Customers are identified by their telephone numbers. Each time a customer makes a purchase or visits a store, Fleet Feet wants to keep a record of the sales contact.

Data Flow Diagrams: See Attachments A and B *(Figures 12.12 and 12.13).*

Output Reports: Attachment C *(Figure 12.15)* shows the desired report format. All pages are numbered as well as dated. Print a maximum of 50 lines per page.

Database Layout: See Attachment D *(Figure 12.22)* for SQL tables.

Input Data: An index is prebuilt that retrieves contacts sequentially by telephone number.

Processing Requirements: Attachment E *(Figure 12.16)* shows the pseudocode to print the contacts by telephone number.

Figure 12.22 SQL table developed during the database design phase of the systems process. This table follows Oracle SQL.

```
create table contacts
         (
         TELEPHONE_NUMBER          NUMBER(10) UNIQUE KEY,
         CONTACT_NAME              CHAR (20),
         FIRST_CONTACT             DATE,
         LAST_CONTACT              DATE,
         NUMBER_CONTACTS           NUMBER(4)
         )
```

The screen designs specify data input requirements that allow programs to collect, validate, and store data on disk or tape. Screen designs should specify data type (numeric, alphanumeric, or others that the DBMS supports) and editing requirements. For example, the main menu screen in Figure 12.14 (page 400) demands an entry of a 1, 2, 3, 4, 5, 6, or 9.

Program definitions and module descriptions finish the specification. The program definition is a verbal description of the goal and purpose of the program (Figure 12.15, page 401). Module descriptions may use pseudocode to outline the functions, purpose, control structures, coupling, and cohesion of each module, as shown in Figure 12.6 (page 395).

CASE AND WRITING SPECIFICATIONS

We have seen how CASE tools help prototype reports, screen entities, relationships, and attributes. Relational database managers allow us to turn our theoretical files into physical realities. Most CASE tools can help us to write specifications.

Our context data flow diagram led us to decompose it and successive diagrams into primitive processes, the point at which further DFD decomposition is unnecessary. At this level, our primitive process eventually becomes a program or a module of a larger program. The DFD "knows" what inputs the primitive process will receive, what outputs it will produce, and what data stores (files) it can reference or update. All is in place, except for the rules that the process must follow in order to translate incoming data to outgoing data. The CASE software also "knows" the parent process and the tree branching that we followed to get to this level.

The specification is a textual description of the process that contains very explicit details. Some CASE software uses pseudocode for the details, while others use decision tables or another of the tools reserved for this part of design (structure charts, flowcharts, Warnier-Orr charts, and so on).

If the CASE tool user picks a structure chart as the tool of preference, the software has a "structure chart editor." Much like a word processing editor such as WordPerfect or a CAD package such as AutoCAD, the structure chart editor allows you to create and maintain the structure chart. The charts that the user creates represent the architecture of the way that the program works. They show the modules that accomplish the tasks and the hierarchy among the modules.

You can define the interfaces or parameters between the modules by listing the data that the module requires and the data that it produces. For the parameters, you would build a table that shows the name of the parameter, its type (numeric, string, real, date, or other), usage (input, output, input/output, or return), and scope (global or local):

Caller: Calculate_Tax
Callee: Tax_Refund

Parameter	Type	Usage	Scope
gross_income	real	input	global
tax_amount	real	return	local
refund_amount	real	return	global

Some CASE systems also allow you to specify how the parameter is passed: by value or by name. The difference between the two types of parameters centers on the parameter's usage. A value parameter lets a function or procedure use the data passed, but not change it. A variable parameter permits the procedure or function to alter the parameter's value; the "new" value is carried back once the function or procedure terminates.

Besides drawing the structure chart, the software will let you specify loop or case constructs plus decision and sequence structures similar to those found in programming languages. For each module, you can describe the circumstances for accessing the module with these structures.

With the design finished, the CASE software will check the the structure chart, modules, and interfaces to make sure they are correct. Among the types of consistency checks are

- Define a module in one place, but allow access in other places
- Ensure that parent modules are correctly specified
- Ensure that child modules are fully defined
- Ensure that the number of parameters passed and received are the same
- Match parameter type, order, and number between parent and child modules

With precise module specification, the conversion to code will lead to fewer errors.

The CASE software now has the entire design definition stored in its data dictionary. We end the use of CASE and design by having the software print the design model. The software draws the structure chart and puts together a package of all the details, ready for a programmer to convert it into code.

THE DESIGN WALKTHROUGH

The design phase also includes a walkthrough, which provides a careful scrutiny of system design and program specifications. During this phase, the analyst, the programmer (or the leader of a programming team), another analyst, and sometimes a user critique specifications in an effort to locate module errors, check for completeness, and ensure clarity for eventual

WORKING WITH PEOPLE
A PROGRAM SPECIFICATIONS WALKTHROUGH

Steve Osterday glanced at the clock as he walked back to his desk: 2:47 P.M. It had only taken 47 minutes to walkthrough his specifications for Fleet Feet's program on posting payments to the database. After his sixteenth walkthrough in two years, Steve now felt comfortable with the process, but he still remembered his first walkthrough, an ordeal that had dragged on for more than three hours. Returning to his cubicle today, Steve propped his feet on his desk and glanced up at his old wall plaque: IN GOD WE TRUST: EVERYTHING ELSE WE WALKTHROUGH.

That first walkthrough had ripped apart his design for a sales analysis system that was supposed to print a report comparing the current year's sales to last year's, including subtotals for a variety of categories and computations of averages and standard deviations. Determined not to get caught with a glaring error on his first major project, Steve had spent two days preparing for the walkthrough. What a wonderful design it had seemed to be at the time.

But what a shock when the walkthrough team tore it to shreds! His database design did not permit the system to capture last year's sales statistics for comparison to the current year's, and some fields were too small to store sales amounts (Steve had provided for six digits, whereas seven were occasionally needed). His critics had found an alarming number of typographical mistakes in his specifications, and they had asked Steve to "translate" some sentences that Steve thought were perfectly clear.

When the first walkthrough concluded, no one said, "Good job, Steve," or even, "Not bad for a beginner." Though Steve had anticipated praise, he felt that he could have handled criticism; but the eerie silence depressed him.

It took over a year (and six more design walkthroughs) for him to appreciate that silence. Walkthroughs are supposed to improve systems, not to evaluate analysts. Evaluations tend to be clouded by emotional judgments that can distract from the pure issue of quality. He had gradually learned not to take objective criticism personally.

Reflecting on two years of walkthrough experience, Steve marveled at how much he'd gained from them professionally. His bag of tricks had grown as he picked up useful design and programming techniques from other analysts and programmers. His writing style had improved, too. A short business communications course had taught him how to replace clumsy passive sentences with crisp active ones. What a revelation that English language skills had benefited his career almost as much as his mastery of C and COBOL!

Steve's reputation and confidence grew as he eliminated most major errors from his designs. No longer did users worry that his lack of business experience would result in complex, hard-to-understand systems.

Suddenly, the new analyst, Sue Aimes, barged into Steve's cubicle, interrupting his reverie. "Steve," she cried. "I just went though my first walkthrough on the new accounts payable system, and the review team shot it full of holes."

Steve smiled. "That usually happens."

"Maybe, but that's not what bothered me."

"What bothered you?" Steve thought he knew.

"No one said anything. It was weird. They must think I'm an idiot. Hey! Why are you laughing?"

"Let me tell you about my first walkthrough," began Steve. "You think *you* fouled up!"

programming. In many ways, the design walkthrough is similar to a wedding rehearsal; it's time to practice and fix what appears wrong before the actual event must take place.

Module errors can arise from improper linkage, incomplete refinement, or violation of the single entry, purpose, exit, and control structure rules. No matter how insignificant a detail might seem, it can eventually create nagging problems in a complex computer system.

Remember that a walkthrough aims at detecting, not correcting, problems. At the conclusion of the walkthrough, the analyst may study problems, make necessary corrections, then release revised system and program specifications for further review. If the team finds a lot of errors, it may schedule a second walkthrough. In any event, since a walkthrough should not attempt to evaluate an analyst's performance, but instead is meant to ensure quality, no formal summary goes to management.

The walkthrough is not a test of the software, but an occasion to trap errors before they arise. Testing is an important part of all software development, but takes place during the third phase, implementation.

DESIGN REVIEW

Once the analyst has corrected any mistakes or added any missing materials, then management, users, and the analyst conduct a formal **design review**. During this presentation of the system, everyone studies the program specifications. Analysts should take great care during this review because it provides the last chance for all concerned to verify that the system will meet the organization's needs within acceptable cost and schedule limits.

In some cases, last-minute changes may delay the system's progress to development, but making changes now will cost less than making them after programming begins.

If everyone agrees that the system should proceed to development, a memo formally authorizes the next step (Figure 12.23). The sooner management circulates it, the better. Otherwise, false rumors can begin alarming staff and stimulating resistance rather than cooperation.

If the organization has suffered financial setbacks since it first requested analysis, has altered its priorities, or has developed more pressing needs, it may still (although very

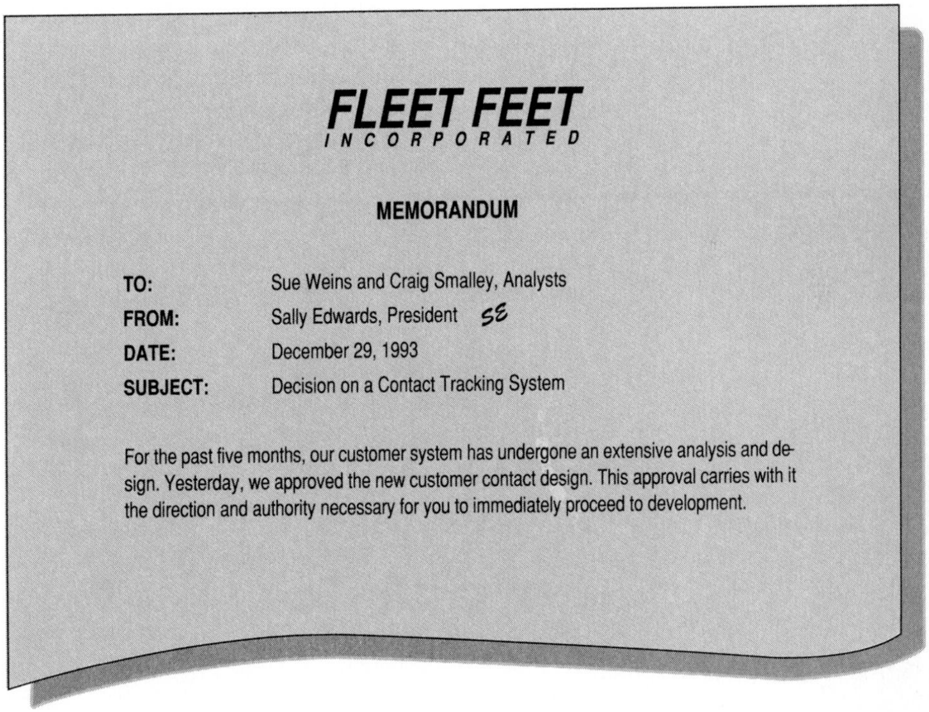

Figure 12.23 A memorandum authorizes system development.

FLEET FEET
INCORPORATED

MEMORANDUM

TO: Sue Weins and Craig Smalley, Analysts
FROM: Sally Edwards, President *SE*
DATE: December 29, 1993
SUBJECT: Decision on a Contact Tracking System

For the past five months, our customer system has undergone an extensive analysis and design. Yesterday, we approved the new customer contact design. This approval carries with it the direction and authority necessary for you to immediately proceed to development.

rarely) abandon a new system at this point. In that case, an immediate memorandum should explain why the organization has decided against the system. Again, false rumors can do a lot of damage to staff confidence and morale.

SUMMARY

The last phase of design involves program definition, which defines the programming of the new system.

When developing program specifications, the analyst must divide the program into single entry, single purpose, and single exit modules that reflect the three control structures: sequence, selection, and iteration.

Having divided programs into modules, the analyst establishes their interrelationships. Modules may be further refined into smaller modules, provided that those modules obey the rules of single entry, single purpose, and single exit—and support the three control structures.

If no programming language is standardized, the analyst selects an appropriate one. Some languages—such as C, Pascal, Ada, and COBOL—lend themselves to the structured methodology.

The analyst gathers all data flow diagrams, report layouts, database specifications, screen outlines, program definitions, and module descriptions into program specifications. This package summarizes the entire system design in a single document.

Before program specifications go to programming, the system undergoes a structured walkthrough to locate errors, ensure completeness, and make sure that specifications are clear. After a successful walkthrough, the specifications move to design review; then, if approved, into the hands of programmers.

A memo authorizes development or explains why implementation will not proceed.

KEY TERMS

Program definition
Modules
Control structure
Sequence control structure
Decision control structure
Case control structure
Repetition control structure

Iteration control structure
Coupling
Cohesion
Span of control
Program specification
Design review

QUESTIONS FOR REVIEW AND DISCUSSION

Review Questions

1. What is a module?
2. What are three rules for writing a module?
3. What are the three module control structures?
4. What are four criteria for good module design?
5. What are three widely used programming languages and how are they used?
6. What are the components of a program specification?
7. Why do we have a design review?

Discussion Questions

1. List the events that occur during the module design phase.
2. List the major parts of the program specifications.
3. Who are the participants in a structured walkthrough of the program specifications?
4. Who are the two individuals first associated with the idea of three control structures?
5. During which phases of the systems process were the following developed?
   a. Program narrative
   b. Database design
   c. Input design
   d. Output report design
6. What is the newest part of the program specifications?
7. What are the goals of the structured methodology?

Application Questions

1. From our brief examination of programming languages, which ones appear to be similar in structure or format?
2. List the goals and objectives of a structured walkthrough.
3. Who is responsible for scheduling and conducting the structured walkthrough of the program specifications?

Research Questions

1. Interview a person who is in a management position for some computer installation. Determine if they use the structured methodology and what benefits they observe.
2. *The New York Times* Project is an often cited example of the success of the chief programmer team. Write a short paper describing this concept and its results.

Program Identification and Specifications for View Video's Titles by Category Report

F rank begins this last phase of the design process with a review of all the programs that View Video will need. He makes a list from the data flow diagram he developed during analysis, output reports, and data collection screens (Figure 12.24). For each program, he assigns a name and writes a brief description of the program's purpose.

Program names are coded. Those in the 100 group are reporting, 200 group are for data collection, 300 are involved in BusinessWorks PC, 400 are utility programs, and 500 are menus. Assigning numeric codes provides a quick reminder from the program's name to the general category of the program.

Figure 12.24 The list of programs for View Video's Video Cassette Tape Rental (VCTR) system.

```
PROGRAM LIST

System:      Video Cassette Tape Rental (VCTR)
Analyst:     Frank Pisciotta
Date:        11/30/93

=========    ==================================================
Program      Description
=========    ==================================================
VCTR110      Prints the Alphabetic Customer List.
VCTR120      Prints the Numeric CustomerList.
VCTR130      Prints the Alphabetic Tape List.
VCTR140      Prints the Numeric Tape List.
VCTR150      Prints the Titles by Category report.
VCTR160      Prints the Overdue Customers report.
VCTR170      Prints the Daily Rentals report.
VCTR180      Prints the Viewing History report.
VCTR190      Prints the Customer Selection report.

VCTR210      Check-out.
VCTR220      Return.
VCTR230      Maintain Customers.
VCTR240      Maintain Movies.
VCTR250      Maintain Clerk Initials.
VCTR260      Maintain Director.
VCTR270      Set Security.
VCTR280      Adjust Rental Rates.

VCTR310      Update BusinessWorks PC.

VCTR410      Set Output Screen.
VCTR420      Set Output Printer.

VCTR510      Main Menu.
VCTR520      Report Selection Menu.
```

Frank likes many small programs instead of one giant program that does everything. Large programs (over 20,000 lines) can have too many side effects, where what seems to be a minor change may cause other unforeseen problems with another portion of the program. His main menu program (VCTR510) will thus determine which other program to call. When the called program finishes, it returns control to the VCTR510 at the point where it left off.

Most systems have a primary output. An accounts payable system's primary output is a check to a vendor; for payroll, it is the paycheck; and in an accounts receivable system, it is the statement sent to customers that requests payment. View Video's situation is different; the primary output is the check-out/check-in of a videotape.

The "Titles by Category" (VCTR150) report is a listing of all tapes that View Video owns, grouping them together by category (Comedy, Sci Fi, Drama, and so on). The "Titles" report falls into one of the classic programs: control break. The report presents the data alphabetically within category, with each category starting on a new page.

Frank Pisciotta completes the specifications for the "Titles by Category" reporting program (Figure 12.25). The specifications provide a single document for the programmers. It starts with a background narrative, system data flow diagram, sample output reports, SQL

Figure 12.25
Specifications for the program to print the Titles by Category report. The specifications contain all the information that the programmer will need in order to write the program.

PROGRAM SPECIFICATION

System:	Video Cassette Tape Rental
Name of Program:	VCTR150
Program Type:	Reporting, Titles by Category
Analyst:	Frank Pisciotta
Programmer:	Violet Lim
Date Assigned:	10/05/93
Completion Date:	11/30/93

Background:
The Titles by Category report is one of many management reports for the Video Cassette Rental system. It helps customers pick a tape from the collection, based on the tape's category (Western, Drama, Comedy, and so on).

Data Flow Diagram:
The data flow diagram for the VCTR150 program is found in Attachment #1.

Output Reports:
See Attachment #2 for a sample copy of the report and Attachment #3 for the data dictionary for the report.

Data Base Layout:
See Attachment #4 for a listing of the SQL table required to print the Titles by Category report. All the tables for the VCTR system are not shown.

Input Data Layout:
All data needed to produce the report are found in the database. Will need to create an index on the table to retrieve data by category by title. The index is temporary and the program should drop the index when it ends.

Processing Requirements:
Attachment #5 shows the pseudocode needed to print the report.

Attachment #1 Data flow diagram for VCTR150.

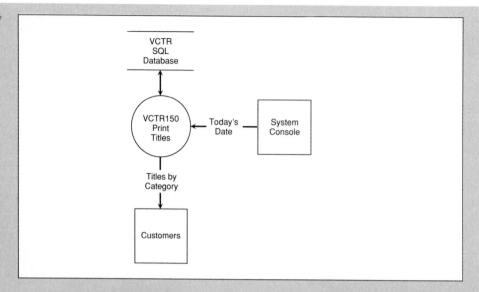

database schema, location of input data, and ends with the pseudocode for the program. Frank's pseudocode does not completely define the program, but is an outline of major activities. He has not spelled out how to print the page headings and control headings, nor advised the programmer on how to number the pages.

Frank is now ready for his design walkthrough. Since he is the only programmer and the only analyst working for View Video, he asks a friend, Pam Enkoji, to participate and (with the Stensaases' approval) promises her 250 free tape rentals once the system is installed. He sends Pam the programming specifications for her to review. Pam meets with Frank at the View Video store and they critique the proposed design. Familiar with BusinessWorks PC (View Video's purchased accounting software), Pam watches Frank's design to see how well it parallels BusinessWorks PC. Convinced that Frank is on the right track, Pam gives Frank a few suggestions, but endorses his overall program specifications.

With the reviewed specifications in hand, Frank conducts a second design review with the owners of View Video. After the review, Mark and Cindy Stensaas give Frank the green light to proceed to development (Figure 12.26). The memorandum is an official acceptance and notifies all parties that the third phase of the new videotape rental system is about to start.

Working With View Video

Frank's program specifications are far from complete. He will need a specification for all the programs in his list.

Attachment #2 Sample of Titles by Category report.

```
Date:   12/20/93              View Video, The Movie People              Page:   1
                                   CATEGORY   LIST

Category: Comedy
=======    ===============================    =========    ===========
Movie                                         Year
Number     Title                              Made         Rating
=======    ===============================    =========    ===========
1234       Arthur                             1981         PG
3145       Airplane II                        1982         PG
3214       All of Me                          1984         PG
4321       Author! Author!                    1982         PG
4589       American Dreamer                   1984         PG
6456       Any Which Way You Can              1980         PG
  *                   *                         *            *
  *                   *                         *            *
  *                   *                         *            *


Category: Drama
=======    ===============================    =========    ===========
Movie                                         Year
Number     Title                              Made         Rating
=======    ===============================    =========    ===========
0123       American Flyers                    1985         PG13
1013       Absence of Malice                  1982         PG
2144       Agnes of God                       1985         PG13
6123       Amadeus                            1984         PG
  *                   *                         *            *
  *                   *                         *            *
  *                   *                         *            *


Category: Musical
=======    ===============================    =========    ===========
Movie                                         Year
Number     Title                              Made         Rating
=======    ===============================    =========    ===========
3123       Annie                              1982         PG
  *                   *                         *            *
  *                   *                         *            *
  *                   *                         *            *
```

Case Study Exercises

Using the format shown, refer to Figure 12.24 and on; write the program specifications for

1. Check-out (VCTR210).
2. Return (VCTR220).
3. Set Security (VCTR270).
4. Maintain Customers (VCTR230).
5. Print Numeric Customer List (VCTR120).

Attachment #3 Data dictionary for tapes by category.

System:	Video Cassette Tape Rental
Name of Report:	Titles by Category
Analyst:	Frank Pisciotta
Date:	11/30/93

Element Name	Length	Data Type	Format
Movie number	4	Numeric	
Title	28	Alphanumeric	
Category	8	Alphanumeric	
Year made	4	Numeric	YYYY
Rating	4	Alphanumeric	

Order of report:	Ascending by category and by movie number.
Subtotals:	None.
Final totals:	None.
Counts:	None.
Frequency:	On demand.
Length:	One page for every 50 titles.
Type of Paper:	8 1/2 by 11 white with perforations.
Distribution:	To owners, store manager, out front.
Security:	None.

Attachment #4 SQL create table required for the report.

```
CREATE TABLE title
    (
    Title                        char(30) not null primary key
        check (Title = initcap(lower(Title))),
    Year_made                    number(04) not null
        check (Year_made between 1900 and 1999),
    Rating                       char(04) not null
        check (rating in ('G   ', 'PG ', 'PG13', 'R   ',
                'X   ', 'None')),
    Category                     char(12) not null
    )
```

Attachment #5
Pseudocode to print the
Titles by Category report.

MAIN DRIVER
1. Perform Opening.
2. Perform Test-Print-Select until end of TITLE.
3. Perform Closing.

Opening.
1. Open Database and printer file.
2. Get today's date from the system console and place it into Heading Date.
3. Select first row from TITLE.
4. Place category in Hold-Category.
5. Print page heading.
6. Print control heading.

Test-Print-Select.
1. If Category not equal to Hold-Category
    Eject paper to top of next page.
    Print page heading.
    Print control heading.
    Place category in Hold-Category.
2. Print detail line.
3. Select next TITLE row.

Closing.
1. Eject paper to top of next page.
2. Close Database and printer file.
3. Terminate program.

Figure 12.26 View
Video's notice to proceed
to development.

||||||| *View* Video |||||||

MEMORANDUM

TO:	Frank Pisciotta
FROM:	Mark and Cindy Stensaas *ms cs*
SUBJECT:	Notification of Design Review Acceptance
DATE:	Dec. 29, 1993

You completed your design of our VCTR (Video Cassette Tape Rental) system last week and we approved it at yesterday's design review meeting. We know that you are a little behind schedule, but expect the system to regain the lost time and make the completion date as planned. We authorize you to proceed on to development as quickly as possible so that we may still reach our spring 1993 installation date.

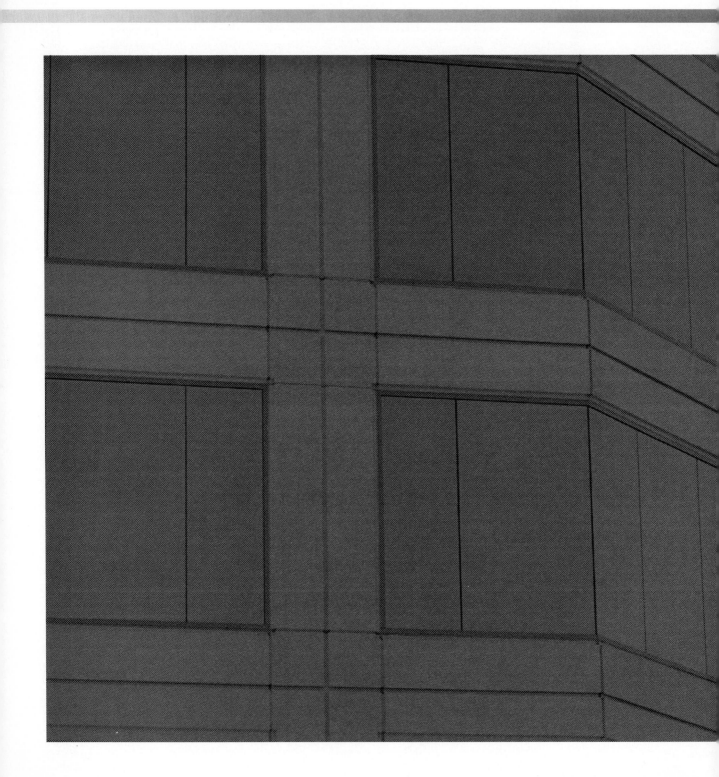

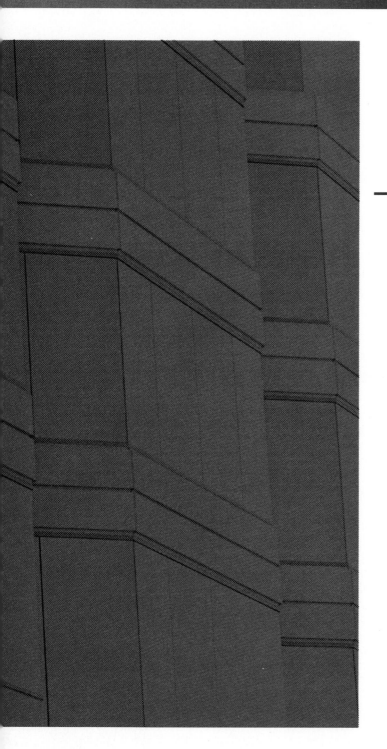

Systems Implementation and Management

Programming, Quality Assurance, and Conversion

GOALS

After reading this chapter, you should be able to do the following:

- Define an implementation plan
- Describe the major activities involved in programming a system
- Develop three examples of programming standards
- Cite the purpose of a program walkthrough
- List the three types of conversions

PREVIEW

Most analysts feel a great sense of accomplishment when they enter the implementation phase. After so much painstaking planning and organization, they relish the day when the system actually begins to take on life. However, as exciting as this prospect is, a few important tasks remain to do.

During the third and final phase of the systems process, programmers will write necessary programs, technical writers will develop appropriate documentation (user and operation staff manuals), and training personnel will create a thorough training program for everyone affected by the new or improved system. In the case of a brand-new system, the analyst must also decide how to convert data from the old to the new system.

During the two earlier phases of the systems process, we focused attention on the organization's goals for the system, taking a perspective not unlike that of an architect drafting plans for a building with its future occupants in mind. But now we must change our perspective to parallel that of the building contractor who will actually erect the structure. At this point, the analyst brings all of his or her skills and tools to bear on coordinating the efforts of programmers, users, and specialists—everyone who plays a vital role in the final crafting of the system.

OVERVIEW OF IMPLEMENTATION

Implementation moves the system closer to realization (Figure 13.1). **Implementation** is the third phase of the systems process, the time where programs are written, tested, and installed. Since an organization has invested a lot of time and effort in getting to this point in the systems process, it usually hesitates to halt at this late date; but on some occasions, that does make sense. For example, after a lengthy analysis and design of a computerized payroll system, a small contractor might ultimately decide to stick with a manual system because the proposed system will not accommodate constant and unforeseen changes in workers' benefit plans that are negotiated by labor unions.

As we have done throughout our study of systems analysis and design, we can level or decompose implementation. The first subtask is scheduling and assignment of tasks. With a schedule settled, the analyst can work with management to assign programmers, determine in what order to write programs, and select a means for testing their accuracy.

After scheduling and assigning tasks, actual programming commences as the code for the system gets written according to the analyst's specifications. This task may consume only a few weeks or several months, depending on the number and complexity of the programs. For example, it might take only a month to program a billing system for a small hardware store, while it might take many years to program an on-line college registration system.

Figure 13.1 Implementation is the third phase in the systems process.

Program testing occurs next. Here, programmers might use real data pulled from a prior system to "test drive" the new system. At the same time as program testing occurs, the organization may begin to convert data from the old to the new system.

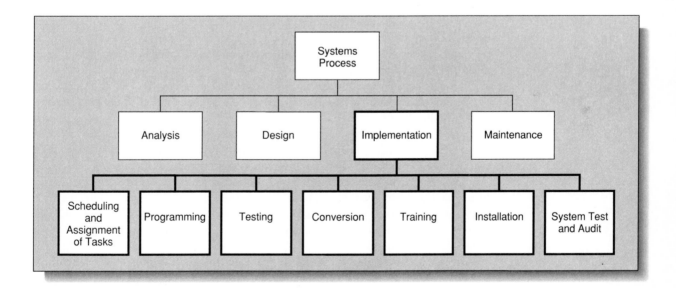

About this time, hardware installation and training of users, management, and computer center operations staff also take place. Outputs from these concurrent tasks include user training manuals, instructions to the computer center staff detailing their anticipated interaction with the system, and the completed and fully tested programs.

Finally, the system must pass a system test (acceptance test) and a review or audit, which is a final evaluation that occurs only after operating the system long enough for users to have gained a familiarity with it. At this point, users, management, and the computer services staff carefully critique the system, citing strengths and weaknesses. They will also compare anticipated and actual budgets and schedules. Results of the audit not only benefit the system in question, they form an important part of the organization's experience with computers, enabling it to make even better decisions in the future.

SCHEDULING AND ASSIGNING TASKS

Analysts begin implementation by writing an **implementation plan** that outlines all forthcoming events, showing activities, times, and events (Figure 13.2. For a review of Gantt charts, see Chapter 5.). If the analyst charged with implementation did not perform the earlier analysis and design, an inspection of the system and its program specifications will help the new person to fully grasp the system's goals and how it will achieve those goals.

To complete an implementation schedule or plan, the analyst fits together all the pieces of the puzzle: facilities, equipment, technical personnel, users, and management. If the system calls for new equipment, the analyst must arrange with the computer center operations staff and supplying vendor for delivery and installation. Quite often, the analyst must indicate to the computer center operations staff if any special alterations are needed within the computer center or a user's work area. When making programming assignments, the analyst must make accurate allowances for computer time, test data, and users' training. Since training sessions usually disrupt normal work schedules, management must remain aware of all schedules and costs.

Two tools help the analyst schedule and plan all of these activities. While a Gantt chart depicts an overall picture of events and each task's schedule, CPM (Critical Path Method) charts provide details about the events (Figure 13.3).

A CPM chart emphasizes tasks critical to maintaining a schedule (Figure 13.4). If a task on the critical path falls behind schedule, the analyst might allocate additional resources to succeeding tasks in order to get implementation back on track.

The CPM chart also reveals "slack" times during which certain events can begin earlier than anticipated, provided that a preceding event has concluded. For example, the CPM chart in Figure 13.5 shows that events 1, 4, 3, 7, 8 and 9 fall on the critical path. Thus, event 2 can start at any time as long as it concludes by day 11. Likewise, event 5 can't start until day 7 and must finish by day 15.

Software is now available that can automatically analyze activities, their relationships, and time factors. This software can calculate the critical path that will ensure completion of the system on schedule, ultimately displaying all of this information in a handy graphic format.

Figure 13.2 Gantt charts depict implementation activities, showing a schedule of weekly activities.

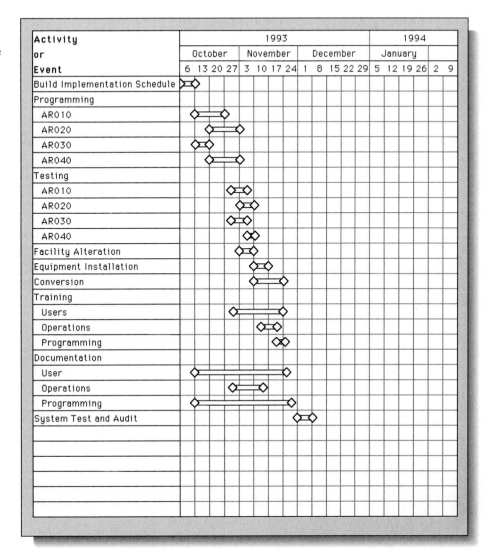

Within the implementation schedule, the analyst includes a personnel chart detailing people assigned to each event (Figure 13.6). The personnel chart for Fleet Feet includes vendors (Holcenberg, a systems engineer from HP, and Fane of Easters Electric), technical staff (Overmiller, Robinson, McAdams and Sprenger), users (Fenolio and Darlington), and management (Valentine and Baker). The lead analyst (Adams-Russell) formally contacts people early on to make sure they understand how and when they fit into the implementation schedule.

Smart analysts invest a lot of time planning for implementation. This type of thoughtful planning can reap many rewards: forestalling hurt user feelings, improving communication, and smoothing the transition to the new system.

Figure 13.3 *(right)* CPM (Critical Path Method) charts graphically represent an implementation schedule. They show precedence relationships of the implementation plan and times required to complete each task.

Event Number	Predecessor Event	Successor Events	Time Required (Days)
1. Build Implementation Plan	none	2	4
2. Programming			
a. AR010	1	3a	12
b. AR020	1	3b	14
c. AR030	1	3c	8
d. AR040	1	3d	13
3. Testing			
a. AR010	2a	5	9
b. AR020	2b	5	10
c. AR030	2c	5	7
d. AR040	2d	5	4
4. Facility Alterations	1	5	9
5. Equipment Installation	4	5	8
6. Conversion Data	3a	7a	9
7. Training			
a. Users	3b	9	24
b. Operations	3a,3b,3c,3d	8	9
c. Management	6	8	3
8. Documentation			
a. User	3	9	22
b. Operations	6	9	15
c. Programming	6	9	29
9. System Test and Audit	8	none	9

Figure 13.4 *(below)* A CPM chart for the implementation phase illustrates inter-relationships between events.

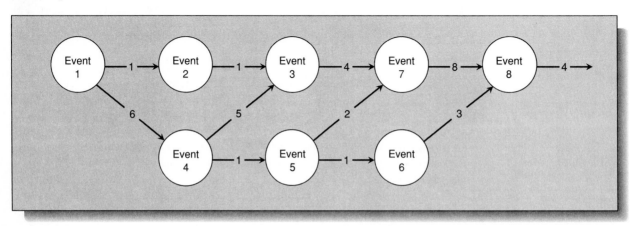

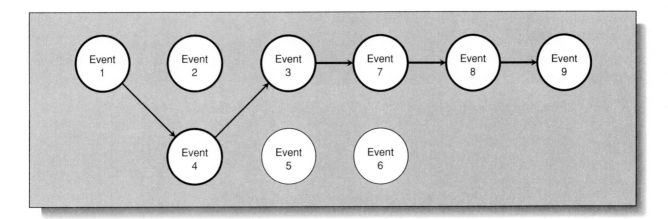

Figure 13.5 *(above)* Critical path (shaded) shows events that, if they exceed their planned time, will delay completion of the overall project by a like amount of time.

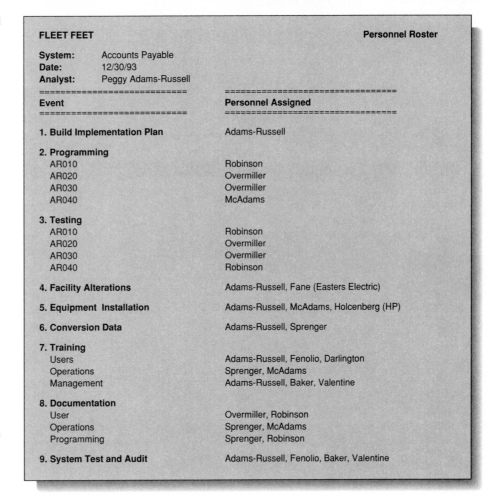

FLEET FEET	Personnel Roster
System: Accounts Payable	
Date: 12/30/93	
Analyst: Peggy Adams-Russell	

Event	Personnel Assigned
1. Build Implementation Plan	Adams-Russell
2. Programming	
AR010	Robinson
AR020	Overmiller
AR030	Overmiller
AR040	McAdams
3. Testing	
AR010	Robinson
AR020	Overmiller
AR030	Overmiller
AR040	Robinson
4. Facility Alterations	Adams-Russell, Fane (Easters Electric)
5. Equipment Installation	Adams-Russell, McAdams, Holcenberg (HP)
6. Conversion Data	Adams-Russell, Sprenger
7. Training	
Users	Adams-Russell, Fenolio, Darlington
Operations	Sprenger, McAdams
Management	Adams-Russell, Baker, Valentine
8. Documentation	
User	Overmiller, Robinson
Operations	Sprenger, McAdams
Programming	Sprenger, Robinson
9. System Test and Audit	Adams-Russell, Fenolio, Baker, Valentine

Figure 13.6 *(right)* A personnel chart for system implementation. Some of the personnel are technical (programming) staff, while others are users.

COPING WITH CHANGE

Franklin Pest Control customer service representative Margot Sue Lyon sat glumly at her desk, worrying about impending job changes. Next Wednesday, the computer department would install a terminal on her desk, putting her "on-line," whatever *that* meant. After all, she hadn't needed any new equipment for 17 years. Why mess with a good thing?

Margot Sue liked her job. Whenever anxious customers whose last names began with A through J called to ask when Franklin would spray their property for gypsy moth caterpillars, she looked up the answer in her manual delivery schedule. Oh, it took a day or two for her to call back with the answer, but no one had ever complained.

Since she cherished the friends she had made with her friendly telephone manner, she was afraid that the new terminal inquiry system would ruin the best part of her job. Her boss claimed that the new computer would make her work more efficient, but she didn't think she could handle additional customers K through M (as management expected her to do). And what about poor Helen Leslie, one of the existing three service representatives that the new system would not need? Was Helen going to lose her job?

Margot Sue also dreaded the terminal that would dominate her desk. What if she accidentally blew it up?

She didn't notice Stan Munoz (Franklin's systems analyst) approach her, carrying two cups of coffee.

Stan smiled. "How 'bout some black coffee? You don't take cream or sugar, do you?"

Margot Sue's eyes brightened. "How did you know that?"

"Oh, Helen told me. You know, she's working in my office now."

"No, I didn't know. What does she do?"

"She's learning all about computers." Unbelievable! Helen had been as upset about the new machines as Margot Sue still was.

"Computers. Ugh!"

Stan laughed. "Hey! If it weren't for computers, I'd be out of a job."

"And because of computers," moaned Margot, "We're all going to lose our jobs!"

Her response seemed to bother Stan. "Where'd you get that idea?" he asked.

"I'm too old for all these changes: Helen's left, I've got to handle new customers I don't know, every day there are new faces in the computer department, and now I get saddled with a stupid machine that I'll never figure out. I won't be able to set a coffee cup on my desk anymore, my parking space has been blocked all week by that truck from IBM, and my granddaughter's getting married on Saturday."

Stan shook his head. "Well, I see I haven't done my job. Sorry about your parking space, but you'll have it back next week. Helen? As I mentioned, she's changed jobs. When I began designing the new order-entry system, I knew one person would have to go. I've spent a lot of time with Helen, so I chose her as our new data-entry control clerk. She loves it. She's taking a night class on computers, with the company paying for tuition, books, and expenses."

"Really!" exclaimed Margot. "Next thing you know, I'll earn a Ph.D."

Stan frowned. "I should have talked to you sooner about all this. I assumed you'd like to see your job streamlined."

"Looks like it'll be streamlined so much I won't make friends on the phone anymore. Everyone's just a number."

"Not at all," said Stan. "Your terminal will help you answer customer questions faster and more accurately. People will appreciate that."

Margot mulled this over. "I guess that's good," she said grudgingly. "But what if I push the wrong button and blow up the computer? Then we'll both lose our jobs."

"You can't blow up the computer, Margot. We'll label all the buttons so you'll know what they're for."

"That doesn't solve my space problem. I won't even have room for a flower vase on this desk."

"The terminal will sit on its own table. And eventually, you can throw these away." Stan pointed to several thick binders that held three years' worth of customer data.

"I didn't know that."

"Like I said, that's *my* fault. I'm so preoccupied with the new system, I've neglected the people that it will affect the most. Sorry I didn't think about talking to you sooner."

Margot laughed deeply. "That's okay. Did I tell you my granddaughter's getting married on Saturday?"

PROGRAMMING A STRUCTURED SYSTEM

After completing the implementation plan, the analyst turns his or her attention to programming. This phase of implementation requires the implementation plan, the computer service staff's expertise, and system and program specifications. When programming is finished, we have the completed programs, along with their test results.

Standards

Most computer services departments have adopted **standards**, a set of rules that programmers must follow when writing programs (Figure 13.7). Standards promote a consistent programming style within a department, thereby making it easier for new personnel to maintain all the programs. Many departments demand consistency for file names, record names, variable or data names, and module names. Most departments also dictate strict rules for writing modules.

Actual **program coding** (the writing of the programs themselves) begins with a review of the program specifications, which should reveal program logic (Figure 13.8). During this review, the programmer identifies and numbers all modules. The programmer's structure chart or VTOC should display four modules: opening, testing, closing, and a main driver module called File-Update that causes the system to perform each module in the correct sequence.

Stubs

With the organization's program standards and specifications in mind, the programmers begin coding in the specified language. At Fleet Feet, for example, the programmers first write the main driver, 000-MAIN-DRIVER, then (using the top-down method), 100-OPENING, 200-PROCESSING, and 300-CLOSING. If they wish to concentrate on one specific module at a time, programmers can **stub** other modules, writing an abbreviated version of the program to facilitate programming. Stubs set aside those modules that will be written later (Figure 13.9, page 434). For example, while writing the 100-OPENING module, a programmer might stub the others by inserting a temporary statement such as a COBOL statement DISPLAY "300-TESTING" or a C statement prints (\n,'Testing function 300') at the point where the fully developed module will eventually appear.

Stubbing unwritten modules allows programmers to test related modules before completing the entire program. Stubbing also permits programmers to write modules in any order, a benefit when a particular individual (such as a busy user or database administrator) is unavailable to assist the programmer.

Because stubbing can free programmers to write the most critical modules first, it can strengthen program implementation and permit testing at intermediate stages. This allows users to "test drive" early versions of programs, thereby enhancing their morale as they watch

Figure 13.7 Programming standards may vary from one computer installation to another.

FLEET FEET
INCORPORATED

PROGRAMMING STANDARDS
adopted 12/03/92

1. File names: Use plural words for file names. Examples are MASTERS, TRANSACTIONS, VENDORS, and CUSTOMERS. File names shall be meaningful as to the data they represent.

2. Record names: Use singular version of file names for all record names. Examples are MASTER, TRANSACTION, VENDOR, and CUSTOMER.

3. Data names: Data names should have prefixes of their file names. Examples are VENDOR-NUMBER or CUSTOMER-LAST-NAME. Prefixes for working storage would be prefixes with WS. Examples are WS-PAGE-NUMBER or WS-CUS-TOMER-TOTAL. Use data names that are unique and meaningful. Programs should not have names like PIZZA, RAVIOLI, or HERE.

4. Module names: All module names should have a prefixed module number and all modules should appear in ascending order. Examples are 100-OPENING and 900-END-OF-PAGE.

5. Module rules: All modules will have a single entry, single exit. Module size is less than 24 lines unless it is simple sequence in structure. Modules must follow one of the three control structures: sequence, iteration, or IF-THEN-ELSE.

6. Statements: Every program statement will begin on its own line. No multiple statements should appear on one line.

7. Indention: Indent statements to show their hierarchy. Indention is especially important when writing IF-THEN-ELSE statements.

their system materialize, and it can especially boost morale when schedules suffer setbacks. Stubbing also forces programmers to follow the top-down structured methodology, which insists that they consider the most important aspects of the program before dealing with details.

Bottom-up offers an alternative to top-down program construction. With bottom-up construction, the programmer writes all of lowest level routines first, progressing up the hierarchy of modules toward the main driver (paragraph 000-MAIN-DRIVER in Figure 13.9).

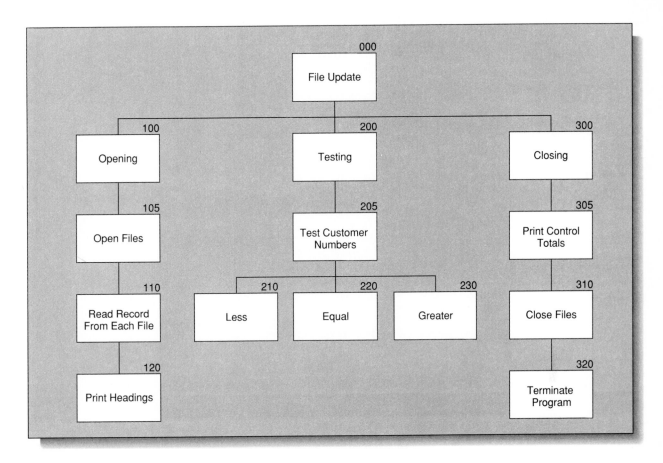

Figure 13.8 Programming specifications should reveal program logic. One can depict program logic with a flowchart, pseudocode, Nassi-Shneidermann diagram, VTOC chart, or as shown with a structure chart.

Bottom-up programming requires the programmer to build a test driver—sometimes called a test harness, test monitor, or exerciser. Such routines, though ultimately replaced by the actual drivers, permit testing of all lower-level modules first. Bear in mind, however, that bottom-up violates the most important to least important hierarchy of the structured methodology.

Program Walkthroughs

Similar in format, scope, and purpose to the walkthroughs we encountered earlier in the systems process, program walkthroughs focus on finding omissions, errors, faulty logic, improper language usage, or otherwise faulty programs. If some small but critical mistake has been made (such as omitting an identifying number on customer statements), the walkthrough should locate the error, which would save the programmer and users a lot of time later on.

After the programmer has written and tested all program modules, the analyst schedules a walkthrough attended by the programmer, analyst, fellow programmers, and perhaps a member of the operations staff if the program requires operator intervention (such as entering dates or changing types of paper in the printer). The **program walkthrough** is a

Figure 13.9 Program stubs allow programmers to set aside some modules while they focus on others. Fleet Feet's update program fully develops the main driver paragraph and the opening module. Stubs are given for paragraphs 210-LESS-THAN, 220-EQUAL-TO, 230-GREATER-THAN, and 900-TOP-OF-PAGE.

```
006000 PROCEDURE DIVISION.
006100
006200 000-MAIN-DRIVER.
006300     PERFORM 100-OPENING.
006400     PERFORM 200-TESTING
006450         UNTIL DONE.
006500     PERFORM 300-CLOSING.
006600
006700 100-OPENING.
006800     OPEN INPUT TRANSACTIONS.
006900     OPEN INPUT CUSTOMERS.
007000     OPEN OUTPUT NEW-CUSTOMERS.
007100     OPEN OUTPUT PRINTERS.
007200     READ CUSTOMERS
007250         AT END MOVE HIGH-VALUES TO CUSTOMER.
007300     READ TRANSACTIONS
007350         AT END MOVE HIGH-VALUES TO TRANSACTION.
007400     PERFORM 900-TOP-OF-PAGE.
007500
007600 200-TESTING.
007710     IF   TRANSACTION-NUMBER IS LESS THAN CUSTOMER-NUMBER
007720         PERFORM 210-LESS-THAN
007730     ELSE
007735         IF   TRANSACTION-NUMBER EQUALS CUSTOMER-NUMBER
007740             PERFORM 220-EQUAL-TO
007750         ELSE
007755             PERFORM 230-GREATER-THAN.
007760
007770 210-LESS-THAN.
007772     DISPLAY "210-LESS-THAN STUB SENTENCE".
007774
007776 220-EQUAL-TO.
007778     DISPLAY "220-EQUAL-TO STUB SENTENCE".
007780
007790 230-GREATER-THAN.
007792     DISPLAY "230-GREATER-THAN STUB SENTENCE".
007800
007900 300-CLOSING.
008000     WRITE PRINTER FROM WS-TOTAL-LINE BEFORE ADVANCING 3 LINES
008100         AT END-OF-PAGE PERFORM 900-TOP-OF-PAGE.
008200     CLOSE TRANSACTIONS.
008300     CLOSE CUSTOMERS.
008400     CLOSE NEW-CUSTOMERS.
008500     CLOSE PRINTERS.
008600     STOP RUN.
009000
009100 900-TOP-OF-PAGE.
009200     DISPLAY "900-TOP-OF-PAGE STUB SENTENCE".
```

peer review of the code to find errors, omissions, faulty logic, or improper language use. Ordinarily, users do not participate in program walkthroughs because they lack sufficient technical knowledge to offer much input. Copies of the program should go to each member well in advance, and if team members are not familiar with the database design, input, or output requirements, they should also receive copies of these materials.

One member of the team takes responsibility for recording all of the errors discovered by the team and reporting them back to the programmer. The team does not correct errors, but leaves that job for the programmer. Some organizations hold walkthroughs before the program is compiled, while others do so afterward. If the team spots few errors or if the errors are not significant, a second walkthrough is not necessary. If the team finds a lot of errors or if the errors are serious, the analyst will probably want to walkthrough the programs again.

Faulty logic may even appear in syntactically correct programs. For example, if a programmer neglected to couple an IF statement with an ELSE statement to accommodate the negative case, the program may fail whenever it encounters a negative situation. In an accounts receivable system, a payment arriving late may determine that a finance charge should be applied to the customer's account, but a test for late payment date without an ELSE statement may cause *all* payments to be considered late, resulting in incorrect finance charges to all customers' accounts.

We refer to an otherwise syntactically correct program that does not perform correctly all the time as having improper language usage. Suppose a programmer has indicated a MOVE statement, but the receiving field is too small to receive the result. In such a situation, truncation occurs. A large city on the West Coast apparently lost over a million dollars in traffic fines when a COBOL program's PICTURE clause could only hold a number up to 999,999.99 and the programmer did not use COBOL's "ON SIZE ERROR" clause to catch field overflows. When the collected amount of fines exceeded this amount, the million portion disappeared, leaving only the hundred thousands on the report. Unfortunately, the cross-footing control on the report, which might have helped spot the error, could only show numbers up to 999,999.99, too. County auditors discovered the discrepancy when hand-kept totals didn't correspond with computer totals. Ironically, a manual back-up saved the day.

USING CASE TO ASSIST PROGRAMMING

Our CASE tool has helped us through analysis and design. We are now entering the implementation and maintenance phases of the systems life cycle and the analyst's needs here are quite different. Instead of helping us with the logical elements of our system, the CASE tool must help us with the physical aspects, creating software, software quality assurance, documentation, schedules, and personnel requirements. We are moving from upper or front-end CASE to lower or back-end CASE. Not all CASE tools help us in all aspects of the system life cycle.

A few CASE systems can generate software. During analysis and design, we told the CASE tool the specifications of a system, the rules that the application must follow (module design), the reports or screens it needs (input/output design), and the relational tables required (during database design). Some CASE tools can take these specifications and generate COBOL or C source language programs. While not the most elegant nor necessarily following structured methodology, the programs will compile and run without errors. If changes are needed, programmers can modify the generated code and make it perform new or additional tasks.

Managing a large software implementation project is also in the realm of a few CASE tools. Assistance here comes in the form of creating and maintaining schedules with Gantt, CPM, and PERTcharts. The CASE tool user describes the events, time spans, and personnel assignments, and the scheduler takes care of monitoring the process. As the project works through the implementation phase, the CASE tool is updated, keeping the staff appraised of progress.

Providing an implementation environment is yet another part of some CASE tools. Environment-based systems come with the following:

- Program editors to help write the source programs
- Program builders to automate the process of compiling a program that is composed of many different source files
- Debuggers to watch complex systems as they operate, allowing developers to place breakpoints and to trace variable values and function calls
- Test data generators that provide data to help prove program correctness
- Record and playback features to allow a developer to recreate steps taken to get to a certain error state
- Analyzers that provide information regarding the structure of a program and automatically recompile a program if changes are made
- Implementation managers to monitor and control the versions of the software, checking them in and out as they are written
- Electronic mail to promote communication among developers

CASE environment-based systems, such as HP's Softbench, provide an integrated environment for the developer. They can accelerate the implementation process while reducing errors. For example, one user reports an overall effort reduction of 40 percent, while defects are reduced by 50 percent.

QUALITY ASSURANCE: TESTING MODULES, MODULE INTEGRATION, AND PROGRAMS

After writing programs and conducting a walkthrough, the system now undergoes rigorous software quality assurance or testing (Figure 13.10). **Software quality assurance** is a check to prove that the program works as expected. Testing should locate any previously undetected errors that might hamper the smooth functioning of the system. Unlike the classroom approach in which we test and retest programs until they work, in the business world we actually try to make the programs fail and then continue testing until we cannot deliberately make them fail any more. Testing is such an important step that some programming groups report spending 30 percent of their implementation time and budget on it.

When a new system replaces an old one, the organization can extract data from the old system to test them on the new. Such data usually exist in sufficient volume to provide ample

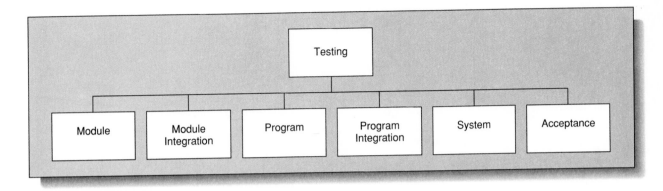

Figure 13.10 Testing involves six phases that lead to a completed system. The first three phases of testing find errors in modules, module coupling, and programs.

testing, and they can create a realistic running environment that ensures eventual system success. However, most students and beginning analysts use artificial data, building a mock data file specifically designed to test all conceivable future situations that the system will encounter. Though artificial, such data can more carefully test each module in every program. Unfortunately, it can be hard to create a sufficient volume of mock data, thus preventing the adequate testing of the system under a normal load of work.

A computer program or special software can also produce test data. Though such programs can generate sufficient volume to test a normal workload, the data may not mirror all the real situations that the system will encounter later.

Another source of test data comes from a library of test data that some organizations maintain. Kept specifically for testing proposed systems, the library is a compendium of data collected from other tests, real data, and artificially created data.

Regardless of the source of test data, the programmers and the analyst will eventually conduct four different types of tests: module testing, module integration, program testing, and program integration. Tests continue until the system can pass acceptance tests by everyone, including end-users.

Testing Modules

Module testing, sometimes called unit testing, centers on validating the correctness of each module. During this type of test, the programmer examines each module separately, actually trying to make it fail.

Once each module survives its test, the programmer evaluates **module integration** to ensure proper module coupling. Faulty module coupling includes failure of one module to access another module. Most vendors supply software to assist in module integration. Such software can tell how many times a module is accessed and tabulate the amount of computer time that the module will consume.

Program tests concentrate on the programs themselves in an attempt to make sure that each program works as it should. With test data in hand, programmers and analysts can test each program individually. The responsibility for conducting these tests falls on the programmers' shoulders. If the programs do not produce intended results, the programmers must repair the programs and continue testing until they do.

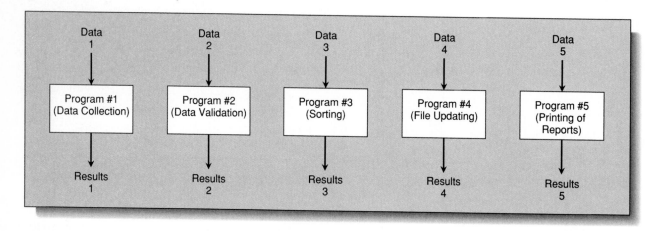

Figure 13.11 During program testing, the analyst and programmers run data through each program and examine the results.

Program integration tests follow next. These tests concentrate on the relationship between programs so that we are sure that the data created by one program are correct relative to the program that follows it. In Figure 13.11, Program #1 will run Data file #1 to produce Results #1, and so on for each program. When Fleet Feet tested its new accounts payable system, it collected data for five programs:

- Program #1 collects raw data
- Program #2 validates them
- Program #3 sorts them into order by vendor number
- Program #4 updates vendor master files
- Program #5 prints a report analyzing payments to vendors

After all errors are cleared up, tests cease until the time comes to evaluate program integration, run the completed system, and recommend acceptance. However, testing never ends because errors and bugs continue to crop up even after an organization has used a new system for a long time.

TYPES OF CONVERSIONS

With programming under way, the analyst can turn his or her attention to the problem of converting from the old to the new system. **Conversion** is the changeover from one system to another. As with so much of the implementation process, smooth conversion from one system to another depends on thorough preparation. The new system may involve installing new equipment, altering facilities to accommodate new hardware, and preparing data— operations that may demand special programs. Furthermore, the analyst must formulate rules for using the new system, including appropriate user responses to errors or exceptions.

GUIDELINES FOR WRITING IF AND GO TO STATEMENTS

Two of the most frequently used commands in many programming languages are the IF and GO TO. Many programmers know how to write them correctly, but these statements can, as we've seen earlier, violate the rules of the structured methodology. Although purists insist that we abolish the GO TO altogether, IF and GO TO can offer valuable alternatives in certain situations in which their use improves an organization's ability to read and maintain its programs.

When writing IF statements:

- Align and indent the IF with its associated ELSE. Place the ELSE on a separate line by itself.

- Some IFs will not need an ELSE. Adding null ELSEs may hamper readability. If you do not write an ELSE, make sure the alignment is proper.

- Write pseudocode *before* writing multiple levels of nested IFs, especially when there are null statements within the nested logic. The pseudocode will help you spot faulty logic.

- Avoid over five levels of nested IFs; they are too hard to read. Three levels are considered ideal as a maximum.

- If you can't avoid more than five levels, consider placing the inner IF statements in another module.

- To improve readability, place parentheses around conditions tested.

- In situations in which the physical span of the IF is too large (when too many statements appear before the ELSE), move these statements to a separate module.

- Place each condition on its own line, linking multiple conditions with an AND or OR.

On the rare occasions when GO TO statements are used:

- Don't use GO TO to move from one module to another.

- Permit GO TO to move downward to the end of the module.

- Using GO TO downward, but not to the end of the module, should not be permitted.

- Permit GO TO in the event of a program abort, allowing it to print a memory dump.

Putting together a good program is as satisfying as putting together a good meal. But keep in mind that although structured methodology should provide the meat and potatoes of the programmer's diet, a few old-fashioned techniques can make the meal tastier, provided that they are used as sparingly as Tabasco sauce.

Most organizations adopt one of three standard methods for converting to a new system from an older one: parallel, phased, or direct. The choice of a conversion method, each of which offers its own peculiar advantages and disadvantages, depends on the particular situation.

Parallel

Parallel conversion requires simultaneous operation of both the old and new systems, with an operator entering data into both systems for processing and comparing the results of each. If both systems produce the same results, then the new system replaces the old one (Figure 13.12). If results do not match, the analyst must repair the new system and continue testing it before conversion can take place.

Raintree Florists used parallel conversion when computerizing their manual billing system with BusinessWorks PC from Manzanita Software. During August, they posted transactions to both systems. When the customer bills matched on September 1, the company immediately dropped its old manual system in favor of the new personal computer system.

Figure 13.12 Parallel conversion requires both the old and new system to operate simultaneously. Cut-over is the actual adoption of the new system and removal of the old system.

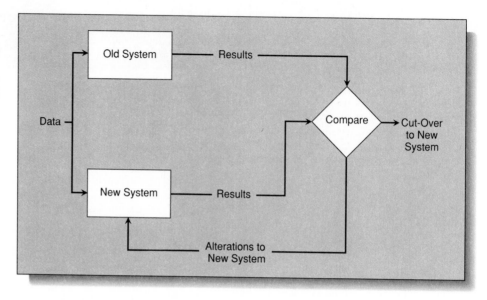

One minor problem instantly became apparent. Raintree discovered that the bills did not show the ZIP code when placed in a window envelope because the printer had positioned it too far to the right. This rather simple error was quickly corrected for the next billing cycle by moving the location of the ZIP code to the left.

Parallel conversion works best when a new system will replace a similar old one. It also offers some security. If the new system fails, the old system can keep on working. However, a danger may arise when an organization, feeling comfortable with an old back-up, takes longer than necessary to accept a new one. Furthermore, the costs of parallel conversion run high because the organization must, in effect, do everything twice.

Phased

With **phased conversion**, an organization gradually replaces the old system with the new one (Figure 13.13). As users become familiar with specific, manageable portions or functions of the new system, they can discard corresponding portions of the old one. Figure 13.13 shows a system with three distinct components: Function #1, Function #2, and Function #3. During Time Frame 1, the organization tests all three functions, but finds only Function #1 satisfactory. Therefore, it converts Function #1 to the new system, but retains Functions #2 and #3 on the old one. In Time Frame 2, the organization tests Functions #2 and #3, which results in the satisfactory conversion of Function #3 to the new system. This process continues until the entirely new system becomes fully operational by the end of Time Frame 4.

The Smiles Dental Clinic used phased conversion to change its accounts receivable system from batch to on-line. It continued its batch system's courier service for picking up charge slips, but began to use the new on-line personal computer system to generate patient reminder notices. Since this function worked fine, Smiles converted it after the first month. The second month, the clinic concentrated on insurance billing, which also worked well, resulting in computerized printing of insurance bills at the end of the second month. The third month, the firm cut-over to the complete patient billing system and finally canceled the unnecessary batch back-up.

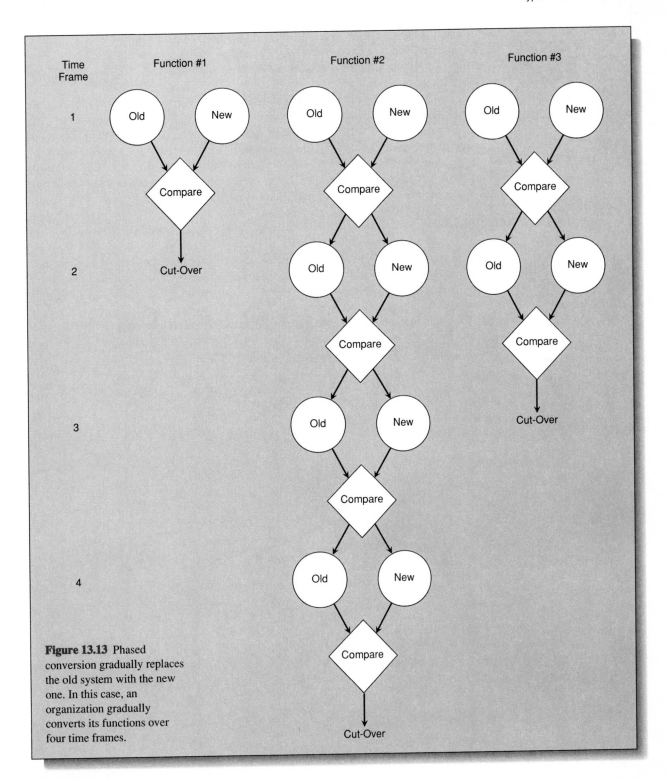

Figure 13.13 Phased conversion gradually replaces the old system with the new one. In this case, an organization gradually converts its functions over four time frames.

Phased conversion costs less than parallel because the organization does not duplicate all data entry, does not process everything twice, and distributes workloads evenly. Since this approach evaluates the system on a module-by-module basis, it reflects the spirit of the structured methodology.

However, phased conversion may confuse users or customers if they simultaneously see results of both systems. Phased conversion also stalls management's appraisal of overall business performance because it cannot easily pull together data spread over two systems. For example, if a company cannot compile its monthly list of customers who haven't paid their bills for over 90 days, it could imperil its positive cash flow, thereby jeopardizing its own good credit with vendors. Delays in spotting certain trends or events that could have profound impacts on the system itself could also create a problem.

Direct

Direct conversion (also called cold-turkey, slash, cut, or inventory) involves immediate changeover from an old to a new system (Figure 13.14). Because it eliminates any backup, this method requires a thoroughly tested new system and poses more hazards than the other two conversion methods.

Mountain Motor Supply, an auto parts retailer, adopted direct conversion when it computerized its manual inventory system. On July 1 (the beginning of its fiscal year), all manual

Figure 13.14 Direct conversion calls for immediate conversion from the old to the new system.

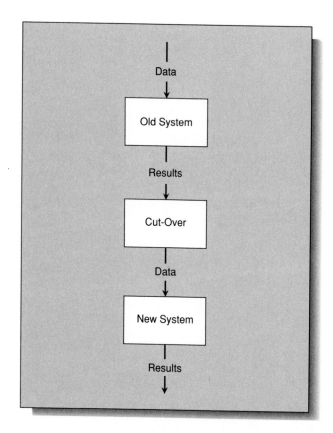

write-ups of customer orders ceased, and clerks began entering customer orders directly on terminals that were located at each parts counter. Sales staff simply entered part numbers, to which the computer immediately responded by printing a sales receipt by using on-line printers that were positioned next to the terminals. When one of the on-line printers suffered a mechanical failure, the spare printer that the analyst kept in reserve replaced the broken one.

Of the three conversion methods, direct conversion costs the least but entails the highest risks. Although it eliminates customer or user confusion over which system produced the results, the lack of sufficient back-up can preclude its use in many applications.

Programs, Facilities, and Procedures

Regardless of the conversion method selected, analysts must convert data files, programs, facilities, and procedures to any new system. Data file conversion becomes necessary whenever new systems transfer data from tape to disk, change from sequential to random file access, or adopt a database environment. The new data files must implement the file design developed earlier in the systems process. Also, the organization must collect and enter new data while dropping data elements that are no longer needed. Sometimes, a special program is written to automatically remove data from the old system and enter them into the new one, thereby speeding the file conversion process.

In one case, the programmers for a small building supply firm that was changing from an MS-DOS PC floppy diskette system to a multiuser UNIX database system wrote programs that could read data from the diskettes directly to magnetic tapes. Other special programs transferred the data from the tapes into the database, while yet another one collected new data required by the new system (such as supplier contacts, ZIP+4 codes, and units of measure), entering them directly into the database. Once all the new data were successfully stored, the company no longer needed these special-purpose programs.

If an application requires a new computer, the organization usually must alter their old programs to get them to run properly. If the new computer fully supports old COBOL programs, then program conversions are probably minimal. However, if old assembly language programs are not compatible with the new system, conversion will be a major undertaking. Some packaged software can assist in certain types of conversions by letting the computer help in the laborious transfer of data from one machine to another.

An organization may also need to alter existing facilities to accommodate a new system. Such alterations might include raised floors, additional electricity and air conditioning, dust and humidity controls, and special lighting. Hardware vendors will readily supply power, space, and air conditioning requirements for their equipment; they may even draw floor plans showing the optimum lay-out.

In some cases, an analyst must also plan for equipment removal. If an organization buys its new computer from a vendor other than the one with which it has done business in the past, the old vendor may not cooperate in this endeavor. In such a situation, an analyst must pay special attention to rapport. For example, one might gain the cooperation of an old vendor with assurances that it will receive consideration for future equipment, even if it failed to land this particular order.

If the organization finds itself with outdated equipment on its hands, it might be able to sell it to a used computer vendor that specializes in older hardware. One college sold its ten-year-old computer to a junk dealer who melted it for the gold, silver, and other metals it contained. Though a used-computer vendor had offered $15,000 for the machine, the junk dealer paid over $25,000 for what had once cost $1 million.

Regardless of how thoroughly a new system is tested, problems will inevitably arise. Pre-printed forms can help users report such problems (Figure 13.15). Forms designed to

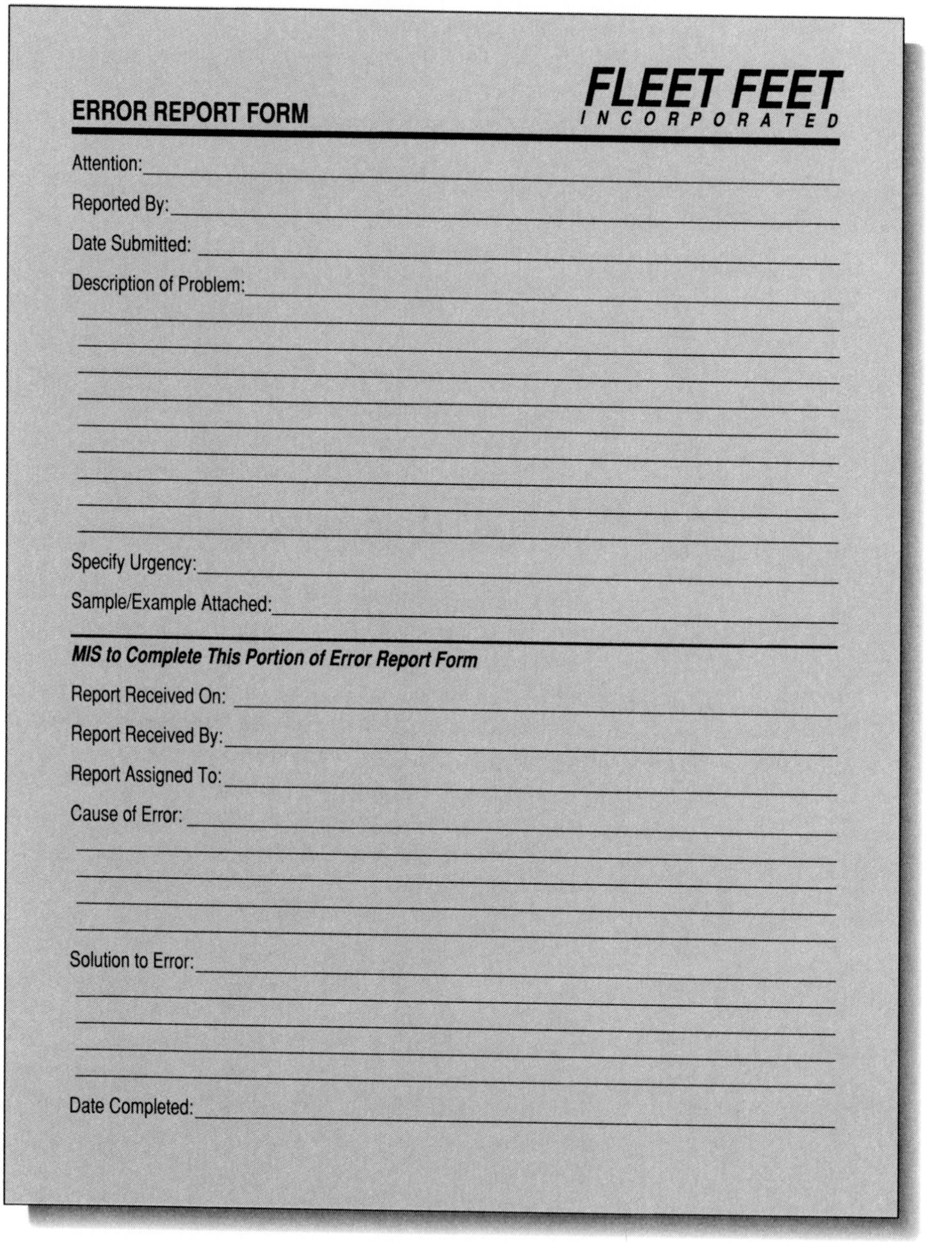

Figure 13.15 A pre-printed form allows users to report problems or errors that they encounter with a new system.

ERROR REPORT FORM

FLEET FEET
INCORPORATED

Attention:_____

Reported By:_____

Date Submitted: _____

Description of Problem:_____

_____

_____

_____

_____

_____

_____

_____

_____

Specify Urgency:_____

Sample/Example Attached:_____

MIS to Complete This Portion of Error Report Form

Report Received On: _____

Report Received By:_____

Report Assigned To: _____

Cause of Error: _____

_____

_____

_____

Solution to Error:_____

_____

_____

_____

Date Completed: _____

report errors must allow space for dates, descriptions, personnel involved, and responses. When a user spots errors or omissions during this stage of the systems process, the analyst should respond rapidly, stopping any problems before they proliferate. This strengthens the user's confidence that the analyst really cares about user needs.

Regardless of the reporting system, frequent (perhaps even weekly) meetings among the analyst, vendors, and users can help prevent problems as well as solve them. For example, the same college that sold its old computer to the junk dealer meets weekly with its new vendor. During these meetings, new problems are analyzed, potential future problems are aired, personnel training is discussed, and future needs are defined. During the second year of the new systems operation, weekly meetings became monthly meetings and eventually ceased altogether in the third year.

SUMMARY

System implementation is the third phase of the systems process. During this time, actual construction of the system occurs.

Analysts begin implementation by framing implementation plans, which outline a schedule and assign tasks to specific groups: users, managers, computer services staff members, and hardware and software vendors.

With the plan in place, the analyst assigns programs to the programming staff. Programmers write the code that is necessary for generating desired reports, collecting necessary data, and manipulating data files. Technical walkthroughs of programs occur before they are actually run on a computer.

Once the code is written, it needs testing to prove that it works as intended. Testing takes the form of trying to make the software *fail*, not to make it work.

While programmers code and test, the analyst decides how to convert from the old to the new system. Conversion methods include parallel, phased, and direct. Parallel conversion requires an existing system to remain in operation while the replacement operates alongside it. Phased conversion involves gradually replacing components of the old with corresponding components of the new system. Direct conversion demands total and immediate change to a new system.

KEY TERMS

Implementation	Module testing
Implementation plan	Module integration
Standards	Program test
Program coding	Program integration
Stub	Conversion
Bottom-up	Parallel conversion
Program walkthrough	Phased conversion
Software quality assurance	Direct conversion

QUESTIONS FOR REVIEW AND DISCUSSION

Review Questions

1. What are the major activities involved in programming a system?
2. What are three examples of programming standards?
3. Who participates in a program walkthrough?
4. What is an implementation plan?
5. What are three types of conversions?

Discussion Questions

1. What items does the analyst have to juggle and schedule when the implementation plan is built?
2. What three factors do program walkthroughs try to find?
3. List the items that must be converted for the new system.
4. List three common standards that a systems department may choose to employ.
5. What are two synonyms for direct conversion?
6. List the two types of program construction.
7. Define a program stub.

Application Questions

1. Could Mountain Motor Supply have used the parallel method of conversion? Explain.
2. Write a data dictionary definition of systems documentation.

Research Questions

1. What are the differences between top-down and bottom-up implementation of a program or system?
2. What did Stan Munoz learn as a result of his talk with Margot Sue Lyon?

Analyst Frank Pisciotta looked forward to the implementation phase of View Video's Video Cassette Tape Rental (VCTR) system. Formally authorized to proceed by management's memorandum, he began by reviewing the design phase documentation and building an implementation schedule (Figure 13.16). After scheduling all vital tasks, he assigned them either to staff or to outside vendors (Figure 13.17, page 450).

Considering the number of required programs and the fact that the programs were needed in three months, Frank elects to contract all of the programming to a professional acquaintance, Kathy Van Hoaf. Kathy is the owner of a software implementation firm, Data Management (DM), that Frank used in the past. DM's work for him has consistently met his high standards and Frank expects no less for View Video.

Data Management has its own networked PC-based system using Oracle and SQL, so their implementation efforts will match View Video's system. They use the C programming language, following top-down design and structured methodology. All of these techniques will work well with the system that Frank is putting together.

DM uses BusinessWorks PC as their accounting software, so they are aware of the interface that needs to occur between the two software systems. Charging $40 per hour for equipment use, programming, software quality assurance and documentation, Data Management agrees to deliver the software in 12 weeks at an estimated cost of $6,200—well within Frank's budget and schedule requirements. Aware of the approaching New Year's holiday, Frank chooses to wait until early January to start the software implementation stage with Data Management.

After evaluating the responses to the request for proposal (RFP) on bar-code readers and comparing all respondents with the evaluation form, Telzon was the winning vendor. They will deliver the new bar-code readers that Frank ordered by the end of January. Frank asks Telzon to deliver the reader to Data Management (instead of to View Video) so that DM may use them in their software implementation and testing.

Maasta Construction performed all of View Video's store remodeling, electrical, plumbing, heating, and air conditioning work. Frank asks them to estimate costs for minor electrical alterations and network cabling to the customer service check-out/check-in counter and office areas. Maasta reports that they can complete both the remodeling and the new tasks by the end of February; the cost is estimated to be under $1,000.

Since View Video is not replacing an existing computerized system and since no rental or customer data exist, Frank's conversion efforts are minimal. He does have the Stensaases start entering data into BusinessWorks PC, such as:

- Suppliers into their accounts receivable system
- Customers into their accounts receivable system
- Employees into the payroll system
- Their chart of accounts into the general ledger system

With BusinessWorks PC installed and a chart of accounts created, the Stensaases will start their new business with up-to-date financial records in the computer, which is a good way to start any type of business.

Figure 13.16 Frank Pisciotta's implementation schedule for View Video's Video Cassette Tape Rental (VCTR) system.

View Video
System: VCTR - Implementation Schedule
Analyst: Frank Pisciotta
Date: Dec. 29, 1993

Activity or Event	Week of									
	01/11	01/18	01/25	02/01	02/08	02/15	02/22	03/01	03/08	03/15
1. Build Implementation Schedule	=====									
2. Programming										
VCTR110		============								
VCTR120		============								
VCTR130		============								
VCTR140				============						
VCTR150				============						
VCTR160				============						
VCTR170						============				
VCTR180						============				
VCTR190						============				
VCTR210				============						
VCTR220				============						
VCTR230				============						
VCTR240						============				
VCTR250						============				
VCTR260						============				
VCTR270		============								
VCTR280				============						
VCTR310						============				
VCTR410		============								
VCTR420				============						
VCTR510		=========================								
VCTR520						============				
3. Testing										
VCTR110				=====						
VCTR120				=====						
VCTR130				=====						
VCTR140						======				
VCTR150						======				
VCTR160						======				
VCTR170								======		
VCTR180								======		
VCTR190								======		

(continued)

As a test of Data Management's operations, Frank decides to follow closely the implementation of the "Numeric Customer List" (VCTR110). He asks Data Management for copies of the C language program and their software quality assurance test plan.

Frank reads the C program to assure himself that it is written following the structured methodology: top-down with single entry, exit, and purpose functions. He looks for GO TO statements in the program and finds none.

Figure 13.16 *(continued)*

Activity or Event	Week of									
	01/11	01/18	01/25	02/01	02/08	02/15	02/22	03/01	03/08	03/15
VCTR210						======				
VCTR220							======			
VCTR230						======				
VCTR240								======		
VCTR250								======		
VCTR260								======		
VCTR270				======						
VCTR280						======				
VCTR310								======		
VCTR410				======						
VCTR420						======				
VCTR510						=========================				
VCTR520							==========			
4. Facility Alterations					========					
5. Equipment Installation						========				
6. Data Conversion										
7. Training										
Users						===================================				
Operations								=======		
Management									=========	
8. Documentation										
Management							========			
Users			===============================							
Operations				===============						
Programming			=================================							
9. System Test and Audit									=======	

Frank asks Kathy about her software quality assurance (SQA) team and their testing plans. The manager of SQA, Dave Williams, outlines his testing methodology, which is based on a "break it" philosophy. Dave's team tries to break the software or make it fail—they do not to try to make it work. If they can't break it, then their assumption is that the software works as planned. Dave reports to Frank that he uses a variety of tests, including module, module integration, and program. He tests to make sure that the software does not fail under high-volume stresses, function key usage is consistent, colors match, words are spelled correctly, prompts line up, and many other measures. Dave tells Frank that he will not release the software to View Video until he knows it is ready.

One of Dave and Kathy's business rules is that *Dave* decides when to release software to the customer. This rather unique rule gives the SQA team power over the programming team and pushes them to write the most error-free code that they possibly can. Kathy tries never to break this rule; as a result, she is very confident of the software when it leaves Data Management's control and is placed in a user environment. In fact, DM customers report far fewer errors than the industry's average. Those that are located, Dave takes personally and works with the SQA team to solve as soon as possible.

Figure 13.17 Personnel assignments for View Video's VCTR system include tasks and schedules for individuals from View Video, Data Management (DM), Telzon, and Maasta Construction.

View Video
System : VCTR - Personnel Assignments
Analyst : Frank Pisciotta
Date : Dec. 30, 1993

Event	Personnel Assigned	
1. **Build** Implementation Schedule	Pisciotta	
2. **Programming**		
VCTR110	Clauson	(DM)
VCTR120	Danielson	(DM)
VCTR130	Cochran	(DM)
VCTR140	Clauson	(DM)
VCTR150	Danielson	(DM)
VCTR160	Cochran	(DM)
VCTR170	Clauson	(DM)
VCTR180	Danielson	(DM)
VCTR190	Clauson	(DM)
VCTR210	Hammer	(DM)
VCTR220	Herod	(DM)
VCTR230	Lawver	(DM)
VCTR240	Hammer	(DM)
VCTR250	Herod	(DM)
VCTR260	Lawver	(DM)
VCTR270	LeFevre	(DM)
VCTR280	LeFevre	(DM)
VCTR310	LeFevre	(DM)
VCTR410	May	(DM)
VCTR420	May	(DM)
VCTR510	VanHoof	(DM)
VCTR520	May	(DM)
3. **Testing**		
VCTR110	Swan	(DM)
VCTR120	Richter	(DM)
VCTR130	Beard	(DM)
VCTR140	Swan	(DM)
VCTR150	Richter	(DM)
VCTR160	Beard	(DM)
VCTR170	Swan	(DM)
VCTR180	Richter	(DM)

(continued)

Working With View Video

Contracting out the programming of a project to another organization is gaining in popularity. However, this does pose special problems that an organization using only its own programming group would not face.

In this situation, Frank does not have a choice; he must contract the programming and testing. If Frank had to write and test the software by himself, View Video would have to wait a full year to place its system in operation. The Stensaases would not accept such a lengthy delay.

Figure 13.17 *(continued)*

```
=============================        =================================
Event                                 Personnel Assigned
=============================        =================================
      VCTR190                         Beard              (DM)
      VCTR210                         Gibbons            (DM)
      VCTR220                         Ford               (DM)
      VCTR230                         Jennings           (DM)
      VCTR240                         Gibbons            (DM)
      VCTR250                         Ford               (DM)
      VCTR260                         Jennings           (DM)
      VCTR270                         Williams           (DM)
      VCTR280                         Williams           (DM)
      VCTR310                         Williams           (DM)
      VCTR410                         Sadler             (DM)
      VCTR420                         Sadler             (DM)
      VCTR510                         Lim                (DM)
      VCTR520                         Sadler             (DM)
```

4. **Facility Alterations**   Pisciotta, Brown (Maasta Construction)

5. **Equipment Installation**   Pisciotta, Engberg (Telzon)

6. **Data Conversion**   Pisciotta, Williams (DM)

7. **Training**
    Users   Pisciotta, Darlington, M. and C. Stensaas
    Operations   Mahan (DM), M. and C. Stensaas
    Management   Pisciotta, M. and C. Stensaas

8. **Documentation**
    Management   Pisciotta, M. and C. Stensaas
    Users   Pisciotta, Mahan (DM)
    Operations   Mahan (DM)
    Programming   Mahan (DM), Williams (DM)

9. **System Test and Audit**   Pisciotta, M. and C. Stensaas

Case Study Exercises

Answer the following questions regarding Frank's decision to contract View Video's programming:

1. Who will provide software repairs or enhancements to the View Video system?
2. How will Frank monitor the progress of Data Management as its programmers go through the program implementation and testing processes?
3. Does software from an outside organization come with any warranties?
4. Who owns the software after it has been developed?
5. What guarantees can Frank impose on Data Management to make sure that it does not resell the new program to other businesses?
6. What should happen to the source code (the original C language program) after Data Management finishes its work?
7. Will Data Management provide any program documentation that explains how the software fits together?

Testing and Training

GOALS

After reading this chapter, you should be able to do the following:

- Describe the three types of tests that a new system must pass
- Identify the individuals who should be trained to use the new system
- Describe system documentation and its use
- List the components of a training manual
- Cite the six stages of certification
- Discuss system maintenance

PREVIEW

By this point in the implementation phase of the systems life cycle, the analyst has put together an implementation plan, ordered programs to be written and tested, and chosen a conversion method. Now comes system testing, training, documentation, and maintenance. Programmers, analysts, and occasionally managers of the computer services department conduct a variety of tests before a system gets up and running because they want a final product that ultimately satisfies both users and their managers.

Testing helps ensure that the system will achieve its goals. Earlier, each module underwent a variety of tests, using sample data, until all modules appeared sound, their linkages proper, and each program correct. If problems and errors arose, programmers worked to correct them. At this juncture, program integration, system, and acceptance tests also occur.

Before, during, and after testing, the analyst or a technical writer composes documentation and training materials. Users, management, and operations personnel need specific instructions, usually in the form of reference and user manuals, that enable them to get the most out of the system and the information that it will generate.

Yet a final test checks the system one last time to make sure it works as intended and before users actually begin running it.

Provisions for maintenance of the system need lots of advanced planning. Maintenance involves not only the upkeep of the system, but a means of updating it if additional problems develop or users feel the need for improvements or modifications.

The systems process now enters the phase in which the new or improved system begins full operation, perhaps finding a few unwanted problems or errors. In many respects, this phase parallels a family that is moving into a new home: the family (user) expects that the new living quarters (system) will be prettier and more accommodating, but only after they've lived in it for awhile will its conveniences and flaws show. After some time, the family may need to make repairs or alterations as appliances break, faucets leak, and children strain the living quarters. So it is with a new computer system. To continue fulfilling user needs, a system must accommodate change. A static system eventually falls apart.

TESTING

Figure 14.1 Testing has six components. Earlier, we followed testing through the first three phases (module, module integration, and program). Now, the analyst looks at the last three that lead to a completed system.

Automobile buyers will not only examine the specifications and appearance of their intended purchase, but they'll actually take the car out for a drive to see how it runs on the road. Similarly, when all components of a computer system are assembled, the analyst puts them through their paces with an assortment of tests (Figure 14.1). Portions of the system—modules, module integration or coupling, and entire programs—underwent testing earlier, but now all the pieces come together with program integration, system proving, and acceptance testing. Tests run the gamut from the simplicity of whether or not a terminal is positioned conveniently, to the sophistication of processing millions of records.

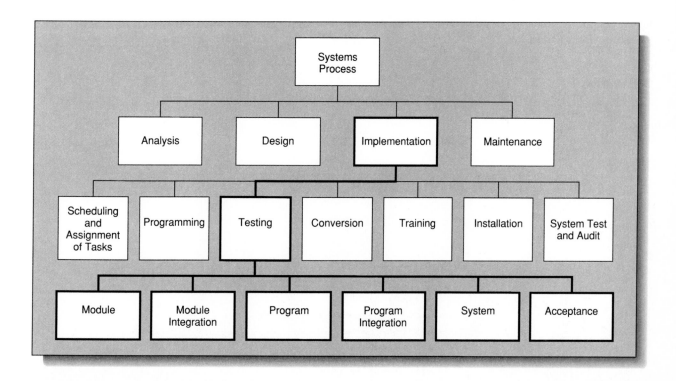

Program Integration

After each program passes its own test, its linkage to the other programs is scrutinized with a **program integration test** (Figure 14.2), also known as a string or link test. Program integration tests ensure that the programs work together as intended. If an error appears, the analyst and programmer will isolate and correct it. For example, even though each of Fleet Feet's individual programs ran perfectly, Program #1 (which collects raw data) did not place the purchase date in the file that Program #2 used for validation purposes. Thus, the program integration test caught an error that the earlier module and module integration tests missed.

When conducting this sort of test, the analyst may devise test data, which is the collection of data that the analyst uses when proving the system's accuracy. As discussed in Chapter 13, good testing actually tries to force the system to fail. In an accounts payable system, for example, failures might occur by having users enter improper account numbers or purchase amounts, credit amounts that are debits, or alphabetic data that should appear numeric. Such failures help programmers enhance procedures for error detection and correction. They allow analysts to observe certain system behaviors, such as the time it takes to process a transaction, response time, and the impact of a new system on the organization's other computer demands.

Another source of test data is the user. The data that the analyst makes up are highly controlled, with each set of data trying to prove or disprove a particular situation. User test data may catch situations that the analyst missed or could not have anticipated.

Figure 14.2 Program integration, string, or link tests unite all programs to ensure that data files created by one program are compatible with the next.

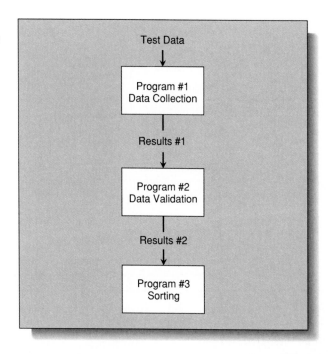

System Test

During a **system test**, users enter data (Figure 14.3) and observe the results. System tests have users verify that the system performs as they expect it to. Unlike program integration tests in which the analyst or programmer provided data, system tests require that users provide their own data. Once the data are entered and processed by the system, users can authenticate the accuracy of the system by matching these results with any known results.

   Multiuser systems, such as Fleet Feet's AP system, have special needs. Multiple users can cause such problems as having two or more simultaneous requests for the same data.

Figure 14.3 A system test occurs after individual program and string tests have been run. This test makes sure that all components of the system properly function as a unit by actually trying to force the system to fail.

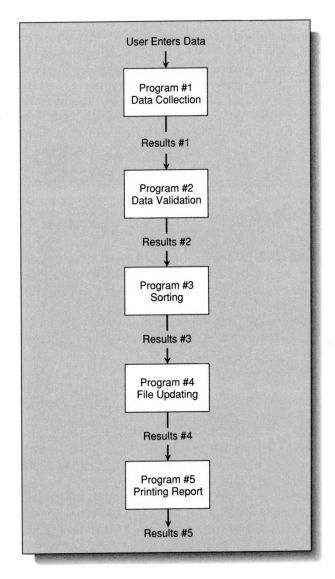

The system must respond properly to this situation. This is an important, complicated, and time-consuming phase of testing, one that is often mistakenly overlooked or skipped.

The real purpose of the system test is to discover whether users can understand and operate the system with success. Theoretically, a "typical" user should be picked for system tests. However, the "typical" user is a myth. Good typical users are those that already know the system's purpose, are motivated to want the new system, and generally know what needs to happen. System tests are a real and practical way of proving the system.

Acceptance Tests

Assuming that users find no major problems with its accuracy, the system passes through a final **acceptance test**. This last test confirms that the system meets the original goals, objectives, and requirements established during analysis. Like the system test, the responsibility for acceptance tests falls on the shoulders of users and management. If the system fulfills all requirements, it is finally acceptable and ready for operation.

The official acceptance of the system by the user usually takes the form of a signed document. Much like a contract, this document is a signal that the user is ready to take over and place the system into operation. In Fleet Feet's circumstances, the owner reviewed the final system against what was expected and determined that it suited her needs very well.

TRAINING

As the new system nears completion, the analyst must finish the materials necessary for **training** users and the computer center's staff. During training, all parties learn how the system operates. The new system may bring major changes to users' jobs—introducing new colleagues and equipment, altering work hours, and affecting even the most minute aspects of the workplace. As we saw earlier, changes can cause anger, resistance, and even sabotage of the new system. Therefore, analysts should consider training not only as a way of introducing a system to its users, but also as a way of minimizing problems associated with change.

Methods

Though diverse training methods exist, the most popular techniques involve training users, their managers, and the computer center staff separately. No one enjoys sitting through a training program when only a small portion of it directly concerns his or her job. Fleet Feet's owners chose to hold two separate training sessions: one for the owners and the other for the clerical staff. Each group learned how to incorporate the new system into its daily routines, with the clerical staff focusing on the inputs required of them and the outputs they would see, and everyone learning more about the mechanics of actually operating the new system.

The analyst may choose to train one user thoroughly; this user will, in turn, train others. Delegating training chores to users helps build confidence because it makes them "experts." This personal confidence rapidly stimulates success for other nonexperts. Users are more familiar than analysts with ongoing staff needs, capabilities, problems, and fears, and they often respond more quickly and cheerfully to someone facing the same challenges. However, users are generally not trained as trainers; despite good intentions and relationships, they may not do it well.

Some large organizations maintain special training departments. For example, the State of California employs an entire separate department to train state employees on IBM hardware and software—not only offering basic instruction to programmers, operators, analysts, and management, but also following up with special classes to upgrade skills. Such in-house programs have become quite popular because an organization can tailor them to its specific needs.

Since vendors routinely provide a certain amount of training, sometimes free of charge, analysts should consider inserting a training clause in the original purchase agreement for hardware or software. Though vendors may not know the inner working of a specific organization, they do understand their own equipment and how it relates to equipment from rival manufacturers. Vendors may send representatives to the organization or they may invite the organization to send people to regularly scheduled classes. Large cities throughout the country support many such vendor schools.

Colleges and vocational training programs can supplement training with courses that closely parallel vendor and large in-house training courses. Colleges also offer more generalized education—especially in word processing, spreadsheets, and database managers for personal computers—which students can use to broaden their education.

Specialists, consultants, and others outside the organization offer seminars on database management, structured methodology, testing, data communication, time management, and other up-to-the-minute topics. New video, audio, multimedia, and computer-assisted instruction training materials have also begun to proliferate. For example, Boeing Computer Services offers a whole videotaped series of classes, ranging from introductory material to an in-depth look at data communication.

Management

Regardless of the training mix that an analyst selects, he or she must anticipate the need for different levels of training required by three groups of people: users, their managers, and the computer services staff. Because managers need a broad view of how the new system will help them more productively fulfill their responsibilities, their training focuses heavily on the information that the system produces. For example, if the new system has an inquiry facility, managers may want to know how to properly interrogate the database with such pressing questions as, "Which customers' debts within ZIP code 10305 have fallen more than 90 days behind in payment?" Of course, managers seldom need to know technical details about entering or updating data or processing transactions.

Small group seminars or one-on-one sessions usually work best for the training of supervisors and managers. Placing managers in a general training session with subordinates who possess a high level of technical knowledge may cause the managers to avoid asking questions that could make them look ignorant. In some instances, management training should take place away from the office, eliminating interruptions from the daily crisis situations so common in active organizations.

Users and Operations Staff

User training must obviously differ somewhat from management training. Not only do users need to know how to interrogate the database, they must also know in detail how to enter data, respond to error messages, and call up routines that will print reports.

Since the computer center's operations staff concerns itself with the system's operation, operators must learn how to run various programs and back-up the system. They must also know about such mundane matters as which paper the printer will use for various programs, on which disk drive the database resides, and how to increase or decrease database file sizes.

Intensive classes, frequently lasting many days, often work best for users. In the case of a particularly complex system, it may take months for them to feel fully comfortable with it.

Any user training program should begin with a system overview, which includes the system's goals and objectives, its mode of operation (batch, on-line, real-time, or distributed), and the kinds of data that the system will collect, store, and report. Then, more specific training can take place that is customized to suit individual users.

Once the system goes into full operation, the analyst should conduct a follow-up training session aimed at more seasoned users, management, and operators who have become familiar enough with the system to pose more sophisticated questions. While such follow-up training may not fall within the analyst's realm of responsibility, it can reap huge rewards. For example, after Fleet Feet's staff worked with their new system for a few months, they learned that they could enter abbreviations of discount codes, which sped data entry by 5 percent—quite a savings over a full year.

DOCUMENTATION

To support training, analysts or technical writers develop written materials that describe the new system both generally and specifically. Such written materials are called **documentation**. As with training, a system may need a variety of manuals tailored to individual management, user, and staff needs.

Throughout analysis, design, and implementation, the analyst should have accumulated a great deal of documentation—report formats, screen formats, database and file layouts, program and module descriptions, and schedules—all of which form the basis for final documentation. Just as success on college exams comes more surely from steady, ongoing study rather than last-minute all-night cramming, forward-thinking analysts compose all early documentation with its ultimate use in mind. Since so many analysts have neglected writing skills in favor of technical expertise, they can find documentation quite frustrating, especially if they've put it off until the last minute.

Traditionally, analysts have treated documentation almost as an afterthought—a tendency that has, unfortunately, filtered down to users. Though management often realizes its value, documentation sometimes lacks quality and completeness because people forget that writing good manuals takes skill, time, and expense. Modern structured methodology

WORKING WITH TECHNOLOGY

EYE-CARE GUIDELINES FOR PC/TERMINAL USERS

Even though the computer profession is a fairly safe one, unanticipated health problems can result from working at a terminal for extended periods of time. Although furniture manufacturers have developed posture-sensitive chairs and new keyboard configurations ease hand pains, few people think about practical ways to reduce eye strain.

Eye-related problems include unexplained headaches, fuzzy vision, and excessive eye fatigue. Unfortunately, as users squirm around to reduce these problems, other physical injuries—such as neck cramps, sore shoulders, and lower-back pain—can result.

Any user experiencing sudden, severe, or unusual eye problems should consult with an eye doctor immediately. In many cases, however, a few preventative measures may help:

1. Take a break from eye strain by resting your eyes for a few minutes every hour or so. Vary your workload so that you get up from the computer now and then. If you can't leave your computer, look out a window or into the distance occasionally. If possible, apply a cold compress as soon as eye strain begins.

2. Go to a furniture store and try out different chairs; an amazing variety of ergonomically designed chairs exist. Tell the salesperson about any physical problems you're experiencing (such as back strain) and ask for advice. Remember that the chair that looks the nicest or feels the best for a few minutes may not be the most comfortable one after several hours of sitting on it. Make sure that the chair you pick provides firm back support and allows your feet to stay flat on the floor.

3. Position your screen away from windows and doors. Light coming from behind you causes unwanted glare and reflections on your screen. If you can, install overhead lighting, which will further cut down on glare.

4. Carefully select your computer screen. Green and amber screens produce the least amount of eyestrain for most people. If you can't afford a new screen, try experimenting with the contrast on the terminal you have—make sure it isn't too bright or too dark for your current surroundings.

5. Use a 9- to 14-inch screen whenever possible. Although smaller sizes are standard on most machines, they increase eye strain. Larger screens may encourage excessive eye movement.

6. Similar to good TV-viewing habits, try not to get too close to your screen. A 24-inch viewing distance is ideal. Also, your eye level should be about 4 inches below the middle of the screen.

7. If you wear contact lenses and your eyes become more irritated than usual, consult your eye doctor about ways to alleviate this. Many contact wearers find that using eye drops or switching from contacts to glasses for a few hours a day makes their eyes feel better.

Even though most computer-related eye problems are relatively minor, the physical pain they cause can reduce productivity, make you more error-prone, and increase irritability. It is a good idea to investigate simple ways of reducing eye strain before eye problems seriously affect your health—and your work.

emphasizes strong documentation, insisting that it get developed in an ongoing way. Good documentation:

- Encourages clearer communication among all participants involved with the system
- Protects the system when personnel advance, transfer, or leave
- Represents long-term money savings because it reduces the cost of training
- Eases system maintenance by centralizing materials describing the system
- Provides a permanent reference regarding the system

Of course, good documentation is not easy to write. Some of the major documentation problems include:

- Incomplete or badly written documents
- Out-of-date or inaccurate documents that do not reflect the evolution of the system
- Unnecessarily technical or jargon-filled manuals that display little sensitivity to readers' comprehension levels
- The reluctance of technically trained analysts and programmers to invest time in writing something easily accessible to all levels of users

Fortunately, modern technology has eased this task. Word processors, text editors, and personal computers can speed the writing process and allow analysts or technical writers to store and update manuals quite easily. A number of word processing and ancillary packages actually flag improper grammar, convoluted sentences, and passive verbs—as well as common misspellings. They measure reading levels (by grade) and note repeated words used in the same sentence or paragraph. They can also count words and compute average sentence and average word lengths. The trick is that these helpful features must be *used*; an alarming number of otherwise impressive documentation contain obvious misspellings and other common errors that could be easily noticed and corrected.

Management

The amount of detail contained in a given document depends on the intended audience. **Management documentation** is the written material for supervisors, team leaders, and other people in administration. This documentation needs the least detail, but should include:

- An overview of the system
- System goals and objectives
- Examples of key reports that enhance decision-making
- Final versus budgeted costs
- Final versus proposed schedule

Such documents should be business-like and jargon-free, stressing the value of the new system in terms of the organization's productivity.

User

User documentation seldom contains much more technical detail than management documentation, and it should employ the same clear and concise writing style that avoids computer jargon. The user documentation includes all necessary written materials that teach people how to use the system. The user manual must offer all the information that users need in order to perform their jobs satisfactorily. The screen menu selector in Figure 14.4 shows the sequence of steps that a user must take to activate the payroll system, process a transaction, and terminate the program. The explanation briefly overviews the system and describes exactly what the user can expect to see on the terminal screen. In this application, similar advice would cover each of the 19 entries that are permitted by the payroll system.

Figure 14.4 User documentation consists of a manual that guides the user through the system. A terminal screen selection menu program for Fleet Feet also appears.

USER INSTRUCTIONS

SYSTEM:	PR (Payroll)	Page 1 of 1
DATE:	Dec. 93	
APPLICATION:	Payroll Menu	
COMMENTS:	The payroll system consists of many programs that provide a range of useful payroll and related accounting functions. You select which program you want to use from the "menu" that is displayed on the screen.	

RUN INSTRUCTIONS

1. To use the payroll menu, type PR.

2. The terminal screen will show:

```
PAYROLL MENU                                        Fleet Feet
PR010                                               12/30/93

    FILE MAINTENANCE
        1.  Employee
        2.  Payroll Time Processing
        3.  Job
        4.  Company
        5.  Payroll Control
        6.  Payroll General Ledger Account
        7.  Deductions and Earnings Codes
        8.  State or City Tax Codes
    REPORTS
        10. Payroll Checks
        11. Payroll Distribution
        12. Payroll History
        13. Employee History
        14. Quarterly Payroll
        15. W-2 Forms
        16. Check Reconciliation
        17. Job Distribution
    PROCESSING
        20. Clear Employee Totals
        21. Calculate Payroll

        [..]  PLEASE SELECT
```

3. Type the number of the application you want, followed by a RETURN. For example, if you print year-end W-2s, type 15 and press RETURN. If you enter anything other than a correct selection, an error message is generated and the screen will wait for a correct entry.

4. To terminate the program, press the END key on your keyboard.

Though analysts or specially trained technical writers actually write users' manuals, users should participate in developing and testing them by reviewing preliminary drafts.

In some situations, the analyst may even design on-line documentation with "help" screens. Users can call up such screens when they become lost or need extra guidance. Such help screens are activated by a dedicated function key and should be context-sensitive, changing the help message whenever the user's location in the program changes.

Program

Because programmers need the most detailed information about the system, **program documentation** includes the greatest amount of technical material (Figure 14.5). Program documentation includes:

- Program narrative
- Design specifications such as pseudocode, VTOC, or a Warnier-Orr diagram
- Written materials explaining the details of each module in the program
- Source code
- Test plan
- Screen, report, and database schemas

Figure 14.5 Program documentation provides great technical detail about the system.

ITEM	DESCRIPTION
Program Title	Most programs are named. Some systems adopt naming and numbering systems, such as AR040. This pattern identifies the program as belonging to the accounts receivable system and is the 40th program in the system.
Abstract	Background information about the program: 1. General description or narrative of the program, stating its purpose and where it fits in the system 2. Date written 3. Author of the program 4. Hardware requirements: printers, disks, tapes, or terminals
Revisions	Similar to the abstract, this is a list of the changes made to the program: 1. Date of change. 2. Name of programmer making the change 3. Description of the change 4. Person authorizing or requesting the change
System Logic	A structure chart, flowchart, or pseudocode that visually shows the location of this program relative to the entire system.
Layouts	Definitions of the reports produced, database requirements, and input screen designs. This portion of the program documentation comes from the design phase of the system.
Module Definitions	Expanded system logic describing each module, its function, data inputs to the module, and data outputs from the module. If the module validates data, the validation requirements are listed. If the module selects another module, the rules for the selection are listed.
Program Listing	A copy of the program, showing all program statements, a list of variables, cross-reference listing of data names, and module names. A copy of the original—as well as most recent—versions of the program should be kept. This listing should be produced by the language compiler.
Test Data	A copy of the data used to test the original and revised versions of the program. Notes showing expected results of the test are also included in this portion of a program's documentation.

Strong program documentation permits other programmers to learn the system quickly. This can be critical in an environment with a high rate of programming staff turnover or in which a program inevitably needs modification.

Furthermore, the body of every program should contain comments explaining the purpose of each module (Figure 14.6). Comments should begin every program's procedural section, explaining the function of the module, any subtle programming practices, and complex logic. Well-constructed comments assist maintenance programmers by reducing the need to read every program line-by-line when making enhancements. Although extreme by some standards, one software house in New England actually tries to have one comment line for every three computer instructions.

When working in a CASE or fourth-generation language environment, this documentation is a part of the software—almost a by-product. The advantage of CASE and 4GL is that a change made to any portion of the system automatically updates the normal documentation the software produces with no extra effort by the analyst or programmer.

Operations

Operations documentation tells the computer center staff how to run the programs. Without such directions, the staff can only guess about such requirements as disk space, back-up, frequency of operation, and arrangement of printed reports. Newer on-line terminal-based systems reduce the amount of operations documentation because they allow each user to control relevant programs.

Computer center personnel need a **run manual** for each system. Such a run manual contains all the useful facts on how to operate the program, handle errors, acquire special forms, access files that the system requires, and handle security provisions. The run manual should include:

1. System function and purpose
2. System flowchart (or similar tool) detailing each program in the system
3. All error conditions and operator responses

Figure 14.6 Comments explaining the module's purpose, data requirements, and results should accompany every program.

```
008000   *
008010   *
008020   * MAIN SELECTOR MODULE
008030   *
008040   * THIS MODULE COMPARES THE INCOMING CODE NUMBER TO DETERMINE
008050   * WHAT SHOULD HAPPEN TO THIS TRANSACTION.  THE CODE NUMBER
008060   * HAS ALREADY BEEN VALIDATED TO BE A 1, 2, 3, OR 4.
008070   *
008080   *
008090   500-SELECTION.
008100      EVALUATE CODE-NUMBER
008110          WHEN "1" PERFORM 510-ADD
008120          WHEN "2" PERFORM 520-CHANGE-TELEPHONE-NUMBER
008130          WHEN "3" PERFORM 530-CHANGE-CREDIT-LIMIT
008140          WHEN OTHER PERFORM 540-CHANGE-STATUS
008150      END-EVALUATE.
```

4. Program run information:

a. Special forms requirements for the printer

b. Names of databases or files required by the program

c. Hardware assignments, including which disks or tapes to use and where these tapes or disks should reside

d. Disposition instructions: who should receive the printed reports, where tapes should go after running the program, and bursting requirements of printed reports

5. Security: who can use the system when, including log-on procedures and passwords

For easy reference, the run manual should fit in a three-ring binder that's kept next to the computer operator's console.

Operator instructions range from lengthy ones for a complex program to short ones for a simpler program. Note that the payroll W-2 operator instructions in Figure 14.7 lead the operator step by step through the running of the program, including such major items as terminating the program and such minor ones as what to do if the printer runs out of paper.

USING CASE TO GENERATE DOCUMENTATION

Our look at documentation has revealed a variety of written materials that are constructed for many different users. If you think back to the beginning of our look at CASE tools and make a list of all the material that the CASE user enters, you will find a complete description of the system.

The CASE tool now needs to bring together all this material into system documentation. The documentation reports could be any or all of the following:

- Narrative descriptions
- Data flow diagrams at all levels
- Data model diagrams
- State transition diagrams
- Screen designs
- Report designs
- Database schemas
- Module specifications, including business rules
- Test data
- A variety of charts, including structure charts, entity relationship diagrams, decision tables, and so on
- Data dictionary of entities, including data type, length, edit rules, default values, and so on
- Time and personnel assignments

Figure 14.7 Operator instructions tell the computer operator how to start up and run a specific program.

FLEET FEET
INCORPORATED

OPERATOR INSTRUCTIONS

SYSTEM:	PR (Payroll)	Page 1 of 1
DATE:	Dec. 93	
APPLICATION:	PRINT YEAR-END W-2 FORMS	
PAPER REQUIREMENTS:	W-2 Forms	
COMMENTS:	This program should be run at the end of the year. It provides all the necessary income and withholding information on a standard W-2 form.	

RUN INSTRUCTIONS

1. Log onto the system with the master payroll user code and select option #10 from the Payroll menu. All data files will be present under this user code.

2. Mount the forms in the printer. The form must be mounted so that printing begins with the first form on the page. (A subtotal will be printed every 42nd form. The IRS requires that this subtotal be at the bottom of the page. The program will comply with this requirement if the forms are mounted correctly.)

3. Print as many alignment forms as necessary for the forms to be properly aligned. One full page will print for each alignment; this is three W-2 forms.

4. Enter the starting and ending employee numbers.

5. You may exit this program by pressing the END key while positioned for entry of the starting employee number. The program then returns to the Payroll menu.

Data-Entry Specifications

Item #	Description	Required	Format
1	Starting Employee Number	Yes	9(6)
2	Ending Employee Number	No	9(6)

6. If the printer runs out of forms while printing, put some plain paper on the printer and let it finish the run. Then go back and find the last subtotal form that was successfully printed and reprint all W-2 forms from that point on. (The W-2 forms are printed in employee number order.)

7. W-2 forms will not be printed for nonemployees.

8. When the W-2s are finished, they should be burst and delivered to the controller's office.

Producing the reports is often as simple as pointing and clicking with a mouse. Starting from a table of contents where icons represent fragments of the system, you assemble the documentation into sections. Since the CASE data dictionary is always up-to-date, you are assured of getting the most recent materials.

Analyst Paul Brewer dreaded writing system documentation. When he chose a career in computer science, he had not realized that he'd have to write reference manuals. If he had, he would have tried to stay awake in his only college English composition class. Learning about C was so much simpler and more elegant. But Paul had no choice today. The system he had designed would be operational soon, and it needed a good user's manual. Paul forced himself to call up the word processor on his personal computer.

Four days later, he put the finishing touches on the data-entry user's manual. It was a tedious and frustrating task, but he felt pretty good about the final product. At least his word processor had made revisions a snap. Finally, he hit the print button.

Later that day, data-entry supervisor Jackie Kortright thanked Paul profusely for the manual and began eagerly devouring it, but it didn't take her long to heave a deep sigh. Having majored in English, Jackie was appalled by Paul's writing style. Unnecessarily complex words, passive constructions, vague technical concepts without adequate concrete examples, and mystifying computer jargon made it almost incomprehensible.

She began circling offensive words, sentences, and even whole paragraphs:

1. "Ascertain whether there are sufficient paper resources for the report to be printed."
2. "Does the payroll package have a check-printing capacity?"
3. "Calculation of the expected profit and payback is accomplished in the TRND module of the program."
4. "Care should be exercised by operating personnel to adequately ensure that they not press CLEAR DISPLAY when INSERT CHARACTER is meant to be pressed."
5. "CONTROL and BK should be pressed in the event that operating personnel desire to initiate backspace without erasing."
6. "It is a requirement that user department managers functionally authorize the F-Spec to show their approval."
7. "The project engineer possesses the requisite data verification responsibility."
8. "Analysts must review all changes. If they are substantial, he will need confirmation from his supervisor."

Jackie rewrote each, using an active writing style and eliminating unnecessarily long words and jargon:

1. "Check the paper supply before printing a report."

Once the documentation is assembled, you can have the CASE software system print or download it to a popular word processing package (such as WordPerfect) or desktop publishing system (such as PageMaker). Once downloaded, you can format it to follow company standards. CASE documentation is never the whole answer. Organizations will want to add special formats, rules, or other inclusions not built into the CASE system.

Often, the CASE documentation software produces the documentation according to some script or standard, such as DOD-STD-2167A. Some organizations have developed their own private scripts and the CASE software can also produce it in this format.

The best part of CASE documentation is updating. When changes (additions, deletions, or enhancements) are made to the under-pinnings of the system, the CASE software can regenerate the documentation, keeping it current. This fact alone is often the reason why people buy CASE systems.

2. "Can this payroll package print checks?"
3. "The TRND module calculates expected profit and payback."
4. "When you press INSERT CHARACTER, be careful not to press CLEAR DISPLAY by mistake."
5. "To backspace without erasing, press CONTROL and BK."
6. "F-Spec approval requires the user department manager's initials."
7. "The project engineer verifies the data."
8. "Analysts must review all changes. Substantial changes require review by supervisors."

At first, Jackie debated whether or not to show Paul what she'd done. Finally, she decided that he deserved her honest reaction to the manual. Surprisingly, Paul loved her editing. "I wish I could have done that myself," he confessed.

"I can teach you some simple tricks," Jackie offered.

"Great! And I can translate the jargon for you."

Concerned that her editing might hamper Paul's schedule, Jackie volunteered to work overtime to help him, provided that he taught her how to use his personal computer's word processor. They struck a deal: She'd edit the whole manual and devise a list of general writing rules if Paul would let her use the word processor to write a magazine article that she was doing. After editing the report, Jackie made the following suggestions:

1. Avoid unnecessarily long words and replace most formal words with informal equivalents. For example, change "The quadruped effected its descent from the arboreal habitat in order to ingest sustenance" to "The squirrel climbed down from the tree to eat."
2. Try to write actively rather than passively. Change "The ball was hit by Bill" to "Bill hit the ball."
3. Illustrate important principles and concepts with concrete analogies, examples, anecdotes, and cases. Most people have a hard time picturing abstractions in their heads. For example, compare sequential to random access by writing, "Readers access novels sequentially, while they access dictionaries randomly."
4. Eliminate as much jargon as possible. Choose common words and analogies to daily experiences that can closely approximate computer concepts. Don't use abbreviations such as CPU, ALU, or DFD.
5. Avoid sexist language and terms. Instead of writing "The programmer should use his judgment," write "The programmer should use his or her judgment" or "Programmers should use their judgments." Similarly, avoid racist, ageist, or stereotypical language of any kind.

Paul, eager to continue improving his writing skills, asked Jackie to recommend a book about clear writing. The next day, she brought him a gift: William Strunk and E.B. White's classic little writing book, *The Elements of Style*. After only four weeks, Paul actually found himself looking forward to writing.

QUALITY ASSURANCE

Quality assurance (QA) tries to remove errors from the system before they are discovered by the user. Error-prone systems are unacceptable and all efforts to remove errors are a must.

Certification

Another kind of test, **certification**, involves an independent audit of the system (instead of by the user, analyst, or programmer). Proving of the system by an outsider helps build a bridge between the analyst or programmer and the ultimate user of the system. The independent tester is called the certifier and it is this person's job to once again verify the accuracy of the system.

When the analyst submits a system for certification, he or she must feel confident that it is complete and ready to run. The submission should include:

- A listing of all programs
- Complete documentation
- All preliminary test results

If, after reviewing the programs, documentation, and earlier tests, the certifier does not clearly understand the system, the analyst and/or programmer will supply explanations or revisions. Open, honest communication becomes crucial here because misunderstandings can cost time and money, plus valuable confidence and rapport.

Runthrough

The certification process progresses through six stages that are similar to those that the analyst followed while testing the system (Figure 14.8). Stage 1, runthrough, familiarizes the certifier with the system and helps determine its degree of "user friendliness." During this stage, the certifier:

- Identifies all files involved
- Identifies all indexes and key fields
- Ensures that screen displays have a logical layout, contain no spelling errors, and include menu options that are clear and consistent
- Checks all subscreens or menus
- Examines error messages

Figure 14.8 Software quality assurance's certification phase has six stages. The system must pass all six stages before it is released to users.

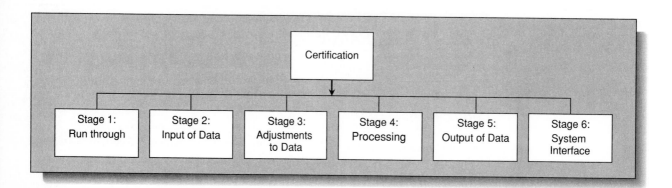

	Certification	

| Stage 1: Run through | Stage 2: Input of Data | Stage 3: Adjustments to Data | Stage 4: Processing | Stage 5: Output of Data | Stage 6: System Interface |

Input of Data

During the input of data (Stage 2), the certifier checks that data get properly validated and written to the correct database. During this stage, certification examines both the field and record levels. At the field level, the certifier:

- Determines whether options displayed at each field work correctly
- Checks that error messages match field errors
- Identifies high and low ranges for each field

  Numeric fields: lowest and highest values

                          alpha characters not allowed

                          values outside range

                          allow zero entry

  Alphanumeric: lowest and highest values

                          numeric allowed

                          values outside ranges

                          allow space entry
- Looks to see if required fields receive data
- Checks the cursor position after error messages
- Tests for field length and truncation
- Makes sure error messages are cleared after field error
- Sees that function keys work correctly
- Ensures that date boundaries are met (month numbers 1–12, day numbers 1–28, 29, 30, or 31 as appropriate)
- Tests numeric edits (dollar signs and commas)
- Determines whether default values work as intended

At the record level, the certifier verifies that:

- Files are updated properly
- End-of-file conditions function as intended
- Record insertions and additions are correct
- Insertions to indexes are done
- Linkages are made to other files
- File sizes and allocations fall within reasonable boundaries

Adjustments to Data

Stage 3, adjustments to data, concerns examination of the records stored in the database. If a record requires adjustment, it must be recalled, fixed, and rewritten to the database. During this stage, the certifier:

- Makes sure that the system writes all fields back to the correct location in the database
- Looks to see that illogical adjustments are not allowed
- Ensures that all other files affected by the adjusted record get adjusted properly
- Tests that a logically deleted record is not redisplayed after deletion and that its data are properly erased

Program Specifications

Stage 4 of certification, processing, takes the input data according to the program specifications. At this stage, the certifier scrutinizes:

- First and last record processing
- Volume processing
- Arithmetic calculations
- The opening and closing of files
- Nonprocessing of deleted records
- Restart capability
- Error messages
- Sort and merge functions
- Record reads and writes

Output of Data

During Stage 5, output of data, the certifier checks the output of the program. Output can take the form of files, lists, or tapes, each with its own specialized checklist:

- Files

  All files are updated

  Pointers and linkages are correct

  File allocations are reasonable

  Files are closed properly
- Lists

  Columns align

  Data in columns match headings

Reports contain titles

Page numbers, dates, and other heading materials are proper

Totals are correct

- Tapes

Data are written to tape

Data can be read back from tape

Format of data is proper

Multireel files work as intended

Multifile reels function correctly

Tape drive is released on time

System Interface

Stage 6, system interface, ensures that the program works smoothly with all other programs as a system. In this stage, the certifier:

- Checks that the software runs harmoniously with other programs that may use records created or modified by the new program
- Determines which other programs in the system operate with the new program

If a system passes Stage 6, it receives its certification and can go into actual operation.

At any of the six stages, a certifier can return a program to the analyst or programmer for repair. The certifier must take care not only to list errors, but also to suggest improvements that might make the program more effective. After correction, the revised program goes through the entire six-stage cycle again.

Certification implements the structured methodology and its aim of producing high-quality, error-free software that is delivered to the user on time. Skipping it could result in erosion of user confidence, as well as tarnishing the reputations of both the analyst and the organization.

A West Coast software firm certifies all software they write before releasing it to their users. Since adopting certification testing, reported errors have dropped almost 90 percent. The firm also pays certifiers the same rate as programmers, so there is no stigma associated with working as a certifier versus working as a programmer. The certification team also specifies when the software is ready for release to customers. In most organizations, this would be a programmer's responsibility; instead, this firm places it in the hands of the certification group. They know that good software quality assurance causes users to become confident of the software and trust it upon receipt.

Instead of having their own in-house certification staff, some organizations rely on outside specialists. Many of the large accounting, auditing, and consulting firms (such as Peat Marwick) sell software quality assurance certification software and services. They claim to offer an independent and unbiased alternative to in-house quality assurance.

SYSTEMS MAINTENANCE

All systems experience some type of change: users find errors, develop new needs, or want enhancements. Most systems spend the bulk, perhaps 80 to 90 percent, of their life in the maintenance mode.

New computer products come on the market and organizations become more and more sophisticated. Though structured methodology may lead to more error-free systems, all good systems evolve as they are used. Maintenance allows changes and modifications to the system in order to repair errors, enhance current functions, and add new capabilities.

Maintenance facilitates system evolution. For example, Fleet Feet decided two months after installation of their new accounts payable system that they needed a new report showing the recent purchases by vendor. To produce this report, Fleet Feet needed to modify its system. Fortunately, the system was already collecting the purchases from vendors, so the analyst only needed to order a single program to print the new information.

Fourth-generation language and CASE-developed software typically exhibit lower maintenance rates than software developed through other methodologies. This fact, coupled with the more rapid development of software, are often the reasons that organizations opt for 4GLs over 3GLs such as COBOL, C, PL/I, or Ada.

To permit and encourage changes, the analyst can develop a form such as the one found in Figure 14.9. Such a form provides space for the description of an error or desired change and its anticipated effect on the organization. The bottom portion of the form indicates what action, if any, the computer services department takes, including all personnel assigned to the task.

Such a form can help inspire change because it encourages users to think about improvements that will really benefit the organization's productivity or profitability. Forms also force users to frame desired changes in specific language, listing benefits and urgency so that they can be prioritized more easily.

A completed form would go to the computer services department for its action. Most departments employ some type of system for managing changes, and the best ones operate like a mini-systems study, using analysis, design, and development techniques. Analysis targets costs, benefits, and impacts on the existing system; design focuses on how to make the change; and development concentrates on implementing it. Some simple changes may take only a few days; complex ones may take months.

Figure 14.9 Change or error request form for Fleet Feet.

CHANGE OR ERROR REQUEST FORM

FLEET FEET
I N C O R P O R A T E D

Attention:_____

Request Submitted By: _____

Date Submitted: _____

Type of Request: Error, Change, Enhancement
                                      (Circle proper request type)

Description of Request: _____
_____
_____
_____

List Benefits or Justification (savings of time, expense, personnel, and so on):
_____
_____
_____
_____

Specify Urgency:_____
Sample/Example Attached:_____

For Use By MIS Only

Report Received On: _____

Report Received By:_____

Report Assigned To: _____

Cause of Request: _____
_____

Solution to Request: _____
_____

Date Completed:_____

SUMMARY

During the two previous phases of the systems process, the analyst functioned like an architect—analyzing and designing the system. Now, the analyst works like a contractor—coordinating and managing the actual building of the system.

Program, system, and acceptance testing commences when programs are complete. While some testing can occur during coding, complete system testing cannot start until the completion of all programs. The analyst must test the whole package of programs by linking them together. Testing, which aims to detect errors, can involve artificial or real data input by users and computer services staff.

Certification provides a final test that is conducted by an outsider. This person can more readily look at the system the way that a new user would. The certifier checks programs for accuracy, completeness, and ease of use.

Users, managers, and computer operations staff must learn the new system. Each group demands tailored training: managers want to know about the information that the system will provide, users must master the system, and operators must keep the system running efficiently. All three groups need different types of documentation. A complete set of all materials makes up the documentation, which resides in the computer services department for reference.

KEY TERMS

Program integration test
System test
Acceptance test
Training
Documentation
Management documentation

User documentation
Program documentation
Operations documentation
Run manual
Certification

QUESTIONS FOR REVIEW AND DISCUSSION

Review Questions

1. What are the three types of tests that a new system must pass?
2. Which individuals should be trained to use the new system?
3. Why do we have system documentation?
4. What do we mean by system maintenance?
5. What are the components of a training manual?
6. What are the six stages of certification?

Discussion Questions

1. List three sources of data that are usually available for testing.
2. Who needs training when the new system is developed?
3. List sources of training.
4. List two advantages of documentation.
5. What are the two types of maintenance activities that can be expected of every system?
6. List three sources of training materials.
7. List three elements of a run manual.
8. List the elements found in a user's manual.

Application Questions

1. Match users with the appropriate level of training:

   a. Computer operator   1. Low level

   b. President of the company   2. Medium level

   c. Data-entry operator   3. High level

   d. Manager of data entry

   e. Manager of computing services

2. Write a data dictionary definition of systems documentation.
3. Write a data dictionary definition of a run manual.

Research Questions

1. What might happen when a system is not accepted?
2. List three examples of what might cause a system to be rejected.
3. What percentage of time has been spent maintaining nonstructured systems?
4. Who performs certification?

Testing, Training, and Documentation of View Video's Video Cassette Tape Rental System

Analyst Frank Pisciotta divides View Video's test plan for its new videotape cassette tape rental into two phases. Phase 1 of the test plan tests each individual module and program with sample data from Mark and Cindy Stensaas. Phase 2 links all programs together for system and acceptance testing.

Since Frank will enter this data himself to see how the computerized system processes it, he asks Mark and Cindy for a hundred or so transactions of all types: adding new customers, changing customers, deleting customers, check-outs, and returns. Mark agrees to provide the data in a few days.

When the data arrive, Frank activates phase 1 of the test plan. Following his list of programs (Figure 12.24, page 416), he starts the main menu program (VCTR510) and calls up the customer file maintenance routine (VCTR230) on his personal computer. He enters data about himself, starting with his customer identification (telephone) number: 4155556735. Then he notes what the system does to it. After entering the data, Frank uses the SQL*Plus portion of the Oracle SQL software. This utility package allows a novice user to query the database while in an interactive mode. The user can then write SELECT, UPDATE, DELETE, and any other SQL statement and the system will perform the desired task. Frank writes the SQL statement:

```
SELECT *
    from customers
    where customer_id = 5571000
```

and looks at the results. Before calling up SQL*Plus, Frank notes the results so that he can compare the two for any discrepancies. Sure enough, all the data Frank entered about himself are there and the other fields are set to their initial or default values. Frank continues entering data, changing and deleting customers, and tests each entry with SQL*Plus.

When he locates a few faulty portions of the programs, Frank notes them so that Data Management can get them corrected. Frank deletes all the data from the database and continues testing for many hours. The errors he finds fall into two basic categories: those derived from his own design mistakes and those that a programmer introduced.

Frank repeats his process for all the programs in the system until he feels that he has uncovered all possible errors. When all the errors have been removed, phase 1 ends.

Phase 2 of Frank's testing plan requires all the programs to work together. Before performing this test, Frank again calculates results manually so that he can compare them with the automated version. As in phase 1, he purges the data from the database and activates the main menu program. He then enters customers, movies (VCTR240), check-outs (VCTR210), and returns (VCTR220). He requests a variety of reports (VCTR110 through VCTR170). To Frank's delight, the hand calculations match the computer-generated ones, so he can stop testing. His new video cassette tape rental system is ready for user training.

Training poses few problems. All of the people that Mark and Cindy hired have used personal computers before. All Frank has to do is teach them how to run the new VCTR system.

To prepare for training, Frank decides to review user's and operations manuals for each module of the system (Figures 14.10 and 14.11). Since the new system is on-line, it needs

Figure 14.10 Video Cassette Tape Rental operator instructions tell the user or operator how to start up and run the Print Alphabetic Customer List report.

|||||||| *View* Video ||||||||

OPERATOR INSTRUCTIONS

SYSTEM: VCTR

DATE: Dec. 93

PROGRAM: VCTR110 — Print Alphabetic Customer List

PAPER REQUIREMENTS: Regular 8 1/2" by 11"

COMMENTS: The Customer List is one of the primary reports from the VCTR system. The data are proprietary and are not for distribution outside of the organization.

RUN INSTRUCTIONS

1. This application may be run on request and will generate the report.

2. Customer list requires read-only access of the VCTR database. It will run regardless of other users running VCTR applications. If other users are accessing the VCTR database, there is no reason to notify them to log off or delay any other activities.

3. Mount the paper when requested. Type "DONE" when the paper is in position.

4. If you wish to print a sample report with X's for alignment purposes, answer "Y" to the question "PRINT ALIGNMENT?" You may continue to request an alignment form until you are satisfied that alignment is correct. When you wish to proceed with the actual printing, answer "N" to the alignment form question.

5. Enter the first and last customer names, as requested. Press the "Esc" key to abort the report run.

6. The report is now printed.

7. When printing is complete, tear off the report and mount the normal paper in the printer.

8. Burst the report and deliver it to the office.

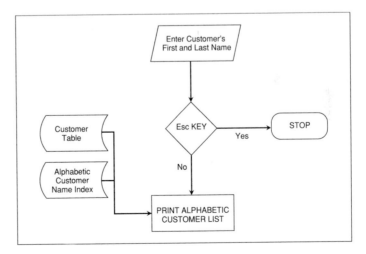

Figure 14.11 User documentation for the Video Cassette Tape Rental Alphabetic Customer List report guides the user through the system.

|||||||| *View* Video ||||||||

USER INSTRUCTIONS

SYSTEM: VCTR

DATE: Dec. 93

PROGRAM: VCTR110 — Print Alphabetic Customer List

PAPER REQUIREMENTS: Regular 8 1/2" by 11"

COMMENTS: Use the option off the main menu to print the list of customers alphabetically by customer last name. The data are proprietary and are not for distribution outside of the organization.

RUN INSTRUCTIONS:

1. To run the application, select this option from the reports menu off of the main menu.

2. The screen will prompt for starting and ending values for the report.

3. Use the up or down arrow keys, Home, or End keys to select the report desired. When selected, press the Enter or Return key. A second window pops up, allowing the user to select a specific group of items to include in the report. If all the items are desired, press the F4 key.

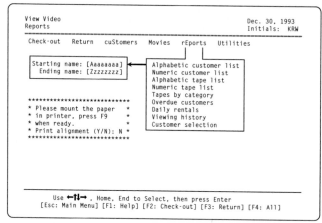

4. Enter the data as follows:

Starting name	Up to 8 letters. Example: Aaaaaaaa
Ending name	Up to 8 letters. Example: Zzzzzzzz

Function keys

Esc	Returns to main menu.
F1	Provides context-sensitive help on the field where cursor is positioned.
F2	Abandons transaction and calls up the check-out a tape function.
F3	Abandons transaction and calls up the return a tape function.
F4	Prints all data in the selected category.

(continued)

Figure 14.11 *(continued)*

5. Mount paper in the printer after the message: "Please mount the paper in the printer." Load the blank paper into the printer and when it is positioned, press the F9 key.

6. After you press F9, you are prompted: "Print alignment?" Answer Y to print a dummy form filled out with 9s and X's to verify the alignment of the paper in the printer. Answer N to bypass printing of a dummy form. When the dummy form is printed (if you request one), the alignment question is asked again. Adjust the paper in the printer. You may align as many times as is necessary to position the paper properly.

7. The report is now printed for the range of customers requested.

8. After the report is completed, the screen will display: "Report finished." Remove the paper and put regular paper back in the printer.

9. Burst the report and deliver it to the office.

only brief run manuals. However, the user's manuals—which must provide a lot of detail about data entry, updates, and report printing—are longer.

Frank adds these manuals to all the other materials he developed during earlier phases of the systems process, thus creating complete systems documentation. He collects all documents in a three-ring binder, with separators dividing one from another.

Copies of the manuals go to Mark and Cindy Stensaas ahead of time so that they can prepare for the training seminar for new users. Early one Thursday, Frank arrives at View Video's office with freshly roasted coffee and bagels for everyone. Frank has learned from past training experiences that such thoughtfulness helps break the ice with any staff. Instead of sitting at the terminal and entering the data himself, he asks each user to log onto the personal computer network, call up the VCTR510, and run it. Each person enters sample data about themselves so that they can compare results. Looking over their shoulders, Frank offers advice whenever necessary.

Because everyone catches on quickly, the training session takes only two hours. "Nice system," observes Cindy, and everyone agrees.

At the end of the training session, Frank asks Mark and Cindy out for lunch for Monday. During their meal, Frank summarizes the project, showing them some of the reports that they will see.

Frank feels great because all the time he devoted to planning and training has really paid off. He delivered the Stensaases' system on time and is particularly pleased that the staff seemed to find it easy to learn and use.

Buoyed by his successful training program for the Stensaases and their staff, Frank Pisciotta decides to continue testing the system for a few more hours. He uncovers a few additional but minor errors, which Data Management corrects quickly.

Working With View Video

Frank uses the Oracle SQL*Plus program to confirm that the data he entered were correct. This program lets the user enter any SQL statement; it then executes the statement, showing the results that the statement produces. This method of testing shows Frank what the computer *says* was entered versus what he *thought* he entered.

Programmers and designers test software to make sure it works, whereas software quality assurance people try to break the software, to make it fail. To a software tester, this difference is very important and crucial.

Case Study Exercises

You can perform a variety of tests on software to make sure that it works correctly. Each is aimed at proving or disproving certain aspects of the system and its operation. Some aspects are still not proven. How would you propose to test for the following?

1. Making sure the system works under heavy workload situations.
2. Dropping a customer from the system.
3. Examining the link between Frank's system and the accounting system, BusinessWorks PC.
4. If we delete a movie from the system, we need to remove all the movie star references to that movie.
5. If we delete a movie star from the system, we need to remove all the references of that movie star to specific movies.

Maintaining and Managing the Systems Process

GOALS

After reading this chapter, you should be able to do the following:

- State what takes place during system maintenance
- Outline management's responsibilities within the systems process
- List the components of a final report
- Identify the participants in the system audit
- Describe how a Gantt chart helps manage the systems process
- Name two professional societies, two books, and two training groups that can extend the analyst's education
- List three concepts of the object-oriented philosophy

PREVIEW

During the systems process, the analyst orchestrates a series of events that lead to placing the system into the user's hands (Figure 15.1). During this final phase, the analyst has essentially turned the system over to the user, yet several systems tasks still need performing.

Figure 15.1 Maintenance is the last phase in the systems process. During this phase, users operate the system on a regular basis.

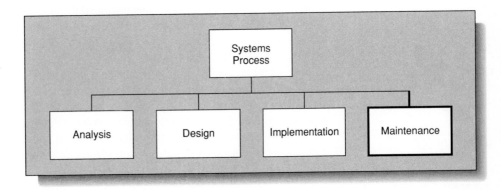

Systems spend the bulk of their lives in maintenance modes. In fact, some organizations report that up to 75 percent of their software development or programming efforts are expended on activities related to maintenance. Minimizing these expenses is essential and high-quality software can help achieve this goal.

Managing the systems process also requires talent and insight concerning people and organizational structures. Part of this chapter will focus on a number of tools that can help management control the systems process stage, from initial user request to installation and operation.

SYSTEMS MAINTENANCE

The system is now ready to enter the maintenance phase. Most systems spend their life here, so it is a crucial aspect of the systems process. This phase represents a culmination of all the systems analyst's efforts up to this point.

Picture a building, perhaps a three-bedroom home, that took 6 months to build, but lasted for 75 years before it was torn down. For 74.5 years, the building was in maintenance mode, occasionally needing repairs, additions, replacement of old appliances, and so on. Maintenance is expensive and many home owners will spend at least as much on maintaining their home as they did in acquiring it. The same is true for software systems; maintenance represents a large expense to an organization. It is important to remember that software does not wear out like a pair of jogging shoes. It keeps doing in the fifth year of life exactly what it did on the first day. What does happen is the organization or business changes, requiring the software to change accordingly.

Operations

Management should try to put the system into operation at the most opportune time. For example, it makes sense to switch to a new payroll system at the beginning of a new year or quarter after all prior federal and state government reports are finished and filed. However, exceptions can exist. For example, the business manager of a power and water company decided to adopt a new on-line payroll system for December's pay period because it could deliver year-end reports much more quickly than the old manual system. While this approach was risky—considering all the problems a new system might experience—W-2 forms went to employees by January 4, a new record for the company. Later, the company uncovered only two errors for its 265 employees! Needless to say, the new system's early success won it a lot of confidence and support.

Organizations usually adopt accounts receivable systems at the end of a month before a new billing cycle begins. A floral shop, for example, decided to switch service bureaus from batch to on-line processing before its May billing cycle. While the batch service bureau had required the florist to close books on the 20th day of the month and then sent customer bills on the 7th working day of the next month, the new on-line system allowed the florist to post charges up to the last working day of the month, with bills getting printed only two hours after the last data entry. During April, the store owner ran the two systems in parallel, going fully to the new one only after verifying results. Parallel operation recorded a rounding error of a penny, which a programmer repaired easily.

Documentation

After notifying all parties of a system's acceptance and operation, the analyst proceeds with one final task: assembling all the materials produced during each step of analysis, design, and implementation (Figure 15.2). If the analysis, design and implementation were thorough, this concluding portion of system documentation is a breeze. Since all necessary materials should exist, the analyst need only arrange them in order, write a table of contents, and bind them together.

In the days of unstructured analysis and design, analysts dreaded this job because the materials were scattered and incomplete. Sometimes the job never got done, especially when the organization assigned the analyst to more "productive" tasks before winding up the current one. However, structured methodology streamlines paperwork by having it completed at appropriate points throughout the systems process, rather than trying to do it all at the end. The widespread use of word processors and personal computers has also aided system documentation by easing the editing and production stages. CASE software also aids the analyst here since it captures all the elements of the system—from analysis through implementation.

Figure 15.2 All the paperwork generated during the systems process forms the complete systems documentation.

A. Systems Analysis
1. User's request
2. Preliminary report with problem description
3. Feasibility study, including budget, schedule, and system DFD
4. Management action

B. Systems Design
1. Assignment roster
2. Screen and printed report formats
3. Data dictionary
4. File layouts or database schema
5. Screen or input layouts
6. Program definitions
7. Program specifications, including narratives and pseudocode, VTOCs, or Warnier-Orr diagrams
8. Management action memorandum

C. Systems Implentation
1. Implementation plan
2. Computer programs
3. Test plan and test data
4. Conversion plan
5. Manuals: management, user, and operations
6. System acceptance notice

Nevertheless, professionals continue to debate the value of extensive documentation. Opponents cite the redundancy of retaining such items as printer spacing chart report layouts once the report begins appearing in its actual form. They contend that documentation should reflect reality, not intentions. However, proponents of careful documentation argue that it provides a historical record that subsequent users and analysts can find quite helpful, especially when replacing the new system with an even newer one. Regardless of the analyst's preferences, system documentation deserves careful thought and tailoring to the needs of each particular system as much as possible.

Audit or Review

After operating the system for awhile, the analyst conducts a final review of the entire systems process, tracking it from preliminary analysis all the way through development. We call this last check of the system to make sure that it meets goals and objectives a **system audit** or review. During the audit, the analyst scrutinizes four aspects of the new system:

1. *Objectives:* Does the new system fulfill the organization's needs?
2. *Costs:* What did the new system cost?
3. *Time:* Was the system delivered on time?
4. *Results:* Did the system meet the original or revised specifications?

To research these four questions, the analyst weighs results of the completed programs, the usefulness of training and operations manuals, the expertise of the computer services staff, users' reactions, and management's observations.

If all objectives, costs, schedules, and results fall within acceptable limits, the analyst and the manager responsible for the system issue an **acceptance notice** to all concerned parties (Figure 15.3). This announcement officially tells all parties involved that the system is finished and operational, and that the analyst will start working on another task. The acceptance memo joins all the other materials to conclude system documentation.

The systems process does not always run smoothly. In some situations, schedules fall badly behind or actual costs greatly exceed estimates, both of which demand a thorough explanation in the system audit. For example, a large Midwestern utility company's energy monitoring system cost twice as much as anticipated. In the eventual system audit, the analyst blamed the overrun on new state regulations that had forced a complete redesign to the system. Fortunately, the analyst had made management aware of the new regulations and everyone was ready for these additional costs.

Rarely does an organization completely reject a system at this late stage because reviews and walkthroughs at earlier stages should have revealed any major deficiencies. If a rejection does occur, the reasons may vary from a change in organizational goals to new ownership. Recently, a software development company did reject a completed international product when the major shareholder sold all her stock to a group of venture capitalists. These new owners decided to focus on domestic American business and withdraw from all business opportunities in Europe, obviating the need for any international products.

Figure 15.3 An acceptance memorandum notifies all involved that the organization has formally adopted the system.

Production Concepts

MEMORANDUM

TO: All Department Managers

FROM: Walt Carson, Sales Manager, Production Concepts

SUBJECT: Acceptance of Sales Analysis System   *WC*

DATE: Nov. 18, 1993

For the past 8 months, we have been developing a new sales analysis system. Yesterday, the system was accepted as complete and is now in full operation. During the last three weeks, we have been using the new system and have found all the reports to meet or exceed our original needs.

The sales analysis by territory reveals some very interesting and valuable data that until now we could not generate: Sales of our new digital/optical record players is very strong in medium-sized metropolitan areas, but weaker than expected in very small and very large cities. We are studying this phenomenon to find out why it is occurring.

Our thanks to all those who worked on this very helpful system. If anyone would like to learn more about the sales analysis system and the data it can provide, please contact me.

Figure 15.3 An acceptance memorandum notifies all involved that the organization has formally adopted the system.

Enhancements

No matter when an organization begins relying on a new system, it can expect user questions, problems, and complaints. They will start immediately and then decrease as users learn the system and analysts learn how to solve the problems (Figure 15.4). As time passes, the need for changes should drop dramatically until the system approaches the end of its life, when it experiences a resurgence of change requests because it can no longer adequately meet users' needs.

Organizations take many different stands on resolving these problems. If a problem is so severe that it stops an organization from operating, then an immediate repair is mandatory. Another stand is to issue "releases" that repair problems and enhance the product by adding new features. Many personal computer software vendors adopt this stand, issuing new versions of their product on a periodic basis, as well as charging a fee for the "upgrade." WordPerfect, Word, Excel, dBASE, Lotus 1-2-3, Windows, and others fall into this release philosophy. CableData, an international software company providing billing systems to the cable television industry, also uses releases to upgrade their software, trying to place a new release every nine months.

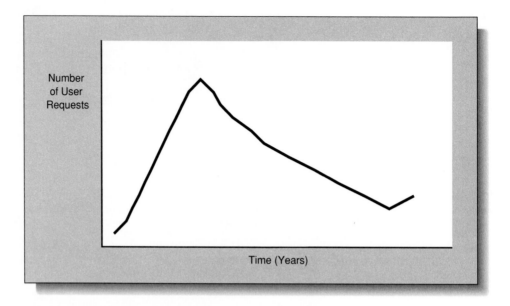

MANAGEMENT ISSUES

Business management activities try to control personnel, funds, inventory, relations with customers or vendors, costs, market share, patents, trade secrets, and a host of other factors that influence the achievement of an organization's goals. Although management may often feel intimidated by the technical aspects of computers, it must nevertheless accept responsibility for its success. Management consulting firms advocate that management solve computer-related problems with the same effective skills and techniques it brings to bear on more traditional areas. A good manager does not have to become a computer scientist in order to successfully manage computer scientists.

Management control is essential throughout the system process. Among other things, control involves scheduling personnel and budgeting funds to achieve maximum results at minimum costs.

Six Keane Rules

Keane, Inc., a New England consulting firm, discovered during its observation of over 350 Fortune 500 companies that six rules can aid in successful management of a computer project:

1. Define the Job in Detail—determine exactly what work must be done and what products must be delivered. Explicitly evaluate the environment and address all assumptions.
2. Get the Right People Involved—involve the appropriate users throughout the project, particularly during planning. Involve the appropriate data processing people.

Ensure that each member of the project team participates in defining his or her own goals.

3. Estimate Time and Costs—develop a detailed estimate of each phase of the development process before undertaking that phase. Estimate the components of a job separately to increase accuracy. Do not estimate what you do not know.

4. Use the 80-Hour Rule—break the job down into "tasks" that require no more than 80 hours to complete. Ensure that each task results in a tangible product. The 80-hour rule provides the framework for scheduling, assigning tasks, identifying problems early, confirming time and cost estimates, and for evaluation of project progress and individual performance.

5. Establish a Change Procedure—recognize that change is an inherent part of systems development. Establish a formal procedure for dealing with these changes and ensure that all parties agree to it in advance.

6. Establish Acceptance Criteria—determine in advance what will constitute an acceptable system. Obtain written user acceptances of products throughout the project so that acceptance is a gradual process, rather than a one-time event at the end.

As Keane suspects, since most good managers apply similar rules to other areas, why not apply them to the management of computer software systems and related functions?

Schedule Overruns

One managerial tool for monitoring the schedule is the familiar Gantt chart. When using a Gantt chart to monitor the systems process, a manager or supervising analyst records actual events as they occur. The Gantt chart in Figure 15.5 shows that management expected the implementation plan to consume one week, and it actually did take one week. Program AR010 was scheduled for two weeks, but only took one week; AR020 should have taken two weeks, but took three; AR030 should have taken one, but took two; and AR040 was scheduled for two weeks and finished on time. Such constant awareness of the actual progress of events helps managers make timely decisions about allocating people and money to solve problems.

Figure 15.5 Gantt charts help management monitor schedules. The black diamonds show the anticipated time required for each activity or event, while the gray diamonds depict the actual completion dates when they differ from the original dates.

Activity or Event	1993			1994	
	October	November	December	January	February
	6 13 20 27	3 10 17 24	1 8 15 22 29	5 12 19 26	2 9 16
Build Implementation Plan					
Programming					
AR010					
AR020					
AR030					
AR040					

People Issues

Managers routinely spend much of their time monitoring and solving problems concerning people and their **interpersonal relationships**, or the differences between people. Such relationships strongly affect the systems process, and problems with them may surface for a variety of reasons, such as philosophical differences, territorial pride, and personality conflicts. Regardless of their origin, such people problems can imperil a new system just as much as hardware or software failures can.

In one large retail business, the lead analyst, who brought ten years of experience to the accounting systems design, could not communicate effectively with the business manager. Because the two had wide **philosophical differences** of how the system should function, they spent more time quarreling and playing politics than they did working together to get the system running. Eventually, the president transferred the analyst to a task completely outside the business manager's sphere of influence, but by then the schedule had slipped three full months.

In another instance, the data processing manager of a wholesale beverage firm suffered from **territorial pride**, an interpersonal problem that arises when one person invades another's area. A young sales manager purchased her own personal computer and learned how to use an electronic spreadsheet, then brought her computer to the office, demanding that the MIS manager provide her with data. The MIS manager fancied himself the organization's top expert on computers and software and resented the young upstart for trying to invade his turf. Although he privately endorsed personal computers, the manager now began publicly attacking them.

The problem became so acute the president of the beverage supply firm had to intercede. Rather than disciplining either party, the president sent the MIS manager to a two-day seminar on creative corporate communication, which tried to teach strong-willed managers how to handle ambitious people. The MIS manager, seeing that he might overcome his territorial pride without relinquishing managerial authority, returned to the situation with new interpersonal insights. He adopted a SQL personal computer product and showed the sales manager how she could download the data she wanted from the corporate mainframe computer.

Personality conflicts arise from differences in individuality or disposition. A financial printing firm experienced a personality conflict when an analyst and a user began loudly arguing over such trivial matters as who should fill the coffee pot in the morning. Their power struggle had less to do with the new computer system than it did with their efforts to impress co-workers. Their two radically different personalities constantly conflicted in other areas, too. Management interceded, threatening to fire each if they didn't begin cooperating with the other. The two still found working together so painfully difficult that one eventually quit.

Cost Containment

Since systems frequently cost more than anticipated, management often must wrestle with **cost containment**. Fortunately, formulas exist for forecasting costs, and these vary from crudely tripling the analyst's expectations to determining expected costs in a more scientific manner.

Two tools to help management control costs are the Program Evaluation and Review Technique (PERT) and the Critical Path Method (CPM). CPM shows relative priorities among tasks; PERT extends CPM by linking optimistic, expected, and pessimistic costs and times with each task (Figure 15.6). Optimistic costs and times assume that all tasks will go exactly as planned, expected costs and times indicate what will probably happen, and pessimistic costs and times provide for the worst possible scenario in which everything conceivable goes wrong. Figure 15.6 also lists event predecessors; that is, what event must precede this event (for example, event A must precede event B, and events C and E must precede event F).

Both PERT and CPM are graphic tools with circles and arrows linking them together in the form of a network. Arrows represent tasks and circles (or nodes) show events (Figure 15.7). Every circle is assigned a number, called the **event identification number**. Arrows are drawn with arrowheads to show precedence between the events. Above each arrow is a letter naming the activity and under the arrow is the expected time to complete the task. Thus, our drawing shows that event A takes 5 days before we can start event B, taking 15 days, or event D, taking 7 days. Event ID numbers mark the completion of the task. Thus, ID number 4 shows that events A and D are finished, taking a total of 12 days.

Just as some mathematicians use × for multiplication while others use *, the use of circles and arrows is not standardized. Some PERT/CPM tools draw the diagrams placing tasks at the nodes. The difference is largely cosmetic and does not alter the theory or usage of the tool.

Figure 15.6 *(right)* Cost, precedence, and time factors required by PERT, a tool for controlling costs.

Figure 15.7 *(below)* CPM chart showing events, tasks, and the critical path.

Task Description ID		Optimistic Cost ($)	Time (Days)	Expected Cost ($)	Time (Days)	Pessimistic Cost ($)	Time (Days)	Event ID Number	Event Predecesser
A	Build Impl. Plan	500	4	1000	5	1500	8	1	Start
B	Program AR010	720	13	900	15	1300	20	2	A
C	Program AR020	1000	10	1100	11	1400	18	3	B
D	Program AR030	2200	5	2500	7	3000	10	4	A
E	Program AR040	680	12	900	16	1150	19	5	D
F	Program AR050	200	4	300	5	500	6	6	C, E
TOTAL		5300	48	6700	59	8850	81		

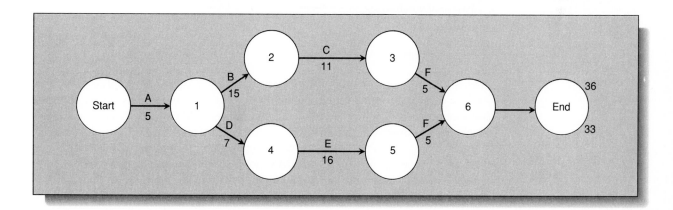

The critical path is the longest pathway through the network. In our example, two pathways exist from start to end:

Path #1: Start - A - B - C - F - End
Path #2: Start - A - D - E - F - End

Using expected times (in days), Path #1 takes 36 days and Path #2 takes 33 days, making Path #1 the critical path.

If management decides that the critical path time is too long, the analyst must shorten the time it takes to perform a task on the critical pathway. The analyst would thus look at tasks A, B, C, or F (not D or E). Shortening can come from hiring more people, spending more money, using another technology, or any other technique that reduces the task's time, but this may drive up the overall cost of the project.

Computing the expected cost and duration of a project uses a standard formula:

$$\text{Expected} = \frac{\text{Optimistic} + 4 \times \text{Expected} + \text{Pessimistic}}{6}$$

Applying this formula for expected cost (refer to Figure 15.6) produces:

$$\text{Expected cost} = \frac{5300 + (4 \times 6700) + 8850}{6} = \frac{5300 + 26800 + 8850}{6}$$

$$= \frac{40950}{6}$$

$$\text{Expected cost} = 6825$$

Knowing the critical path, we use the same formula for expected duration in days:

$$\text{Expected duration} = \frac{48 + (4 \times 59) + 81}{6} = \frac{48 + 236 + 81}{6} = \frac{365}{6}$$

$$\text{Expected duration} = 61$$

With all the permutations and combinations possible in optimistic, expected, and pessimistic costs and time, countless possible pathways exist. Fortunately, many computer software products—such as Harvard's Project Manager, Microsoft's Project, or Claris's MacProject—can assist management in keeping track of all tasks, events, time durations, and costs. Most CPM/PERT software will compute the critical path, costs, and time durations, plus display the network. Some CASE tools also allow the import or export of data. Other formulas consider such factors as the number of modules, number of programs, system complexity, and number of people involved.

Regardless of the cost forecasting approach used, management eventually wants to compare anticipated and actual costs. The closer the two match, the better because accurate forecasting builds confidence not only in an analyst's budgeting ability, but in the resulting system itself.

WORKING IN TEAMS

A buzzword today is "team." All over the world, organizations are breaking their large workforces into small work units or teams. The teams may have names such as "quality circles" or "quality teams," but the concept is the same—working in small groups.

Many software development businesses saw the value of teams years ago, while others are only now recognizing the value of a small group of dedicated people. Highly motivated, they work together to produce a quality product. Team members feel a sense of ownership. They work in a give-and-take environment with a high degree of autonomy, while still striving to attain both their individual and the organization's goals and objectives. Included in the team approach is recognition of both individuals and teams for the contributions they make.

One large West Coast software development group divided its people into three major teams: analysis and programming (AAP), software quality assurance (SQA), and documentation (DOC). Within AAP and SQA, subteams formed that specialized in specific aspects of the software, order entry, money processing, and report areas. The order entry team focused on taking customer orders in a real-time mode. The money team had authority and responsibility for any portion of the software that manipulated money. The report team was responsible for all reports that the system produced for users and management.

All three teams (AAP, SQA, and DOC) had early and constant involvement with the user. They all participated in fact finding, various stages of design and prototyping, and visited users at their places of business to make sure they were still understanding the problem that the software was designed to solve. The use of teams helped the software company deliver their product on time, within budget, and with the features promised.

Like software, teams have their own life cycle. They start with a formative stage in which members get to know each other and recognize individual strengths and weaknesses. The second stage is stormier because conflicts start to arise; the team soon realizes that it needs a conflict resolution methodology. Teams then move into a stage in which they set ground rules and objectives, schedule assignments, delegate decisions, empower one another, and establish regular meetings and communication methods. Now the team is ready to perform the task at hand. Once the task is completed, the team recognizes both individual and group accomplishments. Finally, the team will either disband or refocus on a new task.

Key to the team concept is management support and endorsement. In many organizations, a senior manager becomes the mentor for the team—someone the team can turn to when organizational barriers block the way to success. The mentor "runs interference" for the team so that team members can stay focused on the task at hand.

The West Coast software group mentioned earlier sent their money team to London for four weeks to make sure the software would interface with British currency and banking

After a few years as a programmer, some people tire of the job's repetition: They always seem to encounter the same old problems. When looking at new opportunities, the job of the systems analyst seems to be the next logical step. But after a few years as an analyst, then what? Some remain as analysts, others move to another technical area (such as database administration), and still others opt for management positions. The choice is a hard one, but you may eventually have to face it.

If the choice is management, new skills are a must. The role of a manager is quite different than that of a programmer or an analyst. Fewer technical skills are needed and more organizational, business, budgeting, and personnel skills are mandatory. Yet during all of the programmer/analyst's prior work years, the focus was on honing technical skills. The skill sets are different!

The next position could carry the title of Director of Information Services or Director of Software Development. In this slot, you may have a number of people working for you as analysts, programmers, quality assurance analysts, and technical writers. But what specifically is your job? We could write a job description, but it really doesn't tell the full story; it only lists general duties and responsibilities, which may not have anything to do with your reality. This new job will require you to manage your staff, boss, and users. Here are a few tips that can go a long way to making you a better manager:

Managing Your Staff

1. Treat your staff as you would expect to be treated.
2. Provide ample training opportunities to upgrade your staff's skills.
3. Write performance appraisals that are specific to the individual involved, not general statements. Make sure the appraisals are on time and that compensation adjustments are quick. Double-check that any paperwork is processed by the due date.
4. Support your staff in public meetings; argue with them in private.
5. Send birthday cards and other notes on special occasions such as yearly anniversaries with the company.
6. Acknowledge superior performance by a staff member or team in a public meeting.
7. Give all credit for your department's success to your staff; never take it yourself.
8. When a staff member leaves the team, acknowledge the event and make sure everybody knows about it. This stops all rumors.

procedures. The team lived in a dark, dank hotel in eastern London, working evenings and sleeping during the days. Upon returning, the manager mentor presented team members with antique British medals. The small celebration was a recognition of the money team's success, which helped all teams toward meeting their delivery dates.

THEORIES X, Y, and Z

Although most professional football, basketball, and baseball coaches once played the game themselves, their success as managers depends on their ability to get the job done through others by motivating, organizing, and directing the team's effort. When motivating people to achieve an organization's goals, a good manager must develop a philosophy about why people behave the way they do. Do people work primarily for money? For practice? Love of the subject matter? Or for the simple satisfaction of accomplishing objectives?

9. Run interference for your staff. It is your job to knock down the roadblocks so that they can succeed.

10. Try to hold meetings in your staff members' offices, not your own. It gets you into their workplace and they feel comfortable in their own surroundings.

11. Foster team building with various events that bring everyone together.

12. Hire good people. Place a new hire under the wing of an experienced employee.

13. Make sure the staff is cross-trained. That way, when somebody leaves, is ill, or is on vacation, you are not left scurrying to understand that person's job.

Managing Your Boss

1. Support your boss in public; argue in a private session.

2. Give credit to the boss for your department's success.

3. Keep your boss informed of your staff's tasks and how well they are progressing. Don't make the boss ask for an update.

4. If something is going wrong, tell your boss early. Your boss should not hear of a problem in a public meeting from somebody else.

5. Get your boss to know your staff members and their skills.

6. Invite the boss to your staff meetings as an observer.

Managing Your Users

1. Listen to users; it is their perception that counts the most.

2. Appoint a staff member to serve as the liaison with the user and force all interaction through the liaison.

3. Never promise what you can't deliver, but take it under careful advisement.

4. Let users guide you in the changes they want in software.

5. Publish a list of user requests and their statuses. Periodically update the list.

6. After a meeting with a user, write up the minutes so that you establish control of the results of the meeting and what everyone agreed upon.

7. Don't promise to deliver by a specific date; instead, set delivery by the quarter if possible (for example, 3rd quarter 1994).

8. It is OK to tell a user, "No, we won't do that."

9. When two users disagree, don't get involved. Instead, sit on the sidelines.

10. When you make a mistake, admit it. That ends the discussion.

11. Write the agenda for a users' meeting. That way, you stay in as much control as possible.

In the mid-1950s, Douglas McGregor described the conventional approach to management, which also included some ideas about basic human behavior. He called it "Theory X":

1. The average human being inherently dislikes work and will avoid it if possible.

2. Due to this fact, most people must be persuaded, rewarded, punished, threatened, and controlled to get them to divert their efforts toward the achievement of organizational objectives.

3. People lack ambition, dislike responsibility, prefer direction, and want security above everything else.

A manager who agreed with Theory X would, of course, tend to dictatorially direct and control people toward organizational objectives instead of involving subordinates in the decision-making process.

McGregor argued that such a philosophy was inconsistent with a certain aspect of human nature described by Abraham Maslow's theory of the human needs hierarchy: people can, and often do, work for higher purposes. Since Theory X couldn't account or allow for behavior directed toward the satisfaction of higher personal and social needs, McGregor proposed what he considered a more effective philosophy of human behavior, "Theory Y":

1. People do not inherently dislike work and are not, by nature, passive or resistant to organizational needs.
2. The capacity for assuming responsibility and readiness to direct behavior toward organizational goals is present in everyone.
3. People do not need to be threatened, punished, and controlled in order to get them to be productive.

If a manager accepts Theory Y, he or she should arrange conditions and methods of operation in such a way as to provide opportunities for employees to satisfy their higher personal and social needs, while still pursuing organizational objectives.

In 1981, William Ouchi followed up on McGregor's work with a landmark book, *Theory Z*. In it, he tried to show how American business people can profit from certain Japanese management styles. Unlike either Theory X or Theory Y, Ouchi argued that increased productivity comes from involving workers deeply in the organization's goals by giving them strong incentives to do so. Ouchi pointed out that such a cooperative approach had made Japanese business extremely productive. Unlike American managers, who seemed preoccupied with next year's bottom line, the Japanese took a long-range view of their workers' needs and their organizations' accomplishments. In short, Theory Z recommended:

1. A holistic concern for people, their families, and children, plus their intellectual and moral development.
2. Collective decision-making, with final responsibility still residing with the individual.
3. Trust in individuals to work autonomously (without close supervision).

Of course, no academic theory can solve all the problems that inevitably occur when people work together. Good managers know that subtlety and intuition also play major roles when pinpointing the right people to hire and develop.

As we move through the 1990s, most organizations are moving toward Theory Z types of management styles. Small groups of dedicated individuals have become very successful in developing software. They are given responsibility and authority for the product, working directly with customers. This type of individual freedom was discouraged in the 1950s, but it will make many 1990s companies more creative, efficient, and globally competitive.

WORKING WITH PEOPLE

LEARNING TO LISTEN

Credentials officer Debbie Dean put down her telephone after scheduling an interview with Ed McMillan, analyst for the State Department of Education. She wanted Ed to explain the Department's planned real-time credentialing system, which would replace the manual one that the Department had used to track over 370,500 teachers' credentials. Since Ed had taught tenth-grade mathematics for several years before beginning a new career as a programmer and then an analyst, Debbie had assumed he knew the old credentialing process by heart. It turned out he didn't.

When it came time for the meeting in Ed's office, Debbie hauled along samples of three credential application forms and blank copies of the credentials themselves. Rather than discussing the old system, Ed—eager to display his computer knowledge—launched into a futuristic lecture on new real-time systems. Whenever Debbie tried to interrupt with a question, Ed put her off, saying, "Yeah, I know, I know, but the new CRT 2000...."

"I can't communicate with someone who's talking to himself," snapped Debbie the fifth time Ed interrupted her.

Still fuming, she pick up the materials that she'd *wanted* to discuss, slammed the door to Ed's office, and marched back into her own office. Should she write Ed a nasty memo about the art of listening? Or should she complain to Ed's boss? This new system would involve a lot of interaction with Ed, but she couldn't stand the thought of spending even ten minutes with that pompous jerk. When her phone rang, Debbie was surprised to hear a sheepish Ed McMillan on the other end of the line.

"Oh, it's you," she said, unable to keep the irritation out of her voice.

"You're mad at me," he said.

"I don't get mad," she snapped.

"I'm sorry, Debbie. No matter how hard I try, I can't help coming across like an arrogant know-it-all. I do want to work smoothly with you. Can we talk about it again?"

He sounded so sincere that Debbie paused, then said, "Okay, I guess I have a short fuse lately. Let's do it tomorrow. I'll have everything ready."

The next day, Debbie—who did most of the talking—soon convinced Ed that the credentialing process had changed considerably since he last went through it. For example, applicants now underwent fingerprinting for clearance by local and federal law enforcement agencies. Debbie also felt that mandatory AIDS testing was just around the corner for teachers. About the only similarity between the new and old procedures was the size of the credential, still 8 1/2 × 11".

"I'm afraid I assumed too much," Ed said. "Sorry I wasn't paying attention to what you were trying to tell me. Next time I do that, remind me of this incident—that should keep me humble for awhile."

KEEPING CURRENT

Analysts must constantly strive to keep current with the state-of-the-art. Fortunately, they enjoy a variety of resources for sharpening and updating their skills: professional societies, periodicals, books, seminars, college classes, extension courses, users groups, and training groups.

Professional Societies

Professional societies can provide a rich source of new materials, tools, and contacts. Most of them offer seminars, as well as regular regional and national meetings. The major ones include:

DPMA—Data Processing Management Association (505 Busse Highway, Park Ridge, IL 60068): DPMA promotes and develops inquiries into the fields of data processing and data processing management. It also presents a Data Processor of the Year award. The organization includes numerous local chapters that provide monthly meetings with guest speakers.

ACM—Association for Computing Machinery: ACM dedicates itself to the advancement of the science and art of information processing, the exchange of ideas, and the development of integrity and competence. ACM produces over 50 different periodicals in all areas of computer activities and presents the "Turing Award" to distinguished persons in the profession. Over 100 local chapters offer monthly meetings and guest speakers.

ASM—Association for Systems Management: This society keeps its members abreast of changes in the field of systems management. ASM has five departments: data communications, data processing, management information systems, organization planning, and written communications. The ASM publishes the *Journal of System Management*.

Journals and Periodicals

A multitude of journals and periodicals keep the computer specialist informed. Each of the professional societies just mentioned (DPMA, ACM, and ASM) publishes journals to which both members and nonmembers can subscribe. The following periodicals are a representative sampling and many others are available:

Computerworld (Box 880, 375 Cochituate Road, Framingham, MA 01701): A weekly newspaper, covering a wide range of current events. Special or Extra editions cover a single topic of special interest and appear six times a year. The large number of advertisements from hardware and software suppliers can provide an ongoing awareness of new products.

Infoworld (155 Bovet Road, San Mateo, CA 94402): Advertising themselves as "The Voice of Personal Computing in the Enterprise," this weekly newspaper is full of articles on recent software releases and comparisons of a variety of competing products. Each issue has many columns that cover specific topics in great detail.

The Wall Street Journal (22 Cortland Street, New York, NY 10007): A daily newspaper not specifically written for computer professionals, but containing business news relevant to the field. A weekly column in the second section covers computer-related issues.

Byte (70 Main Street, Petersborough, NH 03458): A monthly magazine for the microcomputer fan. This advertising-laden periodical contains articles relating to hardware and software issues.

Software (Sentry Publishing Company, 1900 West Park Drive, Westborough, MA 01581): A monthly magazine specializing in software issues. Recent topics included CASE, graphical user interfaces, and operating systems.

Special magazines and newspapers have sprung up to serve special interest groups such as Apple Macintoshes (*Macworld*) and IBM PCs and their clones (*PC World*). Still other magazines cover microcomputers, CASE, office automation, data communications, desktop publishing, and database managers.

Books

Books should form part of the analyst's growing reference library. Consult the reference sections at the end of each chapter of this book for a wide range of computer titles. In addition to these, you may want to collect some on the history and philosophy of software development computing and on general management topics:

Structured Analysis and System Specification by Tom DeMarco (Yourdon, Inc.): An easily understandable book, teaching the construction and evaluation of data flow diagrams, data dictionaries, and system modeling. This book is considered the classic work on these topics.

Peopleware: Productive Projects and Teams by Tom DeMarco and Timothy Lister (Dorset House Publishing Company): DeMarco's second classic. A review of research that tells what it takes to develop high-quality software. A "must read" for people involved in any aspect of software development.

Structured Requirements Definition by Ken Orr (Ken Orr and Associates, Inc.): Written by one of the developers of the popular Warnier/Orr diagram, this text expands on diagrams and introduces a new tool: the entity diagram.

The Macintosh Way: The Art of Guerrilla Management by Guy Kawasaki (Scott, Foresman and Company): "Doing the right thing and doing things right" in the development of the Macintosh computer by one of the original team members. A funny, yet serious, look at development from the inside.

On Line Business Computer Applications, 2nd Edition by Alan L. Eliason (Science Research Associates, Inc.): A practical guide to accounting-based computer applications, including sample reports and file layouts.

Using SQL by James R. Groff and Paul N. Weinberg (Osborne/McGraw Hill): One of the easiest to read books on this topic. Covers all the popular implementations of SQL.

Theory Z by William G. Ouchi (Addison-Wesley Publishing Company). Examines the three management systems, explaining why the Japanese approach achieves such high productivity.

In Search of Excellence: Lessons from America's Best-Run Companies by Thomas J. Peters and Robert H. Waterman, Jr. (Harper & Row): Well over 2 million copies of this excellent book have been sold. It focuses on corporate management, analyzing the successful management styles of pace-setting companies. The authors illustrate Ouchi's *Theory Z* concept in action.

Training Groups

A booming training industry aimed at upgrading and advancing the skills of managers, analysts, and programmers developed in the early 1980s and is quite active in the 1990s. The major ones include:

Yourdon, Inc. (1131 Avenue of the Americas, New York, NY): Provides seminars across the country (as well as in Europe and Asia) on CASE, structured analysis, design programming, COBOL, Pascal, C, and Ada.

Ken Orr and Associates (715 East Eighth Street, Topeka, Kansas 66607): Provides seminars on Warnier/Orr and entity diagram tools across the country and at customers' sites. Other seminars focus on structured analysis and design.

DMW Group and Database Design, Inc. (Ann Arbor, MI): Led by James Martin Group, this training company specializes in issues relating to the correct design of large and small databases.

Finally, many vendors also provide training to those buying or leasing their equipment. Colleges (both two- and four-year) and universities also offer degree, certificate, and extension programs that can pack a lot of education into a limited period of time.

OBJECTS AND THE FUTURE OF CASE

As you learn more about the CASE tool at your disposal, you will find that it probably supports several types of diagrams (such as data flow, entity relationship, and state transition) and various styles within each (such as Yourdon/DeMarco or Gane/Sarson for data flow diagrams). A few CASE systems support **object-oriented** analysis and design.

The object paradigm is more than object-oriented programming (OOP). The paradigm is finding its way into object-oriented analysis (OOA), object-oriented design (OOD), and object-oriented database management systems (OODBMS). Some software developers predict that OOP, OOA, OOD, and OODBMS will replace traditional structured programming, analysis, and design. The paradigm focuses on breaking complex problems into smaller ones and examining the data passing between them, just like the structured methodology.

So, is object-oriented paradigm that much different? Yes! The paradigm is bottom-up and not top-down. Object orientation calls for a critical look at the data flows first. Once identified, small procedures are written to make the data flows operate. The procedures are combined into objects, which are then combined into programs. Objects are activated when they receive a message telling them to behave in a stipulated manner. After the object completes the task, it may send a message to one or more other objects.

A key advantage to objects revolves around prototyping. Once an object is defined, it becomes easy to prototype. When in the prototype mode, a designer and the user can study it together to make sure that the object behaves as it is supposed to. The designer and user modify the object's behavior until they are sure it works correctly.

The object-oriented paradigm also affects how we view the system life cycle (Figure 15.8). This object-oriented and CASE view has the phases overlap instead of seeing them as discrete events. An advantage to this seamless approach is that it requires no mind-set change as the system moves between phases. Also, the staff does not have to learn any new applications in order to move ahead in the system.

The object-oriented philosophy will probably become the driving force behind software development in the 1990s. While still evolving, object orientation has produced benefits in many software development organizations already.

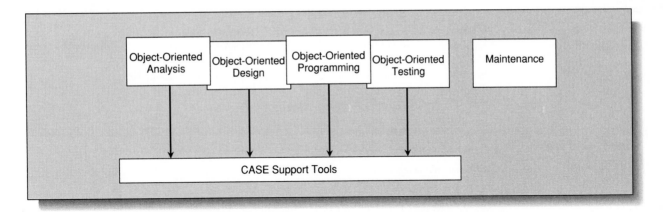

Figure 15.8 The system life cycle as viewed from an object-oriented CASE model.

Object orientation has many goals, but they all aim at producing quality software that meets user needs, is on time, and stays within budget. In addition, this philosophy strives to create reusable software building blocks that are pre-tested and highly reliable. The net gain is to enhance software development productivity.

SUMMARY

System maintenance is the fourth phase and concludes the systems process. All systems begin with a user request and end with the user, management, and the MIS group reviewing the final system. These people should closely examine the system to ensure that it achieves the goals, costs, results, and schedules established by the organization. If the system does so, a notice of acceptance goes to all relevant parties.

Management holds the final responsibility for controlling all software development functions. Good managers do not need technical computer expertise to oversee people, schedules, and funds. Gantt charts help monitor staff progress, while PERT charts assist in tracking schedules. People problems demand sensitivity to human psychology, not computer technicalities.

All analysts must keep up-to-date. A wide variety of books, magazines, organizations, and training groups can keep skills honed.

Today's philosophy of system development is quite different from past years. We now have small teams, highly focused, working together with new concepts such as CASE and object-oriented analysis, design, programming, and databases—testing and building software for the rest of this decade and into the next century.

As you have read in this book, system development is not an art, it is a science. System development involves planning, working with users, listening and communication skills, technical talents, and an ability to cope with constant change. Developing new systems requires constant feedback from users; unfortunately, much of it will seem negative. A good analyst takes information from a variety of sources—both positive and negative—to achieve his or her primary goal: solving users' problems.

KEY TERMS

System audit
Acceptance notice
Management control
Interpersonal relationships
Philosophical differences

Territorial pride
Personality conflicts
Cost containment
Event identification number
Object oriented

QUESTIONS FOR REVIEW AND DISCUSSION

Review Questions

1. What are management's responsibilities within the systems process?
2. How does a Gantt chart helps manage the systems process?
3. What are the components of a final report?
4. What are the names of two professional societies, two books, and two training groups that can extend the analyst's education?
5. Who are the participants in the system audit?
6. What are the three concepts of the object-oriented philosophy?

Discussion Questions

1. What are the four factors that need to be examined during the system review?
2. List the inputs to the system review.
3. List three people-oriented problems that may occur to an analyst.
4. When is a new system placed into operation?
5. Who decides when to place a new system into operation?

Application Questions

1. Why are users included in the system review?
2. Why should a user, and not a member of the computer services staff, sign the notice of acceptance?

Research Questions

1. What do you suppose happens when a system is *not* accepted?
2. List three examples of what might cause a system to be rejected.
3. Find a copy of William Ouchi's book, *Theory Z*, and list the corporate goals of two different businesses.
4. Thomas Peters and Robert Waterman, Jr., list factors affecting the future in their bestselling book, *In Search of Excellence*. Make a list of six of these factors.

Acceptance of View Video's New Rental System

S hortly before the end of the month of parallel operation, Frank calls a final acceptance meeting, sending notices to all concerned after first checking by telephone to determine an agreeable time. When everyone convenes, Cindy Stensaas calls the meeting to order, then briefly reviews the objectives and goals of the new Video Cassette Tape Rental System. She reminds everyone of anticipated costs for the detailed analysis, design, implementation, and aggressive schedule.

Then Frank takes charge, presenting the finished system to Mark and Cindy Stensaas. Frank starts by reporting that the system was delivered with all features stipulated by the study. However, it had a 2-1/2 percent cost overrun and was a month late.

Frank first reviews the feasibility study's objectives and benefits:

- Fast and accurate tape check-out and returns
- Use of a bar-code scanner
- Cross-reference between videotape titles and movie stars, plus selection by category, rating, and year made
- Customized reports that provide vital financial and other business information
- Interface to BusinessWorks PC with possible royalties if the software is ever resold to another video store
- Ownership of the software held by Mark and Cindy Stensaas
- Availability of marketing data on customer preferences and viewing habits

Next, Frank recaps the anticipated costs versus those actually incurred:

Cost Category	Expected	Actual
System Design	$3,500	$3,750
Programming	8,500	8,400
BusinessWorks PC Link	1,200	1,500
Training, Documentation, Installation	900	700
Yearly System Maintenance	800	750
Equipment	5,900	6,200
TOTALS	$20,800	$21,300

Finally, Frank reports on intangible (non-monetary) benefits:

- Better service to customers. It lets them choose movies by rating, category, stars, or year made. This fact alone is drawing customers to View Video because they know they can find the tape they want.
- Useful information for the Stensaases, which helps them manage their operations better.
- View Video's new system outperforms its competitors' systems. Checking tapes in and out is easy and quick, which pleases View Video customers. Some customers have even remarked on the "no-hassle" positive attitude of the system.

- The business data kept by BusinessWorks PC is always up-to-date, which leads to accurate financial statements available at the press of only a few keys.
- View Video utilizes their personal computer network at a high rate and can add new work stations when they want.

Finally, Frank mentions that it took seven instead of six months to deliver the software. Why? Frank's estimate of the link between BusinessWorks PC and the video check-out/check-in software was overly optimistic. The interface took more time and effort than was originally thought. However, it is working as planned, providing the Stensaases with accurate accounting data for their financial reports.

Cindy resumes control of the meeting and asks Mark for his ideas. He reports a high degree of satisfaction with the new system. Commenting last, Mark Stensaas says, "It's simple to use and extremely accurate." He thanks Frank for the considerate way in which he treated everyone at View Video.

In conclusion, Mark and Cindy formally accept the new software. They agree to write a memo that they will both sign (Figure 15.9). Cindy concludes the meeting by congratulating Frank on a job well done. Mark and Cindy walk Frank to the door, wrapping up last-minute details. "Please feel free to call me if any more questions come up," Frank tells them. Then he leaves View Video for the last time—happy that his system is working well for the owners, users, and customers.

Working With View Video

The end of a system implementation project is a time filled with a variety of feelings. To the analyst, it marks a milestone; the project is over and the analyst will probably move onto another endeavor. It is often a time of sadness since close personal relationships developed over the past months, and in some situations even years, will cease. But that is all a part of system development.

Frank Pisciotta delivered the software within reasonable time and budget guidelines and with features as planned. This is a rare occurrence in the software development world, but one that all software developers strive to achieve. We still read about software (such as OS/2 and new versions of dBASE) that fail to make it to market with these three qualities. With good planning and the use of the proper tools—such as CASE—more software than ever before is making it to users on time, within budget, and with features promised.

Case Study Exercises

View Video's cassette rental system will now move completely into the fourth phase of the systems life cycle, maintenance. In fact, the bulk of almost every software's lifetime, perhaps as much as 90 percent, is spent in the maintenance phase.

In day-to-day operation, users will want new features added and repairs made to errors that they may not discover for months. In View Video's case:

Figure 15.9 View Video's notice of acceptance for their new system.

|||||||| *View* Video ||||||||

MEMORANDUM

TO: Frank Pisciotta
                 View Video Staff Members

FROM: Mark and Cindy Stensaas, Owners   *MS CS*

SUBJECT: Notice of Completion of Videotape Cassette Rental System

DATE: March 3, 1994

We have accepted our new on-line videotape check-out/check-in system, which is now in full operation. At yesterday's meeting with Frank Pisciotta, we reviewed the system and it was approved and accepted. We are happy to report that it was completed within the budgeted amount, although it was a month late due to problems with the BusinessWorks PC interface.

As you know, the system has performed very well for the past month and our users are reporting that it exceeds their early expectations. Customers are also telling us that the system is fast and convenient. We are pleased to be the only store in this area to offer them this extra service.

We wish to commend all of you for the excellent cooperation you gave Frank these past seven months. If you encounter any problems with the new videotape system, please let us know immediately.

1. What new features do you think they will want added in the months ahead?
2. What happens when Manzanita Software issues a new release of BusinessWorks PC?
3. Who will take over maintenance activities for View Video?
4. Will Mark and Cindy Stensaas want or need to tie their software to popular word processing, spreadsheet, or database packages?
5. Microsoft Windows made a large impact on the personal computer marketplace in the early 1990s. How would Windows impact View Video's system?

References

Appleton, Daniel S. "The Modern Data Dictionary," *Datamation,* March 1, 1987, 66-68.

Bachman Information Systems. *The Bachman Approach.* Cambridge, MA: Bachman Information Systems, 1988.

Baecker, Ronald M. and Aaron Marcus. *Human Factors and Typography for More Readable Programs.* Reading, MA: Addison-Wesley, 1989.

Baker, F.T. "Chief Programmer Team Management of Production Programming," *IBM Systems Journal,* January 1972.

Barker, Richard. *CASE*Method, Entity Relationship Modelling.* Wokingham, England: Addison-Wesley, 1990.

Booch, Grady. *Object-Oriented Design with Applications.* Redwood City, CA: Benjamin/Cummings, 1991.

Bouldin, Barbara. *Agents of Change: Managing the Introduction of Automated Tools.* Englewood Cliffs, NJ: Prentice-Hall, 1989.

Brand, Stewart. *The Media Lab: Inventing the Future at MIT.* New York: Viking Penguin Inc., 1987.

Brock, Rebecca. *Designing Object-Oriented Software.* Englewood Cliffs, NJ: Prentice-Hall, 1990.

Carr, Robert. "12 Steps to Better Menus," *PC World,* May 1987, 274-281.

Cherin, Allen H. *An Introduction to Optical Fibers.* New York City: McGraw-Hill, 1983.

Class, Colin S. "Making the Change to Structured Code," *Interact,* January 1987, 67-72.

Coad, Peter. *Object-Oriented Analysis.* Englewood Cliffs, NJ: Prentice-Hall, 1990.

Codd, E.F. "A Relational Model of Data for Large Shared Data Banks," *Communications of the ACM,* 13:6.

Connell, John L., and Linda B. Shatner. *Structured Rapid Prototyping: An Evolutionary Approach to Software Development.* Englewood Cliffs, NJ: Prentice-Hall, 1989.

Constantine, L.L. "Objects, Functions, and Program Extensibility," *Computer Language,* January 1990, 34-56.

Date, C.J. *An Introduction to Database Systems, Volume 1, 5th Edition.* Reading, MA: Addison-Wesley, 1990.

Davis, George R. *The Local Network Handbook, 2nd Edition.* New York City: McGraw-Hill, 1987.

DeMarco, Tom. *Structured Analysis and Systems Specifications.* Englewood Cliffs, NJ: Prentice-Hall, 1979.

DeMarco, Tom, and Tim Lister. *Peopleware: Productive Projects and Teams.* New York: Dorset House, 1987.

Deutsch, Michael S. *Software Verification and Validation: Realistic Project Approaches.* Englewood Cliffs, NJ: Prentice-Hall, 1982.

Deutsch, Michael S., ed. *Inside Macintosh, Vol. V.* Reading, MA: Addison-Wesley, 1988.

Dickinson, Brian. *Developing Structured Systems: A Methodology Using Structured Techniques.* New York: Yourdon Press, 1986.

Dijkstra, E. "GO TO Statements Considered Harmful," *Communications of the ACM,* II:147-148.

Duncan, Mark. "Training is CASE Leading Edge," *Computerworld,* January 30, 1989, 81.

Eckols, Steve. *How to Design and Develop Business Systems: A Practical Approach to Analysis, Design and Implementation.* Fresno, CA: Mike Murach and Associates, 1983.

———. *Circular of Information: For the Use of Human Blood and Blood Components.* American Association of Blood Banks, 1991.

Edwards, Perry. *Advanced COBOL.* New York: Macmillan Publishing Company, 1987.

Effron, Joel. *Data Communications Techniques and Technologies.* Belmont, CA: Wadsworth, 1984.

Eliason, Alan L. *Online Business Computer Applications, 2nd Edition.* Chicago: SRA, 1987.

Fitzgerald, Jerry. *Internal Controls for Computerized Information.* Redwood City CA: Fitzgerald and Associates, 1978.

Fitzgerald, Jerry and Ardra F. Fitzgerald. *Designing Controls into Computerized Systems, 2nd Edition.* Redwood City, CA: Fitzgerald and Associates, 1981.

Gane, Chris. *Computer Aided Software Engineering.* New York: McGraw-Hill, 1988.

———. *Rapid System Development.* New York: Prentice-Hall, 1989.

———. *Computer-Aided Software Engineering: The Methodologies, the Products, and the Future.* Englewood Cliffs, NJ: Prentice-Hall, 1990.

Gilb, Tom. *Principles of Software Engineering Management.* Reading, MA: Addison-Wesley, 1988.

Goldberg, Adele, and Kenneth S. Rubin. "Taming Object-Oriented Technology," *Computer Language,* October 1990, 34-45.

Groff, James R. *Using SQL.* Berkeley: Osborne/McGraw-Hill, 1990.

Hansen, Gary W. *Database Processing with Fourth Generation Languages.* Cincinnati: South-Western Publishing Co., 1987.

Hawryszkiewycz, Igor T. *Relational Database Design: An Introduction.* Englewood Cliffs, NJ: Prentice-Hall, 1991.

Heckel, Paul. *Elements of Friendly Software Design.* Berkeley: Sybex, 1991.

Hickman, Craig R., and Michael A. Silva. *Creating Excellence.* New York: New American Library, 1986.

Higgins, David A. *Designing Structured Programs.* Englewood Cliffs, NJ: Prentice-Hall, 1983.

Higgins, David A. *Prototyping and 4GLs.* Topeka, KS: Ken Orr and Associates, 1987.

Hildebrand, Carol. "Managing the Aftermath," *Computerworld,* August 5, 1991, 58.

IBM Corp. "SQL/Data System, Logic," *Reference Manual LY24-5217.* White Plains, NY: IBM Corporation.

Index Technology Corp. *A Guided Tour of Excelerator.* Cambridge, MA: Index Technology Corporation, 1987.

Johnson, Leroy F. and Rodney H. Cooper. *File Techniques for Data Base Organization in COBOL, 2nd Edition.* Englewood Cliffs, NJ: Prentice-Hall, 1986.

Kapur, Gopal K. *Management Forum 1980: How to Effectively Improve Programmer Productivity.* Danville, CA: Kapur & Associates, Inc., 1982.

Kawasaki, Guy. *The Macintosh Way: The Art of Guerrilla Manaagement.* New York: Harper-Collins, 1990.

Keane, Inc. *Principles of Productivity Management in the Development of Computer Applications.* Englewood Cliffs, NJ: Prentice-Hall, 1985.

Kernighan, Bryan W. and P.J. Plauger. *Elements of Programming Style, 2nd Edition.* New York: McGraw-Hill, 1978.

Keuffel, Warren. "Building Essential DFDs," *Computer Language*, April, May, June, and July, 1991.

Kidder, Tracy. *The Soul of a New Machine.* New York: Avon, 1982.

Kroenke, David. *Database Processing, 3rd Edition.* Chicago: SRA, 1988.

Larson, Orlando. "Relational Databases," *Interact,* February 1987, 36-46.

Laurel, Brenda. *The Art of Human Computer Interface.* Reading, MA: Addison-Wesley, 1990.

Laurel, Brenda, ed. *Human Interface Guidelines: The Apple Desktop Interface.* Reading, MA: Addison-Wesley, 1987.

Lorie, R. *SQL and Its Applications.* Englewood Cliffs, NJ: Prentice-Hall, 1990.

Martin, James. *Application Development Without Programmers.* Englewood Cliffs, NJ: Prentice-Hall, 1982.

———. *Strategic Data Planning Methodologies.* Englewood Cliffs, NJ: Prentice-Hall, 1982.

———. *Managing the Database Environment.* Englewood Cliffs, NJ: Prentice-Hall, 1983.

———. *Advanced Database Techniques.* Cambridge, MA: MIT Press, 1986.

———. *The James Martin Productivity Series, Volume 6.* Marblehead, MA: High Productivity Software, 1987.

Martin, James, and Carma L. McClure. *Diagramming Techniques for Analysts and Programmers.* Englewood Cliffs, NJ: Prentice-Hall, 1985.

McClure, Steve. "White Paper: Object Technology: A Key Software Technology for the 90's," International Data Corporation, 1992.

Mellor, Stephen. *Object-Oriented Design.* Englewood Cliffs, NJ: Prentice-Hall, 1987.

Messenheimer, Susan, and Carol Weiszmann. "Quality Software Quest," *Software Magazine*, February 1988, 29-36.

Metzger, Phillip W. *Managing a Programming Project, 2nd Edition.* Englewood Cliffs, NJ: Prentice-Hall, 1981.

Meyer, Bertrand. *Object-Oriented Software Construction, 2nd Edition.* Hertfordshire, England: Prentice-Hall International, 1988.

Morland, D. Verne. "Human Factors Guidelines for Terminal Interface Design," *Communications of the ACM,* July 1983, 484-494.

Myers, Glenford J. *Software Reliability: Principles and Practices.* New York: John Wiley & Sons, Inc., 1976.

———. *Composite Structured Design.* New York: Van Nostrand Reinhold, 1978.

———. *The Art of Software Testing.* New York: John Wiley & Sons, Inc., 1979.

Naisbitt, John. *Megatrends: Ten Directions Transforming Our Lives.* New York: Warner Books, 1988.

Nolan, Daniel. "A New System's Life," *Computerworld,* October 13, 1986, 93-101.

Oracle Corp. *ORACLE.* Belmont, CA: Oracle Corporation, 1987.

Orr, Kenneth T. *Data Structured Systems Development.* Topeka, KS: Ken Orr and Associates, 1984.

———. *How to Evaluate a Methodology.* Topeka, KS: Ken Orr and Associates, 1986.

———. *Structured Systems Development.* Englewood Cliffs, NJ: Prentice-Hall, 1986.

Ouchi, William G. *Theory Z: How American Businesses Can Meet the Japanese Challenge.* New York: Avon, 1982.

Peters, Lawrence J. *Software Design: Methods and Techniques.* Englewood Cliffs, NJ: Prentice-Hall, 1986.

Peters, Thomas J., and Robert H. Waterman, Jr. *In Search of Excellence: Lessons from America's Best-Run Companies.* New York: Harper & Row, 1982.

Polya, Gyorgy. *How to Solve It.* Princeton, N.J: Princeton University Press, 1971.

Rettig, Marc. *Database Programming and Design.* San Francisco: Miller Freeman, Inc., 1991.

Scheier, Robert L. "Cost Justification Is Crucial for IS Projects," *PC Week,* February 11, 1991, 67.

Stamps, David. "Taking an Objective Look," *Datamation,* May 15, 1989.

Strunk, William S., Jr., and E.B. White. *The Elements of Style, 3rd Edition.* New York: MacMillan, 1979.

Swirczek, Linda. *DSSD Program Design Model: A Template for Program Design.* Topeka, KS: Ken Orr and Associates, 1987.

Tog. *Tog on Interface.* Reading, MA: Addison-Wesley, 1991.

Ward, Paul T., and Stephen J. Mellor. *Structured Development for Real Time Systems (3 volumes).* New York: Yourdon Press, 1986.

Warnier, Jean-Dominique. *Logical Construction of Programs.* New York: Van Nostrand Reinhold, 1974.

Weinberg, Gerald. *The Psychology of Computer Programming.* New York: Van Nostrand Reinhold, 1988.

Weinberg, Gerald M. and Daniel P. Freedman. *Handbook of Walkthroughs, Inspections, and Technical Reviews, 3rd Edition.* Boston, MA: Little, Brown and Company, 1978.

Yourdon, Edward. *Techniques of Program Structure and Design.* Englewood Cliffs, NJ: Prentice-Hall, 1975.

———. *Powerhouse Primer: The Universal Development Language.* Ottawa Canada: Cognos Inc., 1985.

———. *Nations at Risk: The Impact of the Computer Revolution.* Englewood Cliffs, NJ: Prentice-Hall, 1986.

———. *Information Engineering Methodology Overview.* Plano, TX: Texas Instruments, 1989.

———. *Modern Structured Analysis.* Englewood Cliffs, NJ: Prentice-Hall, 1989.

———. *Structured Walkthroughs, 4th Edition.* Englewood Cliffs, NJ: Prentice-Hall, 1989.

Yourdon, Edward. *The CASE Report.* Southfield, MI: Nastec Corporation. A newsletter issued six times per year.

Zmud, Robert W., and James F. Cox. "The Implementation Process: A Change Approach," *Management Information Systems Quarterly,* 1979.

Credits

Figure 1.2 Goals and values statement courtesy of McCaw Cellular Communications, Inc. Copyright 1990, all rights reserved; **1.6** courtesy of Manzanita Software Systems, Roseville, CA.

Figure 3.2 Check courtesy of Fleet Feet, Inc., Sacramento, CA.

Figure 4.10 Data flow diagram courtesy of Adamis Distribution.

Figure 5.8 Check courtesy of Fleet Feet, Inc., Sacramento, CA; packing slip courtesy of JBI, a division of Playtex Apparel, Inc.; invoice courtesy of Convert-It!, San Jose, CA.

Figure 7.4 Marketing piece courtesy of R & D Publications, Lawrence, KS.

Figure 8.2 Timecard and purchase requisition courtesy of Fleet Feet, Inc., Sacramento, CA; **8.3** Invoice courtesy of Flowers by Foote, Auburn, CA; **8.4** Photograph courtesy of

American Microsystems, Euless, TX; **8.5** BusinessWorks PC's main menu courtesy of Manzanita Software Systems, Roseville, CA; **8.6** Keystrokes specifications from Microsoft® Windows™ SDK version 3.1, reprinted with permission from Microsoft Corporation; **8.8** Mouse button click specifications from Microsoft® Windows™ SDK version 3.1, reprinted with permission from Microsoft Corporation.

Figure 11.6 Photographs courtesy of Practical Peripherals, a Hayes Company; **11.16** Data flow diagram courtesy of StructSoft, Inc., Bellevue, WA.

Chapter 15, page 486 Six Keane rules reprinted, with permission, from Keane, John F., Marilyn Keane, and Mark Tegan. *Priniciples of Productivity Management in the Development of Computer Applications.* Englewood Cliffs, NJ: Prentice-Hall, 1984.

Working with People and Working with Technology

Page 113 List of rules for more effective meetings adapted from *Computer Decisions,* July 1982; **page 182** adapted from Rand P. Hall, "Seven Ways To Cut Software Maintenance Costs." *Datamation* 33, July 15, 1987; **page 224** "Illustrated Guide to Type" reprinted with permission from *Macworld,* July 1991. Graphic by Arne Hurty. Text by Alex Brown. Copyright © Macworld Communications, Inc. 501 Second St., San Francisco, CA 94107; **page 271** "12 Steps to Better Menus" reprinted with the permission of *PC World,* May 1987. Text by Robert Carr; **page 334** abstracted from Carol Thiel, "Relational DBMS, What's In a Name," *Infosystems,* September 1982; **page 378** abstracted from "Local Area Networks and the Functional Distribution of the Workload," a CAI Whitepaper published in *dataBASE,* 7:3.

Index